Boffey.

PHOTOSYNTHESIS

Molecular Biology and Bioenergetics

Springer-Verlag
Berlin
Heidelberg
New York
London
Paris
Tokyo

Narosa Publishing House
New Delhi
Madras
Bombay

PHOTOSYNTHESIS

Molecular Biology and Bioenergetics

*Proceedings of the International Workshop on
Application of Molecular Biology and Bioenergetics of
Photosynthesis*

Edited by

G. S. SINGHAL, JAMES BARBER, RICHARD A. DILLEY,
GOVINDJEE, ROBERT HASELKORN, PRASANNA MOHANTY

Springer-Verlag

Narosa Publishing House

Narosa Publishing House
6 Community Centre, Panchsheel Park, New Delhi 110 017
35–36 Greams Road, Thousand Lights, Madras 600 006
306 Shiv Centre, D.B.C. Sector 17, K.U. Bazar P.O., New Bombay 400 705

ISBN 81-85198-09-8 Narosa Publishing House New Delhi
ISBN 3-540-50451-6 Springer-Verlag Berlin Heidelberg New York
ISBN 0-387-50451-6 Springer-Verlag New York Berlin Heidelberg

Published by N. K. Mehra for and on behalf of
Narosa Publishing House, 6 Community Centre, Panchsheel Park, New Delhi 110 017
Photocomposed at Taj Services Ltd., E-100 Sector-VI, Noida, (U.P.)
and printed at Rekha Printers Pvt. Ltd., New Delhi 110 020

"Believe nothing
merely because you have been told it,
or because it is traditional,
or because you yourself have imagined it,
Do not believe what your teacher tells you,
merely out of respect for the teacher,
But whatever after due examination and analysis,
you find to be conducive to the good
the benefit,
the welfare of all beings,
that doctrine believe and cling to,
and take it as your guide."

—LORD BUDDHA

Foreword

The Workshop on Applications of Molecular Biology and Bioenergetics of Photosynthesis has been held at a time when plant scientists are searching for new methods of raising the ceiling to yield in economic plants. Yield of a crop plant is the product of interaction between the physiologic efficiency of the plant and the management efficiency of the farmer. The physiologic efficiency is in turn controlled by both the genetic make-up of the plant and the environmental conditions under which the crop is cultivated. This workshop has therefore dealt with a wide range of problems affecting photosynthesis under conditions of optimal growing conditions as well as under different stresses—such as moisture deficit and excess, heat, salinity and micronutrient deficiencies.

I am glad the effects of UV radiation have also been studied since we need more data in this field consequent on the postulated increase in UV radiation arising from the depletion of the ozone layer caused by the excessive use of chlorofluorocarbons. Similarly, an understanding of the molecular and physiological aspects of herbicide resistance is important, since in the tropics and sub-tropics weed infestation is a major factor limiting fertilizer-use efficiency and thereby yield.

The papers on the impact of senescence on different aspects of wheat chloroplasts are exceedingly interesting from the point of view of understanding the changes associated with ageing. I am also glad that photosynthesis in mangrove vegetation has received attention since several countries are endowed with rich mangrove ecosystems.

Countries in the tropics and sub-tropics are faced with shrinking land resources for agriculture and expanding human and animal populations. India, for example, will have to produce annually 230 to 250 million tonnes of food grains, about 700 million tonnes of livestock fodder and over 200 million tonnes of fuel wood by the year 2000. The additional production will have to come from higher productivity per unit of land, water and time. So far, the pathway to yield improvement adopted by plant breeders has been the alteration of the harvest index in favour of grains or other economic parts. The challenge now is to increase the total biological yield. The scope for merely altering the partition pathway has been practically exhausted in crops like wheat and rice.

We need a greater understanding of the molecular aspects of bioenergetics as related to photosynthesis in order to undertake purposeful experiments in genetic engineering. The proceedings of the Workshop on this topic thus constitute a timely and valuable addition to the literature on this subject. We owe a deep debt of gratitude to Professor G. S. Singhal, Professor Prasanna Mohanty and their colleagues for this labour of love. I also congratulate them on the excellence of this book.

The earth has remained a comfortable place for living organisms for a whole 3.5 billion years since life began, despite a 25 per cent increase in the output of heat from the sun. Will we maintain this legacy? This will depend upon our ability to maintain a healthy photosynthetic pathway of development.

(Dr. M. S. Swaminathan)
President, International Union for the
Conservation of Nature and Natural Resources

Preface

Plant productivity depends upon the photosynthetic conversion of the energy of sunlight into free energy stored in the biomass of plants. An intermediate step in this energy conversion process is electron transfer and proton translocation.

Efforts are currently under way, using molecular biology as a tool, to alter the polypeptides that are involved in electron transport and proton translocation. For example, site-specific mutagenesis is being used to replace amino acids thought to be essential for electron transfer and proton translocation. Work of this type will not only help to determine the mechanisms of electron transfer and proton translocation, but will also provide an opportunity to improve plant performance under conditions of stress, such as drought, salt stress and herbicide resistance, which have all been shown to influence photosynthetic electron transport. The latest techniques of molecular genetics and molecular biology have the potential to answer the vital questions about the mechanism of energy transduction and molecular structure of the photosynthetic apparatus.

At present, several research groups are working on projects that are expected to lead to rapid improvement of our understanding of the photosynthetic process and, as a consequence of the application of biotechnology to higher plants, to improved plant performance. From 4 to 8 January 1988 an international workshop on the Application of Molecular Biology and Bioenergetics of Photosynthesis was held at the School of Life Sciences, Jawaharlal Nehru University, New Delhi. This workshop brought together scientists from all over the world, who were using the latest techniques of molecular biology, biochemistry and biophysics for the study of the chemical and physical processes of photosynthetic electron transport and proton translocation. This book consists of the contributions of the participants at this workshop and is a compilation of the state-of-art knowledge on this subject.

—Editors

Acknowledgements

The organizing committee of the International Workshop on Applications of Molecular Biology and Bioenergetics of Photosynthesis would like to express its appreciation for financial and other support received from the following agencies—Department of Biotechnology, Government of India; Department of Science and Technology, Government of India; University Grants Commission, India; Council of Scientific and Industrial Research, India; Jawaharlal Nehru University, India; United Nations Educational Scientific and Cultural Organisation, India; National Science Foundation, USA and India; United States Department of Agriculture, USA; British Council, UK and India; and other agencies who totally or partially funded the delegates from England, Germany, Greece, Holland and Japan. We would like to thank Drs G. Papageorgiou, H. N. Singh, H. S. Raghavendra, U. C. Biswal, V. Krishnan, H. Pakrasi, A. K. Verma and B. C. Tripathy for their oral presentations at the Workshop on various interesting topics related to molecular biology and bioenergetics of photosynthesis. Due to unavoidable circumstances, these presentations could not be included in this book.

We also extend our thanks for presentations in the form of posters to Drs Anbudurai, Behra, Bhardwaj, Bhonsale, Amba, Bose, Bhagwat, Chaturvedi, Choudhury, Khan, Khanna-Chopra, Krishnaswamy, Mathur, Misra, Sainis, Sengupta, Shyam, Panda, Khan and Vivekanandan. We would like to thank Dr Ranjana Paliwal for assisting in the preparation of the index.

Contents

PART 2: Energy Transduction

PART 3: Stress Effects on Photosynthesis Organisms

PART 4: Molecular Structure—Functions, Relationships

PART 1
Genes and Polypeptides of Photosystem II

Genes of the Photosynthetic Apparatus of Higher Plants—Structure, Expression and Strategies for Their Engineering

A. K. TYAGI[1,3], N. Y. KELKAR[1,2], S. KAPOOR[1,2] AND
S. C. MAHESHWARI[1,2]

[1]Unit for Plant Cell and
Molecular Biology and
[2]Department of Botany,
University of Delhi,
Delhi-110007, INDIA

[3]Department of Plant Molecular Biology,
University of Delhi South Campus,
Benito Juarez Road,
New Delhi-110021, INDIA

Summary

Approximately 50 proteins in the form of various multi-subunit supra-molecular complexes are associated with the thylakoid membranes of the chloroplast which carry out various functions of photosynthesis. Gene(s) for these proteins are encoded either in the plastid chromosome or in the nucleus. With the characterisation of the genes for more than 35 polypeptides, it has become clear that the thylakoid membrane is the product of a complex interaction between the nuclear and chloroplast genetic systems. The organisation of chloroplast-encoded genes is complex, involving polycistronic transcription. Expression of both nuclear- and chloroplast-encoded genes is regulated at several steps and light has an important role to play during biogenesis of thylakoids. Nuclear-encoded proteins are produced as precursors on the cytosolic ribosomes with amino-terminus transit-peptides which help in intracellular and intra-organelle targeting of the proteins and get cleaved off during this process. Interesting strategies for manipulation of genes of thylakoid proteins are emerging to study structure/functional relationships in photosynthesis.

Introduction

The light reactions of photosynthesis include capture of light energy, separation of charge, splitting of water, transport of electrons, generation of proton gradient and production of ATP which in higher plants are performed by specialised photosynthetic membranes, commonly known as thylakoids, located in chloroplasts. These membranes are made up of well over 50 proteins, most of them organised into four supra-molecular complexes (PSII, cytochrome b/f complex, PSI, and ATP synthase). Some evidence had already accumulated in the seventies that while certain proteins (e.g. cytochrome f) concerned with photosynthesis are encoded by the chloroplast genome, others (e.g. chlorophyll a/b binding proteins) are encoded by the nuclear genome. However, with the advent of recombinant DNA techniques, it has become evident that the 'dual' control, i.e. by both chloroplast and nuclear genomes, extends to all supra-molecular complexes, which creates a unique situation since biogenesis and the functioning of the photosynthetic membranes depends on a rather complex interaction of the nucleocytoplasmic and chloroplast genetic systems.

As a first step towards understanding the intricacies involved in the expression of individual gene(s) and biogenesis of photosynthetically active membranes, it is imperative to isolate and characterise the genes. Various probes from well characterised genes are being used to elucidate the mechanisms of regulation of their expression at the transcriptional, post-transcriptional and translational levels. It is thus of great interest to know how this regulation occurs in time, space, and stoichiometry; and how products from two different compartments of the cell get organised together. This article gives a bird's-eyeview of the progress made in the above-mentioned areas in recent years by employing recombinant DNA techniques. Also, at the end, we propose certain strategies for engineering the genes of the photosynthetic apparatus. Due to the limited space available, all relevant references could not be included and for details attention is invited to other recent reviews (1–9). It may be further added that although, in a broad sense, the entire chloroplast is generally considered as the photosynthetic apparatus, our attention here will be focused on the components of the thylakoid membranes where primary light reactions actually occur.

Structure and Organisation of Genes for Proteins of the Photosynthetic Membranes

As shown in Table 1, genes for the photosynthetic apparatus proteins are located both in the chloroplast and nuclear genomes. The first indication of the distribution of these genes in two compartments came from early experiments on hybridization with various genotypes and mutants which indicated that whereas with respect to certain characters inheritance followed the Mendelian pattern, in respect of others it did not. In the early seventies, experiments began employing compartment-specific inhibitors of protein synthesis. These experiments were followed by *in vitro* protein synthesis from mRNA isolated from cells—by separating poly A^+mRNA from other RNAs, one could further confirm whether the nucleus or the chloroplast DNA coded for a particular product. In the late seventies, chloroplast genomes (cpDNAs) were characterised from several plants, e.g. spinach and maize. Although some variations do exist, most cpDNAs have turned out to be circular molecules of 120–180 kbp and containing a pair of 10–25 kbp inverted repeats, as illustrated in Fig. 1, which divide the remaining cpDNA into one large and one small single copy region (2). Finally, with pure cpDNA available from several plants as also various restriction endonucleases and recombinant DNA technology, the stage was

TABLE 1

Gene(s) for the Proteins Constituting the Photosynthetic Membranes in Higher Plants[1]

Components of Supra-molecular Complexes	Coding Compartment[2]	Gene Designation[3]	Clones for Nuclear-encoded Genes
PS II Core Complex			
51kDa	Chloroplast	psbB	
44 kDa	Chloroplast	psbC	
34 kDa, D2	Chloroplast	psbD	
32kDa, D1	Chloroplast	psbA	
24 kDa	Chloroplast	psbG	
10 kDa, Phosphoprotein	Chloroplast	psbH	
9 kDa, Cyt b_{559}	Chloroplast	psbE	
7kDa	?		
6.5 kDa	?		
5.5 kDa	?		
5 kDa	?		
4 kDa, Cyt b_{559}	Chloroplast	psbF	
PS II Antenna Proteins			
Chlorophyll a/b binding apoproteins (24–27 kDa)	Nucleus	cab	Genomic
Water-Splitting Complex			
33 kDa	Nucleus		cDNA[4]
23 kDa	Nucleus		cDNA[4]
22 kDa	?		
16 kDa	Nucleus		cDNA[4]
10 kDa	Nucleus		cDNA[4]
Cytochrome b6/f Complex			
Cytochrome f	Chloroplast	petA	
Cytochrome b6	Chloroplast	petB	
Rieske-FeS Protein	Nucleus	petC	cDNA[4]
Subunit IV	Chloroplast	petD	
PS I Core Complex			
68 kDa, Ia	Chloroplast	psaA	
68 kDa, Ib	Chloroplast	psaB	
22 kDa, II	Nucleus		cDNA
19 kDa, III	Nucleus		cDNA
18 kDa, IV	Nucleus ?		
16 kDa, V	Nucleus		cDNA
12 kDa, VI	Nucleus		cDNA
9 kDa, VII-FeS Protein	Chloroplast	psaC	
PS I Antenna Protein			
Chlorophyll a/b binding apoproteins (22–25 kDa)	Nucleus		Genomic

TABLE 1 (*Contd.*)

Components of Supra-molecular Complexes	Coding compartment[2]	Gene Designation[3]	Clones for Nuclear-encoded Genes
Other Components of the Electron Transport Chain			
Plastocyanin	Nucleus	petE	Genomic
Ferredoxin	Nucleus	petF	cDNA[4]
Ferredoxin-NADP$^+$ oxido-reductase	Nucleus		cDNA[4]
CF I, ATP Synthase			
58 kDa, α Subunit	Chloroplast	atpA	
57 kDa, β Subunit	Chloroplast	atpB	
38 kDa, γ Subunit	Nucleus	atpC	cDNA
25 kDa, δ Subunit	Nucleus	atpD	cDNA[4]
14 kDa, ε Subunit	Chloroplast	atpE	
CF$_o$, ATP Synthase			
18 kDa, I Subunit	Chloroplast	atpF	
16 kDa, II Subunit	Nucleus	atpG	cDNA
8 kDa, III Subunit	Chloroplast	atpH	
19 kDa, IV Subunit	Chloroplast	atpI	

[1] To avoid a long list of literature, references have been omitted.

[2] All chloroplast-encoded genes (given here) for photosynthetic apparatus proteins have been sequenced.

[3] Gene designations have been given as reported in the literature.

[4] These cDNAs contain nucleotide sequences encoding the complete precursor protein.

set for detailed characterisation of the chloroplast genome and for localisation of the genes encoded by the plastome on cloned cpDNA fragments. Genes were assigned to various proteins by using the immunological approach, following one of the two preparative strategies for obtaining the proteins: (i) the DNA fragments were cloned and transcription coupled with translation employing cell-free extracts of *E. coli* itself, (ii) cpDNA fragments were used to hybrid select specific mRNAs which were then translated *in vitro* using the reticulocyte system. Many cloned cpDNA fragments have now been sequenced and open reading frames (ORFs) for various polypeptides identified. The deduced sequences have also been compared to partially or completely sequenced polypeptides. In 1986, the complete nucleotide sequences of cpDNAs from a higher plant, tobacco (*Nicotiana tabacum*, 11) and a liverwort (*Marchantia polymorpha*, 12) were reported. Several ORFs have already been identified and efforts are being made to find products for those ORFs whose function is not yet known.

Rapid progress has similarly been made in respect of genes residing in the nucleus. In the early eighties, cDNA libraries from a number of plants

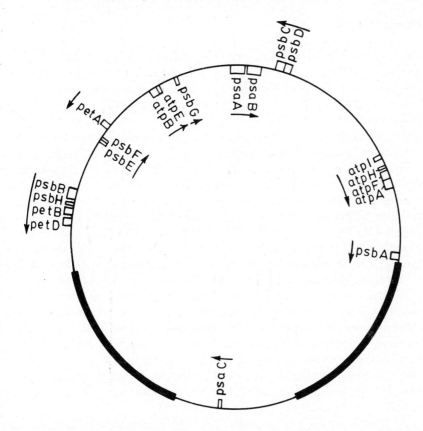

Fig. 1. Location of genes, encoding photosynthetic membrane proteins, on a typical plastid chromosome, e.g. from spinach (10) or tobacco (11). Thick regions of the circle represent inverted repeats.

were made and clones encoding chlorophyll a/b binding polypeptides of PSII were identified. The cDNA clones were then used to isolate genomic clones. More recently, a cDNA library of spinach has been made in expression vectors and clones, specific to several proteins of the photosynthetic membranes, have been isolated by using antisera (13). The cDNA clones have been employed to isolate and characterise genomic clones. Interestingly, all these genes encode precursor polypeptides with amino-terminal transit peptides which help to target these polypeptides into the chloroplast and which are removed during the transport across the chloroplast or thylakoid membranes (5).

Some details about the organisation and characteristics of the genes encoding polypeptides of various supra-molecular complexes and other associated proteins of the photosynthetic membranes are given in Table 1. A general summary follows.

Photosystem II

Photosystem II of the photosynthetic apparatus of higher plants is made up of three smaller complexes, viz. the core complex, the light harvesting complex, and the water-splitting or oxygen evolving complex.

CORE COMPLEX

The core complex is made up of a dozen polypeptides. Genes for eight of these have been localised in the large single copy region of the chloroplast genome. Recently, Nanba and Satoh (14) have made preparations showing P680 reaction centre activity containing three polypeptides, namely, D1, D2 and cyt b559. D1 and D2 proteins show homology and conservation of residues with those of bacterial reaction centre proteins and might be responsible for binding P680, QA, QB, pheophytin and other electron transfer intermediates such as Z (15). D1, the herbicide-binding polypeptide, is encoded by the gene psbA and is one of the first to be localised on chloroplast genome (16). The gene encodes a polypeptide of 353 amino acids; but the latter is known to get post-translationally modified possibly by removal of a stretch of 12–16 amino acids at the carboxy-terminal end (17, 18). Sequencing of this gene from atrazine type herbicide-resistant plant species has shown a change from A to G resulting in replacement of the amino acid serine by glycine at position 264 (19, 20). The gene psbD encodes the polypeptide, D2, of 353 amino acids (21) and, like D1, it is highly hydrophobic with several membrane-traversing spans.

As is well known, a cytochrome, cyt b559, is associated with PS II. This is now known to be made up of two polypeptides of 9 and 4 kDa encoded by genes psbE and psbF. That cytochrome b559 is a product of psbE and psbF genes was predicted after sequencing a fragment containing the gene for psbE which revealed the presence of an ORF for psbF (22). Later research proved that this was indeed true (23). Although the gene(s) for psbE and psbF are highly conserved in spinach, tobacco, *Oenothera*, and wheat, the exact role of cyt b559 remains an enigma (24, 25).

Two genes, psbB and psbC, encode for 508 and 473 amino acid long polypeptides, commonly known as the 51 and 44 kDa polypeptides, which bind chlorophyll a and are possibly part of the antenna pigment system involved in transferring the light energy to P680 (21, 26). Both polypeptides show homology and are predicted to have several spans traversing the membrane. The gene psbC at its 5′-end overlaps the 3′-end of another gene, psbD, by 50 bp. The genes, psbG and psbH, for two other proteins associated with PSII have also been identified and characterised recently. These two genes encode polypeptides of 248 and 73 amino acid residues, respectively (27, 28). The product of psbH is a phosphorylated protein.

LIGHT-HARVESTING COMPLEX

At least four polypeptides of the light-harvesting complex of photosystem II are encoded in the nuclear genome by a family of 3–20 cab genes which can be divided into sub-families on the basis of sequence homology (29–34). Furthermore, these genes have also been localised on individual chromosomes. In pea, chromosome two contains a locus for the cab genes (35), whereas in tomato the genes are distributed on at least 5 chromosomes at 5 different loci: cab-1 (chromosome 2), cab–2 (chromosome 8), cab–3 (chromosome 3), cab–4 (chromosome 7) and cab–5 (chromosome 12) (36). The locus cab–1 contains four genes, arranged in tandem, whereas locus cab–3 contains three genes and a truncated gene (31).

The nucleotide sequences of these genes show an open reading frame with a coding capacity of 264–269 amino acids for the precursor proteins. Of these, the first 33–36 amino acids represent the transit peptides which help in intracellular and possibly also intraorganellar targeting of the mature proteins and which get cleaved off by processing protease(s). Most of these gene(s) contain no intron. But, one gene from *Lemna* has been found to contain an 84 bp intron with certain features of a transposable element (37). Possibly, the intron is not an ancestral feature of the cab genes and has got introduced into the gene later during evolution. Recently, intron-containing cab genes have also been localised in *Petunia* (38) and tomato (39). They encode slightly divergent, type II, polypeptides which are also incorporated into the light-harvesting complex of PSII.

WATER-SPLITTING COMPLEX

At least five polypeptides have been found to be associated with the water-splitting complex of PSII (Table 1, 40). Genes for some of these proteins have been localised in the nucleus (41, 42). Recently, cDNAs encoding the complete precursors for 33, 23, 16 and 10 kDa proteins have been sequenced and found to encode 331, 267, 232 and 140 amino acids which contain amino-terminal transit sequences of 84, 81, 83 and 41 amino acids, respectively (43–45). Of the four polypeptides, three (33, 23, 16 kDa) are located inside the thylakoid lumen; hydropathy plots show their hydrophilic nature. Nonetheless, the transit peptides do not show any similarity with those of others at the primary amino acid sequence level. However, they do share among themselves and with the transit peptide of plastocyanin—which is also located towards the thylakoid lumen—a unique hydrophobic domain, just before the final processing site, and a region which can form an amphipathic β-sheet and possibly includes the site for primary processing (43, 44, 46). Secondary structure predictions for the 10 kDa protein suggest one transmembrane segment with the N–terminus in the thylakoid lumen and a 17 amino acid C-terminal segment outside (45). The transit sequence does not show similarity at the primary sequence level with those of the three extrinsic, luminal proteins of PSII.

Cytochrome b–f Complex

This is the simplest of the supra-molecular complexes associated with photosynthesis and consists of at least four components which are involved in electron flow from PSII to PSI and in cyclic-photophosphorylation. Polypeptides for three of its components, viz. cyt f, cyt b6 and subunit IV are encoded by the genes petA, petB and petD, respectively, located in the large single copy region (47). Gene petA, for cytochrome f has been shown to code for a pre-apoprotein of 320 amino acids in several species except for *Oenothera* where the pre-apoprotein is smaller by two amino acids (see 48 and references cited therein). Comparison of the deduced amino acid sequence with that of the N-terminal sequence of the mature protein has revealed the presence of an N-terminal presequence of 35 amino acids, which gets cleaved off, possibly during transport of the protein across the thylakoid membranes (49, 50). A topological model for the arrangement of this polypeptide has been proposed on the basis of hydropathy analysis and partial proteolytic cleavage data: it appears that there is only a single span of alpha helix traversing the membrane and the bulk of the amino-terminus polypeptide lies in the intrathylakoid space (49, 50).

Cytochrome b6 and subunit IV are encoded by two closely placed genes, petB and petD. Recently, Northern blot, reverse transcription, and S1 nuclease analysis have shown that both petB and petD contain introns (of 753 and 742 bp, respectively) and produce polypeptides of 215 and 160 amino acids, the smaller exons code only two and three residues of the amino terminus (10, 51). Earlier, cytochrome b6 and subunit IV polypeptides have been shown to have homology with the amino and carboxy-terminal ends of mitochondrial cytochrome b, respectively (52). They have several histidine residues which have been implicated in haem binding. The hydropathy plots show five and three membrane spanning regions for cyt b6 and subunit IV, respectively (52).

The Rieske-FeS protein has been found to be encoded in the nucleus (47) and a cDNA has been sequenced recently which encodes a complete precursor polypeptide of 247 amino acids, including also a putative N-terminal transit sequence of 68 amino acids (53). Comparison of the deduced sequence of the mature polypeptide with sequences of the Rieske-FeS proteins of the respiratory electron transport chains of bacterial and mitochondrial origins shows remarkable similarities in secondary structure and the region involved in binding of the Fe_2S_2-cluster. The transit sequence shows some similarity to that of 10 kDa polypeptide of the water-splitting system (45) and the delta subunit of ATP synthase complex (54).

Photosystem I

The core complex of PSI is constituted of at least eight polypeptides (Table 1). So far, genes for only three of these have been characterised and these are located on the chloroplast genome; four others are located in the nucleus and the location of the remaining one is still unknown. Of the three genes located on chloroplasts two, psaA and psaB, encoding 750 and 734 amino acid polypeptides (and which are substantially homologous) are located in the large single-copy region of the chloroplast genome (55, 56). These genes possibly arose by duplication. Their products constitute the photosystem I reaction centre and hydropathy analysis shows several regions which may traverse the membrane. More recently, a third gene, psaC, encoding a 9 kDa polypeptide, has been identified in the small single-copy region of the tobacco plastome (57). The polypeptide possibly functions as the apoprotein for the iron-sulphur centres A and B and shows three regions, rich in cysteine residues, with a potential of traversing the membrane.

The light-harvesting complex of photosystem I contains several polypeptides of 22–25 kDa. Recently, one cDNA and one genomic clone (representing however another gene of the family) have been obtained from tomato (58); these genes have been located on chromosome 5 at a distance of about 7 kbp. The gene represented in the genomic clone contains three small introns of 107, 87 and 89 nucleotides and the encoded polypeptide has 246 amino acid residues. There is more than 55% divergence in the gene for PSI with members of PSII from tomato.

Genes for Other Proteins Associated with Electron Transport

Plastocyanin, a soluble type I copper protein, is a component of the photosynthetic apparatus that transfers electrons from cyt b/f complex to PSI and is located towards the luminal face of the thylakoid membranes. Recently, cDNA and genomic clones for this have been isolated and characterised in spinach (46, 59) and *Silene* (60). The gene, petE, encodes a precursor of 168 amino acids with a transit peptide of 69 residues. The mature polypeptide is hydrophilic. The transit peptide shows structural similarity with corresponding regions of those members of the water-splitting complex which are located on the luminal face (43, 44) as discussed earlier. Available data show that the protein derives from a single copy gene which does not contain any intron (46).

Two other proteins associated with photosynthetic electron transport are ferredoxin-NADP$^+$-oxido-reductase (FNR) and the stromal protein, ferredoxin. Genes for these are located in the nucleus and cDNAs, encoding complete precursor proteins, have been sequenced (61, 62). The ferredoxin cDNA encodes a polypeptide of 146 amino acids comprising 48 amino acids of the transit peptide (61) whereas the cDNA for FNR encodes a polypeptide comprising 55 amino acids of the transit peptide and 314 amino acids of the mature protein (62).

ATP-Synthase

The ATP-generating supra-molecular complex, ATP synthase, is made up of nine subunits, of which four are part of the membrane-bound sub-complex CFo and five of the extrinsic sub-complex CF1. Subunits α, β and ϵ of CFl and I, III, and IV, of CFo are encoded by the genes atpA (507 codons), atpB (498 codons), atpE (134 codons), atpF (184 codons), atpH (81 codons), and atpI (247 codons), respectively, which are located in two clusters (atpB, E; atpI, H, F, A) about 40 kbp away in the large single-copy region of the plastome (63, 64 and references cited therein). The atpF has been found to be split by a single intron in wheat (65), spinach (63), tobacco (66), and pea (64). A comparison of the sequence of the actual protein product of atpF with that deduced from the gene sequence reveals that a 17 amino acid peptide from the N-terminal end gets cleaved off before integration (65). The genes, atpB and atpE, overlap by 4 bp in several species though not in pea (67).

The genes for two subunits γ and δ of CFI and subunit II of CFo are designated as atpC, atpD and atpG, respectively, and these have been found to be located in the nucleus (68, 69). A cDNA encoding the complete precursor (257 amino acids) containing a 70-residue long transit peptide for the δ subunit, has been sequenced recently (54). The transit peptide shows some similarity to that of the Rieske-FeS protein, but differs from those of the luminal proteins.

A remarkable feature of the amino acid sequences deduced from these genes is the high degree of conservation not only among different plant species, but also with corresponding genes from bacterial sources (54, 63, 64).

Regulation of Expression

In this section, we shall briefly touch upon the regulatory aspects of expression of genes for the photosynthetic membranes. As mentioned earlier, these genes are located both in the chloroplast and nucleus demanding a tight, coordinated, expression which is also evident from the incompatibility of nuclear and plastid genomes in certain interspecific hybrids as studied in detail in the genus *Oenothera* (70). Although the molecular basis of various malfunctions (e.g. yellowing of hybrids) is still unknown, it is possible that these may arise either from altered signals, and thus breakdown of communication between chloroplasts and nuclei, or the failure to reassemble functional photosynthetic complexes due to improper juxtaposition. Quite apart from the problem of nuclear–organelle interaction, expression of the various genes is also affected by developmental and environmental stimuli, particularly light. A proper understanding of various control mechanisms has not developed yet, although some insight has been gained by certain recent investigations (4, 8, 71, 72) as summarised below.

Regulation of Genes of Chloroplast DNA

One extremely important finding is that the cpDNA often codes for a large primary transcript having information for more than one gene which is then processed to smaller units (Fig. 1, 11). Such organisation of information is similar to that of operons in bacteria and is reflective of the prokaryotic ancestry of the chloroplast. Structurally also, upstream regulatory sequences of chloroplast genes resemble those from prokaryotes in having the −10 and the −35 canonical sequences (73). At the 3′-end, the precise location of the processing region (i.e. where termination of mRNA may occur) is not clear although several sequences in the non-coding 3′ region show the capacity to form stem and loop structures as in prokaryotes. No poly-adenylation of transcripts takes place in chloroplasts. One reason for the very limited understanding of mechanisms of regulation of expression of chloroplast genes is the non-availability of chloroplast transformation methods. However, as an alternative, *in vitro* transcription systems have recently been developed for chloroplasts and are being employed to transcribe plastid genes (74–76) and hopefully rapid progress will be made in the near future.

We have remarked above that light strongly influences plastid development through stimulation of synthesis of certain proteins. This provides a special opportunity to investigate the control mechanism of gene expression. Recent studies have shown, however, that mRNA for most of the chloroplast-encoded genes are already present in the dark. Their levels are increased by light, but this increase is rather marginal and much less than of the several proteins which rapidly accumulate in light (77,78). By 'run-on' transcription assay, it has been demonstrated that light does not affect relative transcription activities of different plastid genes (79). Thus, light seems to exercise its effect mediated not so much at the transcriptional but at the post-transcriptional or translational levels.

The best opportunity for a study of events associated with the processing of the primary polycistronic transcripts is provided by the psaA: psaB, psbD: psbC, psbB: psbH: petB: petD and atpI: atpH: atpF: atpA operons. An understanding of such control is important since of the multiple polypeptides encoded by a single transcript (4 in case of psbB operon), one or more may accumulate even in the dark and others may need light (78). An intensive investigation into the processing of the transcript of the psbB operon (containing also the psbH, petB and petD genes) has shown a highly complex mechanism. Here, surprisingly, the processing takes place even in the dark, indicating that further controls, including those mediated by light, must operate at the post-transcriptional, translational or post-translational levels (10). However, we know very little about such controls and unravelling them is a task for the future.

Regulation of Nuclear Genes

As mentioned earlier, genomic clones for only three polypeptides, viz. chlorophyll a/b binding polypeptides of PSI and PS II and plastocyanin, have been characterised so far and they contain the typical eukaryotic expression signals in the 5′ –upstream and 3′ –untranslated, downstream, regions (30, 46, 58). Involvement of light in regulation of nuclear-encoded genes became apparent with the studies of Apel and Kloppstech (79a). Subsequently, with the availability of specific gene probes, it was found that both quantity as well as quality of light greatly influence the transcript levels for cab genes, also various members of the gene family are expressed differentially in various organs and cells (see 72, 80).

In an effort to identify the elements involved in such regulation, various 5′ –upstream regions of the cab gene have been tested by transforming a heterologous system with engineered genes in which either deletions have been made in the 5′ –upstream region or various parts of this region have been fused to a marker gene containing its own constitutive type of promoter (81–85). Simpson *et. al.* (82) reported that in pea a 400 bp fragment from the cab gene is responsible for light-regulated and organ-specific expression as is evident from experiments involving fusion between this fragment and a modified bacterial kanamycin resistance gene—both engineered into a Ti plasmid—and subsequently employed for transformation of tobacco. Out of this, a 247 bp (−100 to −347 bp upstream) stretch acts not only as a light-regulated enhancer element but also as tissue-specific silencer (83).

Lamppa *et al.* (81) have reported that a wheat cab gene maintains its light-regulated characteristics even after transfer to the more distantly related tobacco and *Petunia* indicating that the controlling elements are ubiquitous in flowering plants. A 5′ –upstream region of this gene from +31 to −1816 bp has been shown to confer light-regulated and organ-specific expression also on a bacterial chloramphenicol resistance gene. Expression of this gene is, in fact, affected reversibly by red/far-red light implying a role of phytochrome in the regulation of its expression (84). Further experiments have shown that only a 268 bp fragments from −89 to −357 bp in the 5′ –upstream region, which acts as an enhancer, is sufficient to confer the typical regulatory characteristics (85). The work mentioned above has set the stage for more detailed investigations on the mechanism of action of the enhancer elements. Such sequences from the rbcS gene have already been used to identify, by gel-retardation and foot-printing methods, possible trans-acting factor(s) responsible for light-regulation as also organ-specific expression of the genes (86). It is hoped that similar work with the cab gene in the near future will further elucidate the pathway and mechanism of transfer of a stimulus to a gene.

As in the case of cab and rbcS genes, mRNA levels of many other nuclear-encoded genes are greatly increased by illumination. However, the quantity by which mRNA levels increase is often greater than that of the increase in level of proteins for several components of the photosynthetic apparatus—e.g. mRNAs of the delta subunit of ATP synthase, the Rieske iron sulphur protein of the cytochrome b/f complex, and the 33, 23, and 16 kDa polypeptides of the water-splitting system. This indicates again that there must exist novel post-transcriptional control mechanisms for further regulation (78). Such controls may lie at the translational level or even at the level of transport of precursor proteins into the chloroplast (the latter may in turn depend on the availability of receptors) for final localisation in the supra-molecular complexes. All these putative mechanisms of control remain to be elucidated.

Strategies for Engineering Genes for Study of Structure-Function Relationship and for Biotechnology

Availability of structural information about several genes has opened a vast new area for experimentation. Combination of this information with the new recombinant DNA technology, site-specific mutagenesis, and emerging plant gene engineering techniques (see 87, 88) should allow us to unravel many mysteries concerning the structure–function relationship among various components of the photosynthetic apparatus. In principle, genes can now be modified and reinserted into a plant, and one can ask specific questions such as (a) which amino acids serve as attachment sites for ligands, (b) which amino acids or patches of amino acids interact with other polypeptides in a supra-molecular complex, (c) what are the functional roles of different regions of a polypeptide, (d) how does a transit peptide interact with receptor(s) during transport, and (e) which domain of a transit peptide is responsible for compartment-specific localisation of a polypeptide?

To answer these questions about a particular polypeptide by the genetic engineering approach, however, one needs to fulfil certain basic requirements. Firstly, a recipient plant system lacking the wildtype polypeptide is required. For this purpose, one can make use of various nuclear and plastid mutants similar to those available in *Oenothera* (1) or one can create specific defects in the recipient plant cell lines by engineering an 'antisense' copy of the concerned gene (89, 90), thereby creating a phenocopy of the organism. Although, the 'antisense' approach can probably be used to inactivate both nuclear and plastid genes, the technology in respect of nuclear genes is already available. While doing this, one should use only a part of the gene, e.g. the region coding the transit peptide, as an antisense gene so that this part of the gene can inactivate the natural gene, and yet the same gene, modified in the structural region but with a heterologous transit peptide region, can be introduced later at a new third locus and the effects studied without interference with the full mRNA which would otherwise be produced in the recipient plant.

Having obtained a suitable recipient cell line, either a plastid mutant or a phenocopy, e.g. for a nuclear gene, the second requirement is the adoption of a suitable transformation strategy to insert the foreign gene at the desired location in the cell. Ideally, one should be able to design a suitable vector and a method for direct transformation of the chloroplasts. However, such a method is not yet established. One possible way to do it, besides microinjection (91), would be to use isolated chloroplasts for direct gene transfer (92), encapsulate them in liposomes (93), and fuse with plant protoplasts (94). However, even if one were to find a vector for integrative transformation, it is highly probable that this approach may turn out to be fruitless because the chloroplast genome is tightly packed not only with overlapping genes on the same strand, but also genes on both strands (11, Fig. 1) and thus the insertion of a stretch of foreign DNA may inactivate another vital function. One alternative is to develop vectors containing the *ori* region of the plastome (95–97) so that they can independently replicate in the chloroplast and the foreign gene can be expressed.

However, another alternative, and a strategy which can be immediately applied, is to use the standard vectors currently being employed to transform nuclear genomes (88). Should one desire to engineer a gene normally coded by the chloroplast genome, it will be necessary in this approach to fuse a transit peptide sequence region and eukaryotic expression signals of a suitable nuclear gene, such as that of rbcS, cab or from plastocyanin gene, so that the engineered gene product can be automatically transported to the chloroplasts (98). Following this strategy, experiments with cyt f gene employing the expression signal and the transit peptide of rbcS gene from pea and the cyt f⁻ mutant of *Oenothera* as a recipient system have been started (Tyagi *et al.*, unpublished). It appears that foreign genes can indeed be transferred this way and products targeted to the chloroplasts. Further, one can adopt this strategy of genetic engineering to introduce agronomically desirable traits as well, like herbicide resistance, where we are dealing with proteins normally encoded by the chloroplast genome and for whose modifications and engineering there are no suitable methods yet.

Transformation of the nuclear genome can also serve for introducing modified nuclear-encoded genes. Thus, exchange and modification of different parts of a gene, either in the region of the transit peptide or the structural protein, should be able to provide important information about structural–functional aspects.

Conclusion

From the foregoing discussion, it is clear that the newly emerging methods of recombinant DNA technology and molecular biology provide a unique opportunity for understanding the structure, biogenesis, and function of the photosynthetic apparatus of the higher plants. Exploitation of this potential in recent years has already added greatly to our basic knowledge. But much remains to be done, specially on the applied front. So far, genetic engineering of plants capable of harvesting light energy more efficiently has been only a dream, but in the forseeable future one can actually begin to engineer more productive plants.

Acknowledgements

We gratefully thank Professor Dr R. G. Herrmann of the Botanisches Institut, München, and Professor Dr Peter Westhoff of the Botanisches Institut, Düsseldorf, for sending us preprints of their work. Thanks are also due to Dr (Mrs) N. Maheshwari and Mrs Kavita Munjal for valuable advice and assistance.

References

1. Herrmann, R. G., Westhoff, P., Alt, J., Tittgen, J. and Nelson, N. (1985). In: *Molecular Form and Function of the Plant Genome* (van Vloten-Doting, L., Groot, G. S. P. and Hall, T. C., eds.), pp. 233–256, Plenum Publishing Corporation, New York.
2. Palmer, J. D. (1985) *Ann. Rev. Genet.* 19, 325–354
3. Thompson, W. F., Kaufman, L. S. and Watson, J. C. (1985) *Bio Essays* 3, 153–159
4. Anderson, J. M. (1986) *Ann. Rev. Plant Physiol.* 37, 93–136
5. Schmidt, G. W. and Mishkind, M. L. (1986) *Ann. Rev. Biochem.* 55, 879–912
6. Gray, J. C. (1987). In: *Photosynthesis* (Amesz, J., ed.), pp. 319–342, Elsevier, Amsterdam
7. Hoober, J. K. (1987). In: *The Biochemistry of Plants—A Comprehensive Treatise* (Stumpf, P. K. and Conn, E. E., eds.), pp. 1–74, Academic Press, Inc., New York
8. Kuhlemeier, C., Green, P. J. and Chua, N.-H. (1987) *Ann. Rev. Plant Physiol.* 38, 221–257
9. Zurawski, G. and Clegg, M. T. (1987) *Ann. Rev. Plant Physiol.* 38, 391–418
10. Westhoff, P. and Herrmann, R. G. (1988) *Eur. J. Biochem.* 171, 551–564
11. Shinozaki, K., Ohme, M., Tanaka, M., Wakasugi, T., Hayashida, N., Matsubayashi, T., Zaita, N., Chunwangse, J., Obokata, J., Yamaguchi-Shinozaky, K., Ohto, C., Torazawa, K., Meng, B. Y., Sugita, M., Deno, H., Kamogashira, T., Yamada, K., Kusuda, J., Takaiwa, F., Kato, A., Tohdoh, N., Shimada, H. and Sugiura, M. (1986) *EMBO J.* 5, 2043–2049
12. Ohyama, K., Fukuzawa, H., Kohchi, T., Shirai, H., Sano, T., Sano, S., Umesono, K., Shiki, Y., Takeuchi, M., Chang, Z., Aota, S., Inokuchi, H. and Ozeki, H. (1986) *Nature* 322, 572–574
13. Tittgen, J., Hermans, J., Steppuhn, J., Jansen, T., Jansson, C., Andersson, B., Nechushtai, R., Nelson, N. and Herrmann, R. G. (1986) *Mol. Gen. Genet.* 204, 258–265

14. Nanba, O. and Satoh, K. (1987) *Proc. Natl Acad. Sci. USA* 84, 109–112
15. Barber, J. (1987) *Trends Biochem. Sci.* 12, 321–326
16. Bedbrook, J. R., Link, G., Coen, D. M., Bogorad, L. and Rich, A. (1978) *Proc. Natl Acad. Sci. USA* 75, 3060–3064
17. Zurawski, G., Bohnert, H. J., Whitfeld, P. R. and Bottomley, W. (1982) *Proc. Natl Acad. Sci. USA* 79, 7699–7703
18. Marder, J. B., Goloubinoff, P. and Edelman, M. (1984) *J. Biol. Chem.* 259, 3900–3908
19. Hirschberg, J. and McIntosh, L. (1983) *Science* 222, 1346–1349
20. Goloubinoff, P., Edelman, M. and Hallick, R. B. (1984) *Nucleic Acids Res.* 12, 9489–9496
21. Alt, J., Morris, J., Westhoff, P. and Herrmann, R. G. (1984) *Curr. Genet.* 8, 597–606
22. Herrmann, R. G., Alt, J., Schiller, B., Widger, W. R. and Cramer, W. A. (1984) *FEBS Lett.* 176, 239–244
23. Widger, W. R., Cramer, W. A., Hermodson, M. and Herrmann, R. G. (1985) *FEBS Lett.* 191, 186–90
24. Carrillo, N., Seyer, P., Tyagi, A. and Herrmann, R. G. (1986) *Curr. Genet.* 10, 619–624
25. Hird, S. M., Willey, D. L., Dayer, T. A. and Gray, J. C. (1986) *Mol. Gen. Genet.* 203, 95–100
26. Morris, J. and Herrmann, R. G. (1984) *Nucleic Acids Res.* 12, 2837–2850
27. Steinmetz, A. A., Castroviejo, M., Sayre, R. T. and Bogorad, L. (1986) *J. Biol. Chem.* 261, 2485–2488
28. Westhoff, P., Farchaus, J. W. and Herrmann, R. G. (1986) *Curr. Genet.* 11, 165–169
29. Dunsmuir, P., Smith, S. M. and Bedbrook, J. (1983) *J. Mol. Appl. Genet.* 2, 285–300
30. Cashmore, A. R. (1984) *Proc. Natl Acad. Sci. USA* 81, 2960–2964
31. Pichersky, E., Bernatzky, R., Tanksley, S. D., Breidenback, R. B., Kausch, A. P. and Cashmore, A. R. (1985) *Gene* 40, 247–258
32. Lamppa, G. K., Morelli, G. and Chua, N.-H. (1985) *Mol. Cell Biol.* 5, 1370–1378
33. Meyerowitz, E. M. and Pruitt, R. E. (1985) *Science* 229, 1214–1218
34. Dunsmuir, P. (1985) *Nucleic Acids Res.* 13, 2503–2518
35. Polans, N. O., Weeden, N. F. and Thompson, W. F. (1985) *Proc. Natl Acad. Sci. USA* 82, 5083–5087
36. Vallejos, C. E., Tanksley, S. D. and Bernatzky, R. (1986) *Genetics* 112, 93–105
37. Karlin-Neumann, G. A., Kohorn, B. D., Thornber, J. P. and Tobin, E. M. (1985) *J. Mol. Appl. Genet.* 3, 45–61
38. Stayton, M. M., Black, M., Bedbrook, J. and Dunsmuir, P. (1986) *Nucleic Acids Res.* 14, 9781–9796
39. Pichersky, E., Hoffman, N. E., Malik, V. S., Bernatzky, R., Tanksley, S. D., Szabo, L. and Cashmore, A. R. (1987) *Plant Mol. Biol.* 9, 109–120
40. Andersson, B. (1986). In: *Encyclopedia of Plant Physiology*, New Series (Staehelin, L. A. and Arntzen, C. J., eds.), Vol. 19, pp. 447–456, Springer-Verlag, Berlin
41. Westhoff, P., Jansson, C., Klein-Hitpass, L., Berzborn, R., Larsson, C. and Bartlett, S. G. (1985) *Plant Mol. Biol.* 4, 137–146
42. Sheen, J. Y., Sayre, R. T. and Bogorad, L. (1987) Plant Mol. Biol. 9, 217–226
43. Tyagi, A., Hermans, J., Steppuhn, J., Jansson C., Vater, F. and Herrmann, R. G. (1987) *Mol. Gen. Genet.* 207, 288–293
44. Jansen, T., Rother, C., Steppuhn, J., Reinke, H., Beyreuther, K., Jansson C., Andersson, B. and Herrmann, R. G. (1987) *FEBS Lett.* 216, 234–240
45. Lautner, A., Klein, R., Ljungberg, U., Reiländer, H., Bartling, D., Andersson, B., Reinke, H, Beyreuther, K. and Herrmann, R. G. (1988) *J. Biol. Chem.* (in Press)
46. Rother, C., Jansen, T., Tyagi, A., Tittgen, J. and Herrmann, R. G. (1986) *Curr. Genet.* 11, 171–176
47. Alt, J., Westhoff, P., Sears, B. B., Nelson, N., Hurt, E., Hauska, G. and Herrmann, R. G. (1983) *EMBO J.* 2, 979–986
48. Tyagi, A. K. and Herrmann, R. G. (1986) *Curr. Genet.* 10, 481–486

49. Alt, J. and Herrmann, R. G. (1984) *Curr. Genet.* 8, 551–557
50. Willey, D. L., Auffret, A. D. and Gray, J. C. (1984) *Cell* 36, 555–562
51. Tanaka, M., Obokata, J., Chunwongse, J., Shinozaki, K. and Sugiura, M. (1987) *Mol. Gen. Genet.* 209, 427–431
52. Widger, W. R., Cramer, W. A., Herrmann, R. G. and Trebst, A. (1984) *Proc. Natl Acad. Sci. USA* 81, 674–678
53. Steppuhn, J., Rother, C., Hermans, J., Jansen, T., Salnikow, J., Hauska, G. and Herrmann, R. G. (1987) *Mol. Gen. Genet.* 210, 171–177
54. Hermans, J., Rother, C., Bichler, J., Steppuhn, J. and Herrmann, R. G. (1988) *Plant Mol. Biol.* (in Press)
55. Fish, L. E., Kück, U. and Bogorad, L. (1985) *J. Biol. Chem.* 260, 1413–1421
56. Kirsch, W., Seyer, P. and Herrmann, R. G. (1986) *Curr. Genet.* 10, 843–855
57. Hayashida, N., Matsubayashi, T., Shinozaki, K., Sugiura, M., Inoue, K. and Hiyama, T. (1987) *Curr. Genet.* 12, 247–250
58. Pichersky, E., Hoffmann, N. E., Bernatzky, R., Piechulla, B., Tanksley, S. D. and Cashmore, A. R. (1987) *Plant Mol. Biol.* 9, 205–216
59. Herrmann, R. G., Westhoff, P., Alt, J., Winter, P., Tittgen, J., Bisanz, C., Sears, B. B., Nelson, N., Hurt, E., Hauska, G., Viebrock, A. and Sebald, W. (1983). In: *Structure and Function of Plant Genomes* (Ciferri, O. and Dure III, L., eds.), pp. 143–153, Plenum Publishing Corporation, New York
60. Smeekens, S., de Groot, M., van Binsbergen, J. and Weisbeek, P. (1985) *Nature* 317, 456–458
61. Smeekens, S., van Binsbergen, J. and Weisbeek, P. (1985) *Nucleic Acids Res.* 13, 3179–3195
62. Jansen, T., Reiländer, H., Steppuhn, J. and Herrmann, R. G. (1988) *Curr. Genet.* (in Press)
63. Hennig, J. and Herrmann, R. G. (1986) *Mol. Gen. Genet.* 203, 117–128
64. Hudson, G. S., Mason, J. G., Holton, T. A., Koller, B., Cox, G. B., Whitfeld, P. R. and Bottomley, W. (1987) *J. Mol. Biol.* 196, 283–298
65. Bird, C. R., Koller, B., Auffret, A. D., Huttly, A. K., Howe, C. J., Dyer, T. A. and Gray, J. C. (1985) *EMBO J.* 4, 1381–1388
66. Shinozaki, K., Deno, H., Wakasugi, T. and Sugiura, M. (1986) *Curr. Genet.* 10, 421–423
67. Zurawski, G., Bottomley, W. and Whitfeld, P. R. (1986) *Nucleic Acids Res.* 14, 3974
68. Westhoff, P., Nelson, N., Bunemann, H. and Herrmann, R. G. (1981) *Curr. Genet.* 4, 109–120
69. Westhoff, P., Alt, J., Nelson, N. and Herrmann, R. G. (1985) *Mol. Gen. Genet.* 144, 290–299
70. Stubbe, W. (1959) *Z. Vererbungsl.* 90, 288–298
71. Harpster, M. and Apel, K. (1985) *Physiol. Plant.* 64, 147–152
72. Tobin, E. M. and Silverthorne, J. (1985) *Ann. Rev. Plant Physiol.* 36, 569–593
73. Hanley-Bowdoin, L. and Chua, N.-H. (1987) *Trends Biochem. Sci.* 12, 67–70
74. Link, G. (1984) *EMBO J.* 1697–1704
75. Orozco Jr, E. M., Mullet, J. E. and Chua, N.-H. (1985) *Nucleic Acids Res.* 13, 1283–1302
76. Gruissem, W., Greenberg, B. M., Zurawski, G. and Hallick, R. B. (1986) *Methods in Enzymol.* 118, 253–270
77. Klein, R. R. and Mullet, J. E. (1987) *J. Biol. Chem.* 262, 4341–4348
78. Westhoff, P., Grüne, H., Schrubar, H., Oswald, A., Streubel, M., Ljungberg, U. and Herrmann, R. G. (1988). In: *Photosynthetic Light-Harvesting Systems: Structure and Function* (Schees, H. and Schnerdes, S., eds.), Walks de Gruyter Verlag, Berlin (in Press)
79. Deng, X. W. and Gruissem, W. (1987) *Cell* 49, 379–387
79a. Apel, K. and Kloppstech, K. (1978) *Eur. J. Biochem.* 85, 581–588
80. Fluhr, R., Kuhlemeier, C., Nagy, F. and Chua, N.-H. (1986) *Science* 232, 1106–1112

81. Lamppa, G. K., Nagy, F. and Chua, N.-H. (1985) *Nature* 316, 750–752
82. Simpson, J., Timko, M. P., Cashmore, A. R., Schell, J., Van Montagu, M. and Herrerra-Estrella, L. (1985) *EMBO J.* 4, 2723–2729
83. Simpson, J., Schell, J., Van Montagu, M. and Herrera-Estrella, L. (1986) *Nature* 323, 551–554
84. Nagy, F., Kay, S. A., Boutry, M., Hsu, M. Y. and Chua, N.-H. (1986) *EMBO J.* 5, 1119–1124
85. Nagy, F., Boutry, M., Hsu, M. Y., Wong, M. and Chua, N.-H. (1987) *EMBO J.* 6, 2537–2542
86. Green, P. J., Kay, S. A. and Chua, N.-H. (1987) *EMBO J.* 6, 2543–2549
87. Fraley, R. T., Rogers, S. G. and Horsch, R. B. (1986) *CRC Critical Rev. Plant Sci.* 4, 1–46
88. Schell, J. S. (1987) *Science* 237, 1176–1182
89. Green, P. J., Pines, O. and Inouye, M. (1986) *Ann. Rev. Biochem.* 55, 569–597
90. Rothstein, S. J., DiMaio, J., Strand, M. and Rice, D. (1987) *Proc. Natl Acad. Sci. USA* 84, 8439–8443
91. Crossway, A., Hauptli, H., Houck, C. M., Irvine, J. M., Oakes, J. V. and Perani, L. A. (1986) *BioTechniques* 4, 320–334
92. Krüger-Lebus, S. and Potrykus, I. (1987) *Plant Mol. Biol.* Rep. 5, 289–294
93. Giles, K. L., Vaughan, V., Ranch, J. P. and Emery, J. (1980). *In Vitro* 16, 581–584
94. Binding, H., Krumbiegel-Schroeren, G. and Nehls, R. (1986). In: *Differentiation of Protoplasts and of Transformed Plant Cells* (Reinert, J. and Binding, H., eds.), pp. 37–66, Springer-Verlag, Berlin
95. de Hass, J. M., Boot, K. J. M., Haring, M. A., Kool, A. J. and Nijkamp, H. J. J. (1986) *Mol. Gen. Genet.* 202, 48–54
96. Wu, M., Lou, J. K., Chang, D. Y., Chang, C. H. and Nie, Z. Q. (1986) *Proc. Natl Acad. Sci. USA* 83, 6761–6765
97. Gold, B., Carillo, N., Tewari, K. K. and Bogorad, L. (1987) *Proc. Natl Acad. Sci. USA* 84, 194–198
98. Cashmore, A. R., Szabo, L., Timko, M., Kausch, A., Van den Broeck, G., Schreier, P., Bohnert, H., Herrera-Estrella, L., Van Montagu, M. and Schell, J. (1985) *Biotechnology* 3, 803–808

Specific Mutagenesis as a Tool for the Analysis of Structure/Function Relationships in Photosystem II

WIM VERMAAS, SHELLY CARPENTER AND CANDACE BUNCH

Arizona State University
Department of Botany
Tempe AZ 85287-1601, USA

Summary

Mutants of the cyanobacterium *Synechocystis* sp., PCC 6803 have been created in which a gene encoding one of the components of the Photosystem II core complex has been specifically altered. The mutations can be made using any of the following concepts: (a) gene interruption or gene deletion, (b) replacement of part or all of a cyanobacterial Photosystem II gene by the corresponding part of a higher plant gene, and (c) mutagenesis that leads to the specific replacement of a single amino acid residue (site-directed mutagenesis). *Synechocystis* sp., PCC 6803 is particularly suitable for such mutagenesis studies since it is one of the few organisms known that (a) is photoheterotrophic (can be propagated in the absence of Photosystem II activity in the presence of glucose), and (b) is transformable and can incorporate foreign DNA into its genome by homologous recombination. Thus, specific Photosystem II mutants can be made by transformation, using appropriate DNA constructions, and can be subsequently propagated. The methodology for the creation of these mutants will be described along with examples illustrating the use of this approach to answer specific questions regarding the function of Photosystem II components. The analysis on one of the mutains indicates that the C-terminal half of CP47 is probably involved in energy transfer from phycobilisomes to PS II. From the results obtained with site-directed D2 mutants it has been concluded that the Photosystem II donor D is the Tyr-160 residue in this protein.

Introduction

Photosystem II (PS II) is the pigment-protein complex catalysing the light-induced reduction of plastoquinone by water in higher plants, algae and cyanobacteria. The PS II activity depends on precisely balanced interactions among at least six proteins (five of which are in the thylakoid membrane), and between prosthetic groups and these proteins. Without these delicate interactions, highly efficient energy transfer and specific redox reactions in PS II would not be possible. The proteins involved in PS

II activity in cyanobacteria include (a) CP47 and CP43, two proteins that bind a number of chlorophylls that serve as an antenna for the PS II reaction center chlorophyll, (b) D1 and D2, the putative PS II reaction center proteins with apparent molecular masses of 32–34 kDa, (c) cytochrome b-559, of as yet unknown function, and (d) the manganese-stabilising protein (the '33 kDa protein'), an extrinsci protein. CP47, CP43, D1, D2 and cyt b-559 together form a complex, the PS II core complex, that spans the thylakoid membrane. For detailed information regarding aspects of the function and structure of the core complex, the reader is referred to reviews of this area (for example, 1–6). However, a number of important aspects of PS II function and structure (binding sites of prosthetic groups, arrangement of the proteins in the complex, protein complex assembly) have not yet been elucidated satisfactorily.

The analysis of PS II mutants in many instances has contributed to a better understanding of how the PS II complex functions and assembles, and how its synthesis may be regulated (for example, 7-11). However, virtually all mutants that have been characterised were selected after random mutagenesis and in many cases it has not been established in which gene(s) the mutations were located; the matter is complicated further by the fact that pleiotropic effects of a mutation are frequently observed (for example, a mutation in a PS II gene may lead to the disappearance of a number of PS II proteins from the thylakoid 8–10). This causes the unambiguous elucidation of the role of a specific protein by classical means of mutagenesis and mutant analysis to be difficult.

However, the development of molecular-genetic techniques over the last few years has led to the opportunity to induce specific mutations in a single gene in certain organisms without affecting any other gene in the genome. This can be used to specifically modify a gene, introduce it into the organism, and analyse the effect of the mutation on the phenotype. Thus, a specific phenotype can be directly correlated with a mutation in one of the genes.

This paper deals with the introduction of specific mutations into the genes coding for PS II components of the cyanobacterium *Synechocystis* sp., PCC 6803 (to be referred to as *Synechocystis* 6803 in this paper) and the subsequent characterisation of the mutants thus generated. Two factors contribute to this cyanobacterium being very suitable for such work. (a) In the first place, because of its natural transformation and recombination system (12–14) it is possible to simply add the DNA that needs to be inserted into the cyanobacterial genome to the cells. A prerequisite for site-specific integration into the genome is that the DNA to be inserted needs to be flanked by sequences identical to those next to the site where the insertion of foreign DNA should take place. Using similar procedures, specific parts of the cyanobacterial genome can also be deleted, or can be replaced by foreign DNA. (b) *Synechocystis* 6803 is a photoheterotrophic organism, which means that it can be propagated (in the light) in the absence of PS II activity in the presence of a carbon source (glucose in this case) (15). Thus, even PS II mutants which are no longer capable of photosynthetic growth can be generated and propagated for a detailed functional and structural analysis.

In contrast to the *Synechocystis* 6803 system, genetic modification of higher plant PS II genes and subsequent expression of the modified genes in the organism is still very difficult due to the absence of a reliable procedure for chloroplast transformation in higher plants (the genes for all PS II core components in eukaryotes are located on the chloroplast DNA). Because of the large similarity between plant and cyanobacterial PS II, it is anticipated that results obtained with the *Synechocystis* 6803 PS II complex can be extrapolated to provide reliable and precise insight into the function and structure of PS II in higher plants.

Materials and Methods

Cyanobacterial Growth and Transformation

Growth conditions of *Synechocystis* 6803 have been described before (16). Most laboratories that have started to cultivate this cyanobacterium over the last few years have not encountered any major problems. In our hands, the doubling time of wild type *Synechocystis* 6803 in logarithmic phase is 10–12 hours under photoautotrophic conditions. Photoheterotrophically (i.e., in the presence of 5 mM glucose), growth is as fast as under photoautotrophic conditions, but the cell density in stationary phase tends to be about two-fold higher. Problems with propagation of this cyanobacterium can generally be traced to (a) the light intensity (the organism does not grow very well if the light intensity is too high; in our laboratory, two 60 W cool-white fluorescent tubes (Sylvania F96T12/CW/HO), placed 1.3 m above the cells, are used for illumination), (b) the pH of the growth medium (10 mM N-tris(hydroxymethyl)methyl-2-aminoethanesulfonic acid (TES)/NaOH pH=8.2 is added to solid media; liquid media generally do not require this), or (c) the presence of sodium thiosulfate (0.3% sodium thiosulfate in solid media increases the plating efficiency; the effects of thiosulfate on growth in liquid media are minimal in our experience). Of course, the use of proper sterile procedures is critical, especially when the media contain glucose.

Transformation

The transformation procedure has been described (13,16). In contrast to most other bacteria, no pretreatment of *Synechocystis* 6803 is necessary to induce the uptake of foreign DNA. To obtain efficient homologous recombination, the homologous regions flanking the DNA that should be incorporated into the genome can be relatively short. A total of 500 base pairs of the homologous flanking sequence is generally sufficient. If the flanking sequence on one side is relatively short (100 base pairs or less), a long flanking sequence on the other side significantly improves the transformation efficiency.

Results and Discussion

Three mechanisms for directed gene modification will be outlined in this section and a few examples of mutant characterisation will be presented.

Gene Interruption/Deletion

Using this method, a gene can be selectively inactivated in the cyanobacterial genome, and thus a mutant can be created in which the functional genetic information for one protein is no longer present. The general procedure to induce such a gene interruption or deletion in *Synechocystis* is outlined in Fig. 1. In principle, this method can be applied to any gene that has been cloned. Interruption of *psb*B (16,17) or *psb*C (18) (genes coding for CP47 and CP43, respectively), or deletion of *psb*E/F (encoding the two subunits of cyt b-559) (19), both copies of the D2-encoding *psb*D gene (20), or the three *psb*A copies (coding for D1) (21) has revealed that expression of at least one copy of each of these genes is required for PS II activity. Interestingly, interruption or deletion of a specific gene not only leads to the absence of the protein coded for by this gene from the thylakoid membrane, but in many cases also induces the disappearance of other PS II proteins from the thylakoid. Only after interruption of *psb*C can the other PS II core proteins still be detected in the thylakoid membrane, albeit in significantly reduced amounts. Deletion of the two *psb*D copies leads to a loss of all PS II proteins assayed (D1, D2, CP43 and CP47) from the thylakoid (18), whereas after the interruption of *psb*B, CP43 (but not D1 and D2) can still be detected (17,18). Thus, even though the features of the gene interruption/deletion mutants are generally rather 'boring' in a functional sense (no PS II reaction centre activity, except probably for the *psb*C interruption mutant), they may be of interest for the study of the *de novo* assembly of PS II proteins into a complex in the thylakoid membrane. On the basis of the data obtained with the *Synechocystis* mutants, a model for the assembly of the PS II reaction centre in cyanobacteria has been proposed (18): it has been hypothesised that D1 and D2 together create a 'pre-complex' which is stablised by the subsequent incorporation of CP47; CP43 would be incorporated independently of CP47 and further stablise the complex (18).

These gene interruption and deletion experiments provide some valuable (although arguably rather coarse) insight into the role of the

A. **GENE INTERRUPTION:**

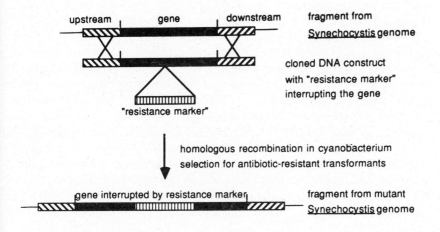

B. **GENE DELETION:**

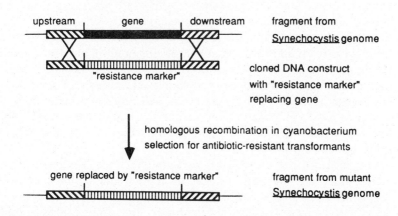

Fig. 1: Procedure to create gene interruption and gene deletion mutants in *Synechocystis*. The gene of interest is cloned out, and is either interrupted (A) or replaced (B) by a DNA fragment that contains a gene that breaks down a specific antibiotic to which *Synechocystis* is sensitive. Upon transformation, this DNA will be integrated into the cyanobacterial genome by homologous recombination. Mutants can be selected by screening for resistance against the antibiotic.

individual PS II core proteins in PS II assembly and function, and provide the basis for more delicate alterations in the PS II genes as described in the next two sections.

Creation of Chimeric Mutants

An interesting opportunity to induce a number of amino acid residue changes in one protein at the same time is the replacement of part or all of a *Synechocystis* PS II gene by the same gene from another organism. The PS II genes code for proteins that are relatively conserved between evolutionary distant species, probably reflecting the precisely balanced interaction between proteins as well as between proteins and prosthetic groups that is a prerequisite for efficient energy capture and electron transport. *Synechocystis* 6803 and spinach share more than 75% amino acid sequence identity for CP47, CP43 and D2 (16,22,23). Replacement of a native *Synechocystis* gene by the homologous gene from a higher plant could conceivably lead to mutants with interesting properties: mutants thus created could be photosynthetically active, but have functional modifications (*e.g.*, in redox kinetics or energy transfer) in comparison to wild type due to an 'imperfect fit' of the higher plant protein in the cyanobacterial PS II complex.

We have generated chimeric *psb*B mutants by transformation of wild-type cyanobacteria with spinach *psb*B coupled (downstream) to a neomycin phosphotransferase gene and surrounded by cyanobacterial DNA from up- and downstream of *psb*B (18). Since expression of the neomycin phosphotransferase gene leads to resistance to kanamycin, transformants can be identified by selection for kanamycin resistance. In kanamycin-resistant transformants recombination will have occurred in the homologous region downstream of *psb*B as well as either in the *psb*B gene (heterologous recombination) or in the region upstream of *psb*B (homologous recombination). The former event will result in a mutant with its 5' end of *psb*B of cyanobacterial origin and its 3' end from spinach, whereas the latter event will lead to a mutant with the entire *psb*B gene being from spinach.

We have not yet identified a *psb*B mutant that is photosynthetically active and that has been proven to contain the entire spinach *psb*B sequence. To address the problem whether spinach *psb*B can support PS II activity in the cyanobacterial core complex, and, if not, which region(s) of *psb*B need to be of cyanobacterial origin, a mutant has been created in which *psb*B has been deleted and replaced by the neomycin phosphotransferase gene (Fig. 2): an *Nco*I site was induced at the start of the *psb*B gene and a 2 kb *Nco*I/*Nco*I cyanobacterial fragment (containing *psb*B (1.5 kb) and 0.5 kb downstream of the gene) was replaced by a 1.2 kb fragment containing the neomycin phosphotransferase gene. This deletion mutant has been subsequently transformed with various constructs containing spinach *psb*B and cyanobacterial DNA from up- and downstream of *psb*B. Thus far, none of these transformation experiments has yielded mutants with functional PS II, suggesting that spinach *psb*B in its entirety may not be able to functionally replace the gene from *Synechocystis* in this

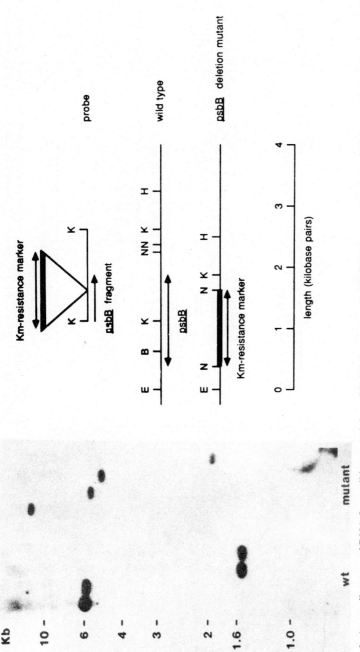

Fig. 2: Autoradiograms of DNA from wild type and the *psb*B deletion mutant (*psb*B-D1) that was size-separated after restriction digestion with *Bam*HI (B), *Eco*RI + *Bam*HI (E/B), *Kpn*I (K), or *Eco*RI + *Kpn* (E/K) and subsequently probed with a *psb*B gene fragment interrupted by the 'Km-resistance marker' (see below). In the deletion mutant the *psb*B gene has been replaced by a DNA fragment containing a gene for neomycin phosphotransferase, leading to kanamycin resistance. This DNA fragment is referred to as 'Km-resistance marker' in the figure. A schematic representation of selected restriction sites in the *psb*B region in wild type and mutant has been included as a reference for the interpretation of the autoradiogram. In wild type and mutant there is a *Bam*HI site 3.3 kb beyond the *Hind*III site indicated in the figure.

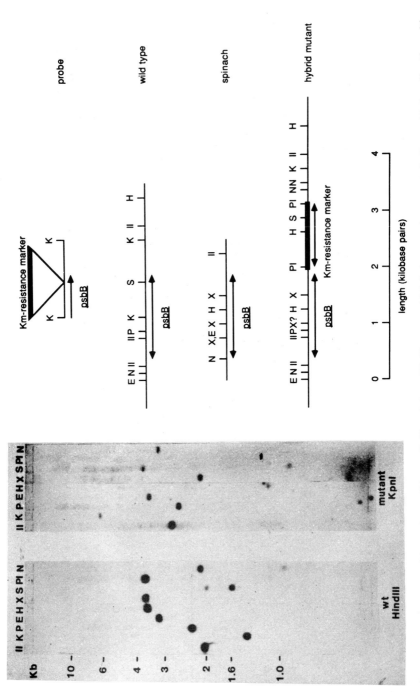

Fig. 3: Autoradiograms of DNA from wild type (left) and the *psbB*-C1 mutant (right) that was size-separated after restriction digestion with *Hind*III (wild type) or *Kpn*I (mutant) and *Hinc*II (II), *Kpn*I (K), *Pvu*II (P), *Eco*RI (E), *Hind*III (H), *Xmn*I (X), *Sma*I (s), *Pst*I (PI) or *Nco*I (N), and subsequently probed with a *psbB* gene fragment interrupted by the 'Km-resistance marker' (see Fig. 2). A schematic representation of the

cyanobacterium. The regions in *psb*B that possibly need to be of cyanobacterial origin for PS II activity are currently being identified by transformation of the non-photoautotrophic mutants (containing spinach *psb*B) with various regions of the cyanobacterial *psb*B gene followed by selection for photoautotrophic transformants.

The original transformation experiments of wild type *Synechocystis* with constructions containing spinach *psb*B have yielded various photoautotrophic mutants. One of these (*psb*B-C1; 'C' stands for chimeric) was selected on the basis of slower photoautotrophic growth rates, and was further characterised. To determine how much of the spinach *psb*B sequence had been incorporated into the cyanobacterial genome and had replaced the native gene sequence, Southern blots of wild type and mutant DNA cut with selected restriction enzymes were compared (Fig. 3). It appears that the recombination site in this mutant is between the bases 563 and 765 of the *psb*B gene (numbering starting at the ATG start codon): the *Pvu*II site at base 563 after the start of cyanobacterial *psb*B is present in the mutant and is absent in spinach *psb*B whereas the *Kpn*I site present at base 765 in cyanobacterial *psb*B and absent in the spinach gene is not detectable in the mutant. The *psb*B gene from *Synechocystis* is 1521 base pairs long; thus, the recombination appears to have occurred close to the middle of the gene. The results of restriction digestions with other appropriate enzymes corroborate this conclusion (Fig. 3).

Although this mutant can grow under photoautotrophic conditions, it was observed that its growth was about two-fold slower than that of wild type under such conditions. We used low temperature fluorescence measurements to test the assumption that this effect could be caused by changes in light-harvesting mechanisms in the mutant. The flourescence emission spectrum with 600 nm (phycobilisome) excitation revealed a relatively high emission at 683 nm (Fig. 4), presumably originating from allophycocyanin (24, and references therein). The emission at 683 nm should not be confused with the (under these conditions much less intense) fluorescence at 685 and 695 nm originating from PS II chlorophylls. The contribution of light absorbed by phycobilisomes towards PS I fluorescence (at 730 nm) is relatively low compared to wild type (Fig. 4). For a further comparison, fluorescence emission spectra from a mutant lacking all of the PS II core components (mutant *psb*DC-D1; deletion of both copies of *psb*D and of *psb*C and from a mutant (*psb*B-I1) in which *psb*B has been interrupted and which lacks CP47, D1 and D2 in the thylakoid membrane (18) have been shown (Fig. 4). It is obvious from these data that the chimeric mutant has a higher allophycocyanin fluorescence as well as a lower contribution of phycobilisomes to PS I flourescence than even the mutant lacking the entire PS II core complex. This implies that the chimeric *psb*B mutant does not have even as much energy transfer from the phycobilisomes to the thylakoid as a mutant lacking PS II entirely. This perhaps unexpected result can be explained by the following rationale. In

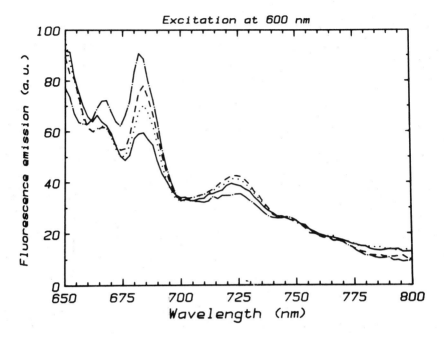

Fig. 4: Fluorescence emission spectrum of intact cells from wild type (———), the chimeric mutant *psb*B-C1 (–·–·), the *psb*DC deletion mutant (*psb*DC-D1; – – – –), and a mutant with an interrupted *psb*B gene (*psb*B-I1; ·····) at 77 K. The excitation wavelength was 600 nm (absorbed mainly by phycobilins). The excitation bandwidth was 18 nm; the emission bandwidth was 3.6 nm. The spectra were normalized at 700 nm and were corrected for wavelength dependence of photomultiplier sensitivity.

the mutant lacking PS II, no phycobilisomes can attach to PS II and may attach—although possibly less specifically—to PS I components instead. This is reflected in the somewhat higher contribution of light absorbed by phycobilisomes to PS I emission compared to wild type. In the chimeric mutant, however, PS II is still present in the thylakoid membrane (18) and the phycobilisome can still attach to the PS II complex, but because of the mutation the native interaction between the phycobilisomes and PS II appears to be disturbed resulting in a low efficiency of energy transfer from the phycobilisome to PS II. This can explain the relatively high allophycocyanin fluorescence and the low PS I emission upon excitation at 600 nm. These results indicate that the C-terminal half of CP47 might play a role in the energy transfer from the phycobilisomes to PS II. This C-terminal half includes a long stretch of rather hydrophilic residues that is relatively poorly conserved between spinach and *Synechocystis* (16). Further studies will be needed to more precisely correlate a region of CP47 with the phenotype observed.

Even though the creation of chimeric mutants generally implies the alteration of a relatively large number of base pairs at the same time (thus making precise correlations between an amino acid residue and a specific function difficult if not impossible), chimeric mutants can be of interest to find correlations between protein regions and a specific function. Subsequently, the amino acid residue(s) that may be specifically involved in this function can be determined by site-directed mutagenesis (see next section). Moreover, as this study shows, the creation of chimeras can lead to mutants with interesting phenotypes that would not have been found otherwise.

Site-directed Mutagenesis

When sufficient data have been gathered on the presumable structure of PS II proteins in relation to their function, and hypotheses can be formulated, residues of (presumable) importance can be mutated and the effects of the mutation on PS II function can be determined.

The development of modern concepts regarding PS II structure and function has greatly benefitted from (a) the availability of a high-resolution structure of the reaction centre from purple bacteria, as determined by X-ray diffraction pattern analysis of crystallized reaction centres, and (b) information on the primary structure of the PS II core proteins from various organisms, as deduced from the DNA sequence of the corresponding genes. The elucidated details of the bacterial reaction centre structure (25–30), coupled with the realisation that there is a significant regional sequence identity between the L and M subunits of the bacterial reaction centres and the D1 and D2 proteins of the PS II complex (31–33), have led to detailed hypotheses regarding the function of various residues in the D1 and D2 proteins. For example, the reaction centre pigment P680 in PS II might be bound by symmetrical His residues in the presumptive 4th helices of D1 and D2 just as the primary donor of the bacterial reaction centre is bound by two 4th helix His residues, one in L and one in a symmetrical position in M.

To test the hypothesis that the D2 His-197 residue from *Synechocystis* (located at a position homologous to the M-subunit His residue that binds a component of the primary donor) is indeed involved in binding P680, this D2 residue was mutated to Tyr. This was accomplished by first making a cyanobacterial mutant in which both copies of *psb*D had been replaced by genes leading to resistance to an antibiotic (using a method similar to that outlined in Fig. 1), and subsequently replacing one of the antibiotic-resistance genes by a copy of *psb*D containing a single-base pair change in the codon encoding the His-197 residue so that it codes for a Tyr residue instead (20,34). The methods for site-directed mutagenesis using (a) derivatives of the single-stranded DNA bacteriophage M13 containing the gene fragment to be mutagenised and (b) oligonucleotides with a single mismatch (compared to the sequence of the gene fragment to be mutagenised) at the site of the intended mutation has been described in detail (35–37). The single-site mutation resulting in a change of His-197 into Tyr was found to lead to a loss of primary PS II activity as well as to a loss of assembly of most of the PS II complex (18.34). The mutant *psb*D fragment was sequenced and the only mutation found was at the desired location. Transformation of this mutant with a 0.5 kb wild type *psb*D fragment containing the codon for the His-197 residue restored the capability of photoautotrophic growth, thus confirming that the phenotype of the mutant indeed was due to the site-directed mutation rather than to some random mutation unrelated to this small region of the *psb*D gene. Using the same methodology, His-214 of D2, the His-residue that appears to be in a homologous position compared to the His residue of M that binds the first quinone-type electron acceptor, Q_A, was changed into Asn. This also led to PS II inactivation and to a lack of PS II assembly (18,34), both of which were reversible upon transformation with a wild-type *psb*D fragment covering the region coding for His-214 (34).

Although these data do not provide direct proof that the mutated His residues are indeed the primary ligands for crucial redox reactants in PS II, they do indicate that these His residues are necessary for proper structure and function of PS II. However, it is obvious that, since the mutation of a single residue creates such a dramatic change in PS II function and stability, the function of these residues is extremely important and most likely these His residues indeed are the primary ligands for the special pair and for Q_A, just as the homologous residues from the M subunit. The dramatic effects of the single-site mutation on PS II structure may also be indications of the importance of prosthetic groups in the proper assembly of PS II: when the prosthetic groups can no longer be bound, protein stability may be drastically decreased. Similar phenomena have been observed in the case of pigment binding to antenna complexes (38–40).

In spite of the good homology between the electron acceptor sides of photosynthetic bacteria and PS II, the electron donor side is functionally and structurally different in the two systems, thus excluding the possibility that the information obtained from bacterial crystal structure analysis can be used to obtain ideas regarding residues possibly involved in electron transport at the donor side of PS II. In this case, the targeting of residues for site-directed mutagenesis experiments is somewhat less straightforward, but even so, two groups (Debus, Barry, Babcock and McIntosh at Michigan State University, and our group at Arizona State University) have shown independently and virtually concomitantly that D, the slow donor to P680 and the entity giving rise to EPR Signal II_s upon oxidation, is likely to be the Tyr-160 residue in D2: mutation of this residue into Phe eliminated Signal II_s (41,42). Also based on the results of recent studies on the effect of tyrosine deuteration on Signal II_s (43), D is now assumed to be Tyr-160 in the D2 protein. The Tyr-160-Phe mutant still grew photoautotrophically, even though at three-or four-fold lower rates. This inhibited growth could not have been caused by a decreased stability of PS II since the number of PS II centres (as measured by the number of binding sites for the PS II-directed herbicide diuron) was not greatly affected. The reason for the decreased growth rate of the mutant lacking D is still unclear.

To make sure that the disappearance of D was specifically correlated with the Tyr-160-Phe mutation, a residue next to Tyr-160, Met-159, was mutated to Asn. In this Met-159–Asn, mutant EPR Signal II_s was still detectable, and the growth rate was virtually normal. Thus, Signal II_s appears to be specifically correlated with the Tyr-160 residue of D2. By analogy, the physiological electron donor to P680, Z, is expected to be the homologous Tyr residue in D1, since Z^+ gives rise to an EPR signal that is of virtually identical shape compared to that of D^+.

To our knowledge, D is the first example in the photosynthetic electron transport chain of an amino acid residue that serves as a redox reactant and remains stable in its oxidized form for an extended period of time. Thus, the D2 protein not only functions as the protein providing the proper binding environment for the prosthetic groups and cofactors for some of the PS II reactions, but also takes part directly in one of the redox reactions. Therefore, according to the biochemical definition, D2 is not merely an enzyme, but at the same time it is a substrate.

Concluding Remarks

The above results indicate that site-directed and chimeric mutants can be helpful in providing a detailed analysis of the role specific proteins and amino acid residues therein play in the various reactions and in the assembly of PS II. However, by no means the creation of mutants and the genetic analysis of such mutants should be the primary aim of the experiments. Once the mutants have been created, a thorough functional and structural analysis can be performed and it is only at this stage that the results will have a significant impact on photosynthesis research as well as on the current views and hypotheses in this field. Nonetheless, in the relatively short time that directed mutagenesis has been applied to PS II, a significant amount of information has been gathered applying these techniques whereas such information could not have been easily obtained otherwise.

Even though this paper has not dealt with any modifications of PS II that would be advantageous for crop production, it is hoped that future information obtained from the cyanobacterial system can be applied to higher plant systems once the proper procedures for modification of genetic information localised in the chloroplast have been developed.

Acknowledgements

This research was supported, in part, by a grant 'from the University Research Fund at Arizona State University.

References

1. Mathis, P. and Rutherford, A. W. (1987) In:*Photosynthesis* (J. Amesz, ed.), pp. 63–96, Elsevier, Amsterdam
2. Vermaas, W. F. J., Pakrasi, H. B. and Arntzen, C. J. (1987) In: *Models in Plant Physiology and Biochemistry*, Vol. I (D. W. Newman and K. G. Wilson, eds.), pp. 9–12, CRC Press, Boca Raton
3. Diner, B. A. (1986) In: *Photosynthesis III, Encyclopedia in Plant Physiology*, New Series, Vol. 19 (L. A. Staehelin and C. J. Arntzen, eds.), pp. 422–436, Springer, Berlin
4. Arntzen, C. J. and Pakrasi, H. B. (1986) In: *Photosynthesis III, Encyclopedia in Plant Physiology*, New Series, Vol. 19 (L. A. Staehelin and C. J. Arntzen, eds.), pp. 457–467, Springer, Berlin
5. Bryant, D. A. (1986) In: *Photosynthetic Picoplankton* (T. Platt and W. K. W. Li, eds.), pp. 423–500, Department of Fisheries and Oceans, Ottawa
6. Satoh, K. (1985) *Photochem. Photobiol.* 42, 845–853
7. Metz, J. G., Pakrasi, H. B., Seibert, M. and Arntzen, C. J. (1986) *FEBS Lett.* 205, 269–274
8. Erickson, J. M., Rahire, M., Malnoe, P., Girard-Bascou, J., Pierre, Y., Bennoun, P. and Rochaix, J. -D. (1986) *EMBO J.* 5, 1745–1754
9. Bennoun, P., Spierer-Herz, M., Erickson, J., Girard-Bascou, J., Delosme, M. and Rochaix, J. -D. (1986) *Plant Mol. Biol.* 6, 151–160
10. Jensen, K. H., Herrin, D. L., Plumley, F. G. and Schmidt, G. W. (1986) *J. Cell Biol.* 103, 1315–1325
11. Leto, K. J., Bell, E. and McIntosh, L. (1985) *EMBO J.* 4, 1645–1653
12. Grigorieva, G. and Shestakov, S. (1982) *FEMS Microbiol. Lett.* 13, 367–370
13. Williams, J. G. K. (1988) *Meth. Enzymol.*, in press
14. Golden, S. S., Brusslan, J. and Haselkorn, R. (1987) *Meth. Enzymol.* 153, 215–231
15. Rippka, R., Deruelles, J., Waterbury, J. B., Herdman, M. and Stanier, R. T. (1979) *J. Gen. Microbiol.* 111, 1–61
16. Vermaas, W. F. J., Williams, J. G. K. and Arntzen, C. J. (1987) *Plant Mol. Biol.* 8, 317–326
17. Vermaas, W. F. J., Williams, J. G. K., Rutherford, A. W., Mathis, P. and Arntzen, C. J. (1986) *Proc. Natl Acad. Sci.* USA 83, 9474–9477
18. Vermaas, W. F. J., Ikeuchi, M. and Inoue, Y. (1988) *Photosynth. Res.*, in press
19. Pakrasi, H. B., Williams, J. G. K. and Arntzen, C. J. (1987) In: *Progress in Photosynthesis Research*, Vol. 4 (J. Biggins, ed.), pp. 813–816, Martinus Nijhoff, Dordrecht
20. Vermaas, W. F. J., Williams, J. G. K., Chisholm, D. A. and Arntzen, C. J. (1987) In : *Progress in Photosynthesis Research*, Vol. 4 (J. Biggins, ed.), pp. 805–808, Martinus Nijhoff, Dordrecht
21. Jansson, C., Debus, R. J., Osiewacz, H. D., Gurevitz, M. and McIntosh, L. (1987) *Plant Physiol.* 85, 1021–1025
22. Chisholm, D. and Williams, J. G. K. (1988) *Plant Mol. Biol.* 10, 293–301
23. Williams, J. G. K. and Chisholm, D. A. (1987) In: *Progress in Photosynthesis Research*, Vol. 4 (J. Biggins, ed.), pp. 809–812, Martinus Nijhoff, Dordrecht
24. Mohanty, P., Hoshina, S. and Fork, D. C. (1985) *Photochem. Photobiol.* 41, 589–596
25. Deisenhofer, J., Epp, O., Miki, K., Huber, R. and Michel, H. (1985) *Nature* 318, 618–624
26. Michel, H., Epp, O. and Deisenhofer, J. (1986) *EMBO J.* 5, 2445–2451
27. Allen, J. P., Feher, G., Yeates, T. O., Rees, D. C., Deisenhofer, J., Michel, H. and Huber, R. (1986) *Proc. Natl Acad. Sci.* USA 83, 8589–8593
28. Chang, C. -H., Tiede, D., Tang, J., Smith, U., Norris, J. and Schiffer, M. (1986) *FEBS Lett* 205, 82–86
29. Allen, J. P., Feher, G., Yeates, T. O., Komiya, H. and Rees, D. C. (1987) *Proc. Natl Acad. Sci.* USA 84, 5730–5734

30. Allen, J. P., Feher, G., Yeates, T. O., Komiya, H. and Rees, D. C. (1987) *Proc. Natl Acad. Sci.* USA 84, 6162–6166
31. Rochaix, J. -D., Dron, M, Rahire, M. and Malnoe, P. (1984) *Plant Mol. Biol.* 3, 363–370
32. Michel, H. and Deisenhofer, J. (1986) In: *Photosynthesis III*, Encyclopedia in Plant Physiology, New Series, Vol. 19 (L. A. Staehelin and C. J. Arntzen, eds.), pp. 371–381, Springer, Berlin
33. Hearst, J. (1986) In: *Photosynthesis III, Encyclopedia in Plant Physiology*, New Series, Vol. 19 (L. A. Staehelin and C. J. Arntzen, eds.), pp. 382–389, Springer, Berlin
34. Vermaas, W. F. J., Williams, J. G. K. and Arntzen, C. J. (1987) *Z. Naturforsch.* 42c, 762–768
35. Zoller, M. J. and Smith, M. (1983) *Meth. Enzymol.* 100, 468–500
36. Sarkar, H. K., Viitanen, P. V., Padan, E., Trumble, W. R., Poonian, M. S., McComas, W. and Kaback, H. R. (1986) *Meth. Enzymol.* 125, 214–230
37. Kunkel, T. A., Roberts, J. D. and Zakour, R. A. (1987) *Meth. Enzymol.* 154, 367–382
38. Bennett, J. (1981) *Eur. J. Biochem.* 118, 61–70
39. Klug, G., Liebetanz, R. and Drews, G. (1986) *Arch. Microbiol.* 146, 284–291
40. Youvan, D. C. and Daldal, F. (eds.) (1986) *Microbial Energy Transduction: Genetics, Structure and Function of Membrane Proteins*, Cold Spring Harbor Laboratory, Cold Spring Harbor
41. Debus, R. J., Barry, B. A., Babcock, G. T. and McIntosh, L. (1988) *Proc. Natl Acad. Sci.* USA 85, 427–430
42. Vermaas, W. F. J., Rutherford, A. W. and Hansson, Ö, (1988) *Proc. Natl Acad. Sci. USA*, in press.
43. Barry, B. A. and Babcock, G. T. (1987) *Proc. Natl Acad. Sci.* USA 84, 7099–7103

Replication of Chloroplast DNA: Replication Origins, Topoisomerase I, and *In Vitro* Replication

BRENT L. NIELSEN, ROBERT MEEKER AND K. K. TEWARI

Department of Molecular Biology and Biochemistry
University of California, Irvine
Irvine. California USA 92717

Summary

Replication origins in pea chloroplast DNA (ctDNA) have been mapped in the 44 kbp *Sal*I fragment A. Restriction digests of pea ctDNA with *Sma*I, *Pst*I and *Sal*I/*Sma*I have allowed orientation of the two D-loops on the restriction map. One D-loop is located in the spacer region between the 16S and 23S rRNA genes, and the second D-loop is located downstream of the 23S rRNA gene, about 6 kbp apart from the first D-loop. Denaturation mapping of two recombinants, pCP12–7 and pCBI-12, which contain the two D-loops, confirmed the location of the D-loops. Denaturation mapping studies also showed that the D-loop present in the rRNA spacer region has a much higher A+T content than the other D-loop.

Using a partially purified pea chloroplast replication system, the supercoiled pCP12–7 and pCB1–12 recombinants served as highly active templates for DNA synthesis. Some smaller subclones of the A+T-rich D-loop region also show activity, including a 1.2 kbp *Bam*HI-*Hind*III clone. A smaller 700 bp *Bam*HI–*Bgl*II clone shows very low activity suggesting that sequences outside the *Bgl*II region are important for replication. Analysis of the DNA synthesised *in vitro* from these templates showed that full-length DNA was made, while other recombinants from other regions of the pea ctDNA showed no significant activity *in vitro*.

Pea chloroplast topoisomerase I was purified 5000-fold to homogeneity, and was found to consist of a single polypeptide of 112,000 daltons. The enzyme catalyses an ATP-independent relaxation of negatively supercoiled DNA. It is resistant to nalidixic acid and novobiocin, and causes a unit change in the linkage number of supercoiled DNA. These properties support the identification of the pea chloroplast enzyme as a prokaryotic-like topoisomerase I.

The role of topoisomerase I in the replication of ctDNA has been established by using a partially purified pea chloroplast replication system which lacks topoisomerase I activity, and recombinant templates which contain the pea ctDNA replication origins. A 2–6-fold stimulation of DNA synthesis resulted when the purified topoisomerase I was added to reactions at 80 mM KC1. When the reactions were carried out at 125 mM KC1, which is completely inhibitory for topoisomerase I but does not affect non-specific DNA polymerase activity, topoisomerase I addition caused no stimulation of activity. Novobiocin, an inhibitor of topoisomerase II, was found to have no effect on the *in vitro* replication of the ctDNA recombinants.

Abbreviation: TPP, thymidine-5′-triphosphate.

Introduction

The chloroplast genome of higher plants consists of a homogeneous population of double-stranded, closed circular DNA molecules of 120–160 kbp (1). The chloroplast DNA (ctDNA) of tobacco (2) and liverwort (3) have been sequenced, and many genes for proteins, rRNAs, and tRNAs, as well as other open reading frames, have been identified and mapped (4–7). Replication of ctDNA has been studied by examining replicative intermediates with the electron microscope (8,9). The replication of pea and maize ctDNAs have been shown to begin by the formation of displacement loops (D-loops) in the supercoiled ctDNA. The two D-loops (700–1100 bp in size) have been mapped about 6 kbp away from each other in pea ctDNA. These two D-loops then expand towards each other, and join to produce Cairns replicative forked structures. These Cairns structures expand bidirectionally to a site $180°$ around the circular molecule from the initiation site, where termination occurs. The daughter molecules then may separate, or replication may continue from a site at or near the termination site by a rolling circle mechanism (10). The D-loop origins of replication in *Euglena gracilis* and *Chlamydomonas reinhardii* ctDNA have been mapped and sequenced (11–14). A crude extract from *C. reinhardii* is abe to carry out *in vitro* replication on a recombinant template-containing the origins of replication of *c. reinhardii* (15). In addition, an *Eco*RI fragment from maize ctDNA, containing sequences homologous to the *C. reinhardii ori*A replication origin, was found to be a highly active template in a heterologous in *vitro* DNA synthesis system (16). While progress has been made in understanding the mechanism of ctDNA replication, and some crude *in vitro* replication systems have been developed, the individual enzymes involved in ctDNA replication have not been extensively studied. Pea chloroplast DNA polymerase has been purified to homogeneity, and shown to play a role in replication (17). Topoisomerase I and II have also been identified in chloroplasts (18, 19), but their role in replication has not been studied. Our goal has been to map and characterise the origins of replication in pea ctDNA, and identify and isolate the proteins involved in replication.

Results and Discussion

Supercoiled pea ctDNA was purified through successive CsCl gradients, as described (9). The restriction map of pea ctDNA is shown in Fig. 1. Supercoiled pea ctDNA molecules were digested with *Sal*I and examined with the electron microscope for D-loop-containing fragments. Fig. 2A shows the *Sal*I fragment A, which was found to contain two D-loops. None of the other *Sal*I fragments were found to contain D-loops, while most of the *Sal*I A fragments contained only one of the D-loops. Several of the *Sal*I

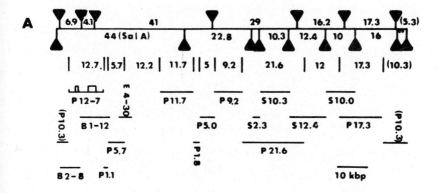

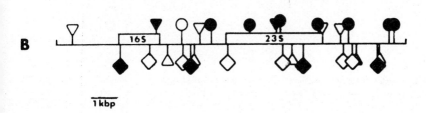

Fig. 1: Restriction map of pea chloroplast DNA (ctDNA) and map of cloned plasmids used in the *in vitro* replication assays. (A) Map of the *Sal*I, *Sma*I, and *Pst*I restriction enzyme recognition sites in the pea ctDNA, oriented with the largest *Sal*I fragment on the left. Beneath the upper restriction map is given the positions of the various ctDNA inserts of the plasmids used in Table II. The plasmid names have been abbreviated thus: pCP12–7 as P 12–7, pCS 10.3 as S 10.3 and pCE 4–30 as E 4–30. (B) A detailed restriction site map of the 12.7 kbp *Pst*I ctDNA fragment in pCP 12–7. The open box symbols (☐) on P 12–7 represent, from left to right, the 16S and 23S rRNA genes. The symbols are: (▲) *Sal*I, (▼) *Sma*I, (|) *Pst*I, (○) *Bam*HI, (▽) *Bgl*II, (◇) *Eco*RI, (●) *Hind*III, (◆) *Pvu*II, and (△) *Sac*I.

fragments were analysed, and a cumulative histogram of D-loop locations is presented in Fig. 3D. The histogram shows a bimodal pattern for the locations of the D-loops in pea ctDNA.

Single- and double-digests of pea ctDNA with *Sma*I and *Sal*I were analysed in order to orient the D-loops within the large *Sal*I A fragment (Fig. 1). A single D-loop was found in a 41 kbp ctDNA fragment (Fig. 2B), while the other D-loop was located in a 4.1 kbp *Sma*I fragment located at the left end of *Sal*I A (Fig. 2C). From the known restriction map of ctDNA (Fig. 1), it is clear that both D-loops are located at the left end of the 44 kbp *Sal*I fragment. This location will predict that both D-loops are contained in the 12.5 kbp *Pst*I fragment, and analysis of *Pst*I-digested ctDNA showed that the two D-loops were actually present in this fragment. Fig. 3 shows cumulative histograms from the analysis of the

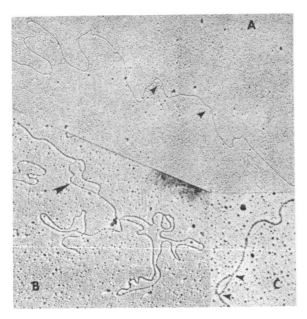

Fig. 2: Electron micrographs of chloroplast replicative regions. (A) A *Sal*I-generated ctDNA fragment containing two D-loop structures. (B) The large ctDNA fragment containing a single D-loop generated by *Sma*I digestion. (C) The small *Sma*I D-loop fragment. The arrows point out the D-loop structures.

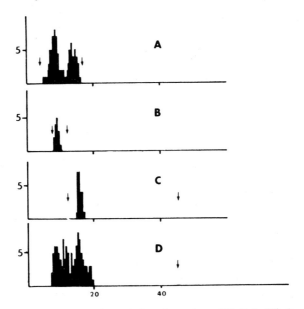

Fig . 3: Cumulative histograms of restriction digest data: (A) *Pst*I, (B) *Sma*I (the small fragment only), (C) *Sal*I/*Sma*I (large fragment only) and (D) *Sal*I. The positions of the fragments are lined up according to their correct position in the restriction map. Arrows in the histograms indicate the restriction sites of each fragment.

different restriction digest analyses, and Table 1 shows the approximate locations of the D-loops calculated from measurements of the molecules with the electron microscope. There is good agreement as to size and location of the D-loops, and the orientation of the D-loops in the pea ctDNA. One D-loop clearly maps in the spacer region between the 16S and 23S rRNA genes, while the other D-loop maps downstream of the 23S rRNA gene.

Using a sightly modified version of the denaturation mapping technique of Kolodner and Tewari (20), we oriented the pea ctDNA restriction map to their denaturation map. Fig. 4 shows the mapping of the cumulative histogram for the *Sal*I fragment A portion of pea ctDNA. The lining up of *Sal* A gives a denaturation peak over one of the six major denaturation regions of Kolodner and Tewari seen at 2.5% total denaturation. This position is further confirmed by mapping of *Sal* B seen in Fig. 4A and B. It can be seen that one D-loop (Fig. 4D) is found directly within the major denaturation region. When denaturation mapping was done they also showed a single denaturation region which maps very near the location of the leftward D-loop, while the other D-loop showed no denaturation under the same conditions. This data suggests that the leftward D-loop is A+T rich, while the rightward D-loop contains significantly higher G+C content.

A partially purified replication system from pea chloroplasts was isolated (16), and the replication activity of various ctDNA recombinants is shown in Table II. The activity against nicked calf thymus DNA served as a positive control for DNA polymerase activity. Supercoiled pBR322 without any ctDNA insert served as a negative control. Of the *Pst*I and other main clones examined, only pCP12–7, pCB1–12, and pCE4–30 showed significant DNA polymerase activity, while the other recombinants showed activity in the same range as pBR322. The use of nicked

TABLE 1

Restriction Map Orientation of D-loops

Restriction Enzyme	Left D-loop (mean)		Right D-loop (mean)	
	[a]Start	End	[a]Start	End
*Sal*I	9 kbp	9.7 kbp	14.2 kbp	15.5 kbp
*Sma*I/*Sal*I (large fragment)	—	—	14.4 kbp	15.6 kbp
*Sma*I, *Sma*I/*Sal*I (small fragment)	8.8 kbp	9.6 kbp	—	—
*Pst*I	8.1 kbp	9.3 kbp	13.4 kbp	14.6 kbp

[a] Distances adjusted as explained in the legend to Fig. 3 (kpb = kilobase pair).

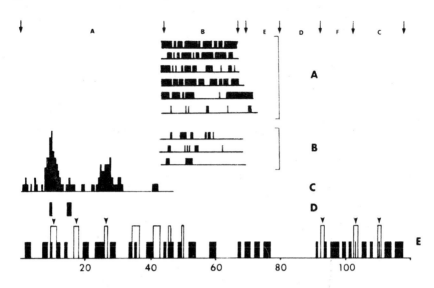

Fig. 4: Denaturation loop mapping of fragment A of the *Sal*I digests of pea ctDNA, presented as a best fit orientation to the denaturation loop map produced by Kolodner and Tewari (20). (A) Map of some representative denaturation looped *Sal*A fragments produced with a 47% formamide hypophase and a 77% formamide spreading buffer solution. (B) *Sal* A molecules denaturation looped at 37% formamide hypophase and 67% formamide spreading buffer. (C) The D-loop positions as determined from Table I. (D) The denaturation loop map of Kolodner and Tewari (20) oriented with the 44 kbp *Sal* A fragment placed at the left. *Symbols:* (▼) *Sal*I restriction sites; in (C) (■) D-loop structurs; in (D) (□) the denaturation loop regions seen at 2.5% denaturation and (■) the denaturation loop regions at 22%; and also in (D) (▼) the six major denaturation loop regions observed at 2.5% denaturation of the intact pea ctDNA molecule. The numbers below (D) indicate size in kbp.

recombinant templates resulted in a complete loss of activity. Both pCP12–7 and pCB1–12 contain the two D-loop regions, while pCE4–30 contains a strong promoter and an A+T-rich region which resembles a replication origin (21), and may be the reason for the activity of this template. Subclones were made from pCP12–7 which contain smaller fragments of the leftward D-loop region, as shown in Fig. 5. The recombinant pCS4.1, which contains the 4.1 kbp *Sma*I D-loop fragment, yields the same level of activity as the pCP12–7 template. Recombinants pCSH2.1 and PCBH1.2, containing a 2.1 kbp *Sma*I-*Hind*III fragment and a 1.2 kbp *Bam*HI-*Hind*III fragment, respectively, showed a 30–50% reduction in activity compared to pCP12–7. The recombinant pCBBg0.7, which contains a 700 bp *Bam*HI-*Bgl*II fragment, showed a reduction to only 16% of the PCP12–7 activity. Sequence determination of the 1.2 kbp *Bam*HI-*Hind*III DNA fragment and further in *vitro* replication studies are required in order to strictly define the sequences involved in replication.

In order to determine whether the incorporation activity of the recombinants as shown in Table II reflects true replication of full-sized

TABLE 2

Chloroplast Replication System Assays[a]

Template[b]	CPM	Template	CPM
– DNA	930	pCP 1.8	7,440
activated calf thymus DNa	99,950	pCP 17.3	4,950
pBR322	3,100	pCP 21.6	3,100
pCB 2–8	7,600	pCP 10.3	9,700
pCB 1–12	35,960	pCP 9.2	5,500
pCP12–7	99,400	pCS 10.0	3,300
pCP 1.1	6,260	pCS 12.4	2,750
pCP 5.7	4,000	pCS 10.3	2,680
pCP 11.7	3,170	pCS 4.1[c]	54,500
pCP 5.0	1,700	pCSH 2.1[c]	28,600
pCP 12.0	1,830	pCBH 1.2[c]	37,300
pCE 4–30	18,520	pCBBg 0.7[c]	8,600

[a] Assays for incorporation of [³H]-TTP on the specified supercoiled plasmid DNA templates (1 μg each), using an enzyme fraction which had been purified through DEAE-cellulose, heparin-Sepharpse, and phosphocellulose columns. This enzyme fraction contains DNA polymerase, DNA primase, RNA polymerase, topoisomerase I and II, single-stranded DNA binding proteins, and other proteins.
[b] The templates used are shown in Fig. 1.
[c] These last four templates are subclones of pCP12–7, and are shown in Fig. 5.
(CPM = counts per minute)

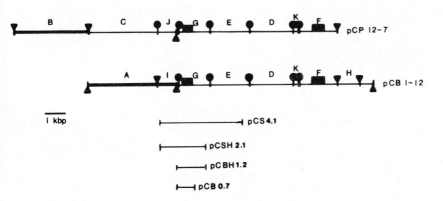

Fig. 5: Restriction map of pCP12–7, pCB1–12, and some subclones. Plasmid pCD12–7 contains the 12.3 kbp *Pst*I fragment from pea ctDNA in pACYC-177, Plasmid pCB1–12 contains the 10.0 kbp *Bam*HI fragment from pea ctDNA in pBR322. The two clones overlap as shown. The thick lines represent the vector DNA, and the filled boxes represent the two D-loops. The four smaller clones at the bottom are subclones of the AT-rich D-loop region, prepared in pUC19, and used as templates as shown in Table II. pCS4.1 contains a 4.1 kbp *Sma*I fragment; pCSH2.1 a 2.1 kbp *Sma*I-*Hind*III fragment; pCBH1.2 a 1.2 kbp *Bam*HI-*Hind*III fragment; and pCBO.7 contains a 700 bp *Bam*HI-*Bgl*II fragment.

DNA molecules, *in vitro* DNA synthesis reactions were carried out using [^{32}P] TTP as the label, with 2 μM cold TTP and 20 μM of the other dNTPs. The products were analysed in a 0.8% alkaline agarose denaturing gel, and the results are shown in Fig. 6. The bands represent single-stranded DNA, with the expected sizes of 17 kb for pCP12–7 and 14.5 kb for pCB1–12. A band of 9 kb with somewhat lower intensity was observed for the pCE4-30 template, while no band could be observed for pCP5.0, one of the templates which showed low activity.

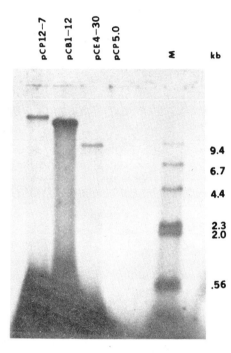

Fig. 6: Alkaline denaturing gel of *in vitro* synthesised DNA. In vitro DNA synthesis was carried out using [^{32}P] -TTP as described, using recombinants pCP12–7, pCB1–12, pCE4–30, and pCP5.0 as templates. The labelled synthesised DNA was analysed in a 0.8% alkaline agarose gel. The pCP12–7 lane shows a band with the correct size of 17 kb, for pCB1–12 a band of 14.5 kb, and for pCE 4–30 a band of 9 kb. In the pCP5.0 lane only a very faint band of 9.4 kb was seen. The marker lane contains *Hind*III-digested lambda DNA, with sizes as shown at the right.

Since the partially purified replication system appears to carry out specific replication of D-loop containing templates, we next proceeded to isolate and identify the individual enzymes present in this fraction. As already mentioned, pea chloroplast DNA polymerase has been purified and characterised (17). Topoisomerase I has been previously identified in chloroplasts (18,19), and is involved in regulating the level of superhelicity of supercoiled DNA. A method for separating topoisomerase I activity from DNA polymerase was developed by eluting a DEAE-cellulose column with a linear salt gradient. Topoisomerase I activity eluted before the DNA polymerase activity, with only slight overlap (Fig. 7). The two activities were separated, and further purification was carried out as illustrated in Fig. 8. This purification procedure led to the 5000-fold purification of pea chloroplast topoisomerase I to a single polypeptide of

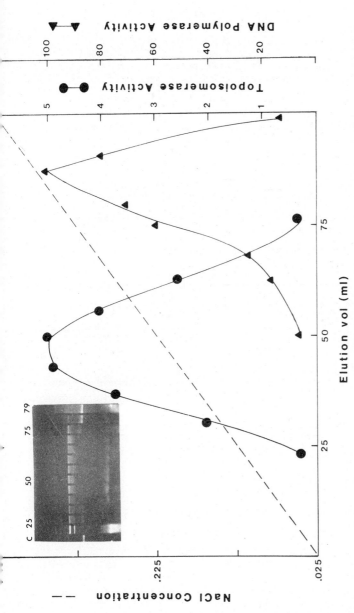

Fig 7: Elution profile of DNA polymerase and topoisomerase I from the DEAE-cellulose column. The column was loaded and washed with buffer A + 0.025 M NaCl, and then eluted with a 100 ml gradient of buffer A containing 0.025 M NaCl to 0.425 M NaCl. The molar salt concentration is indicated by the dashed line. Topoisomerase I activity (●) is indicated as the number of units (defined in Table III) in 5 μl of sample. DNA polymerase activity (▲) was assayed as in (17), and is shown as CPM X 10^{-3}. The inset photograph illustrates agarose gel analysis of topoisomerase I activity, with the lower band being supercoiled plasmid DNA, and the upper band relaxed circular DNA. Lane C is DNA without addition of enzyme. The numbers indicate the elution volume of the column.

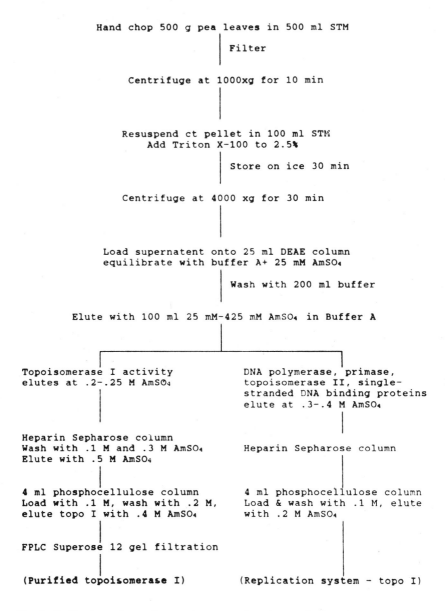

Fig. 8: Purification scheme for topoisomerase I and a replication fraction devoid of topoisomerase I. Chopping was in STM buffer as described (17). Buffer A was used for the DEAE-cellulose, Phosphocellulose, and FPLC columns, and contains 50 mM Tris, pH 8.0, 50 mM 2-mercaptoethanol, 0.2 mM PMSF, and 25% glycerol. Buffer B was used for the heparin-sepharose column, and is the same as buffer A with the addition of 0.1 mM EDTA and 0.1% Triton X-100 (NH₄)₂So₄ was the salt used, and is designated in the figure as $AmSO_4$.

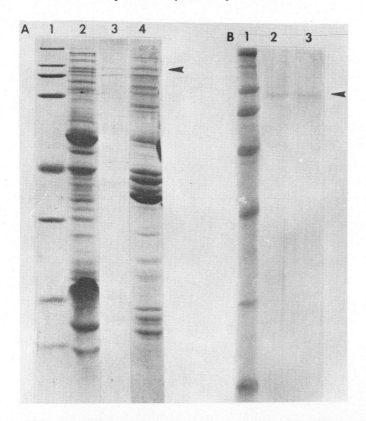

Fig. 9: SDS-polyacrylamide gel of pea chloroplast topoisomerase I purification. Fractions from the various columns were analysed in a 10% polyacrylamide denaturing gel electrophoresed at 120 volts. Panel A. Coomassie blue stained gel. Lane 1, protein molecular weight markers, from top to bottom, myosin (200 kDa), B-galactosidase (116 kDa), phosphorylase b (92.5 kDa), bovine serum albumin (66 kDa), ovalbumin (45 kDa), carbonic anhydrase (31 kDa), soybean trypsin inhibitor (21.5 kDa), and lysozyme (14.4 kDa). Lane 2, DEAE cellulose gradient topoisomerase I fraction. Lane 3, phosphocellulose column fraction. Lane 4, heparin-sepharose column topoisomerase I fraction. Panel B. Silver stained gel. Lane 1, molecular weight markers, same as in A. Lanes 2 and 3, FPLC Superose-12 fractions 14 and 13, respectively. Fraction 14 shows two minor bands slightly larger and smaller than the major 112,000 dalton band, while fraction 13 (lane 3) shows only the 112,000 dalton band.

112,000 daltons (Fig. 9). Gel filtration analysis of the topoisomerase I also gave a calculated molecular weight of the native enzyme of about 112,000 daltons. The purification of the enzyme through the various columns is summarised in Table III.

The topoisomerase I was characterised, and was found to have a broad temperature optimum, with peak activity at 37° C, and irreversible loss of activity at 42°C or higher. The enzyme showed highest activity at 25–50 mM KCl, but the activity decreased at higher salt concentrations, with very

TABLE 3

Purification of Pea Chloroplast Topoisomerase I

Fraction	Protein (mg)	Activity units[a]	Specific Activity	Fold Purification	Recovery %
Lysed chloroplasts	520	26,000[b]	50	—	100
DEAE-cellulose	5.1	25,000	4900	98	96
Heparin-Sepharose	1.32	16,000	12,120	242	61.5
Phosphocellulose	0.05	5,600	112,000	2240	21.5
FPLC Superose-12	0.008	2,000	250,000	5000	7.7

[a] One unit of topoisomerase I is defined as the amount of enzyme required to catalyze relaxation of 50% of the supercoiled DNA molecules from 1.0 μg pBR322 DNA in 30 min at 37°C.
[b] Estimated activity.

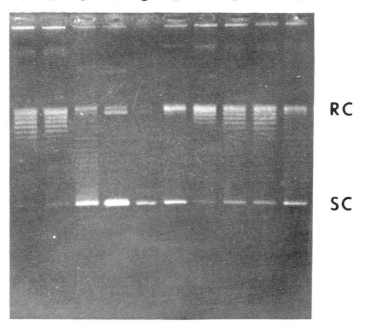

Fig. 10: Salt requirements of pea chloroplast topoisomerase I. Standard reactions were carried out as described in the legend to Fig. 7, at 37°C for 30 min. Lanes 1–5, each with 5 mM $MgCl_2$ and 0, 50, 100, 150, and 200 mM KCl, respectively. Lanes 6–10, each with 50 mM KCL and 0, 5, 10, 15, and 20 mM $MgCl_2$.

little activity at 100 mM KCl, and no activity at higher concentrations (Fig. 10). The enzyme required 5 to 15 mM $MgCl_2$ for optimum activity, and was significantly inhibited by 20 mM $MgCl_2$. The pea chloroplast topoisomerase I was unable to relax positively supercoiled DNA, was not dependent on ATP for activity, and causes a unit change in the linkage number of negatively supercoiled DNA. These properties of the enzyme are similar to those reported for the spinach chloroplast topoisomerase I, and confirm the identification of the chloroplast enzyme as a prokaryotic type I topoisomerase.

The role of the topoisomerase I in replication of ctDNA was studied by adding the purified topoisomerase I to the replication system devoid of topoisomerase I activity (see Fig. 8). Using pCP12–7 and pCB1–12 as templates, the addition of topoisomerase I resulted in a 2–6-fold increase in DNA synthesis activity (Table IV). *E. coli* topoisomerase I was able to cause a similar stimulation of activity, while calf thymus topoisomerase I caused no significant increase in activity. Similar results were obtained by Kornberg and coworkers (22), who found that calf thymus topoisomerase I could not substitute for *E. coli* topoisomerase I in stimulating *in vitro* replication of an *E. coli* replication origin. The prokaryotic and eukaryotic

TABLE 4

Role of Topoisomerase I in Replication of Chloroplast DNA

Template	mM KCl	± topoisomerase I[a]	± novobiocin[b]	CPM
activated calf thymus DNA	80	+/−	−	64,000
				103,000
”	125	+/−	−	6,700
pCP12–7	80	−	−	47,700
”	”	+	−	9,570
”	”	+ 20 units calf thymus topo I	−	44,000
”	”	+ 20 units *E. coli* topoI	−	2,020
”	”	−	+	36,360
”	”	+	+	3,570
”	125	−	−	2,090
”	”	+	−	21,000
pCB1–12	80	−	−	
”	”	+	−	38,000
”	125	+	−	9,000

[a] When added, 10 μl of the purified topoisomerase I (app. 20 units or 0.25 μg) was added to 10 μl (5–10 μg) of the DNA polymerase fraction devoid of topoisomerase I (see Fig. 8), except for the reactions carried out with calf thymus or *E. coli* topo I, as shown.
[b] When present, novobiocin was added to 250 μg/ml, which is inhibitory for topoisomerase II activity.
(CPM = counts per minute)

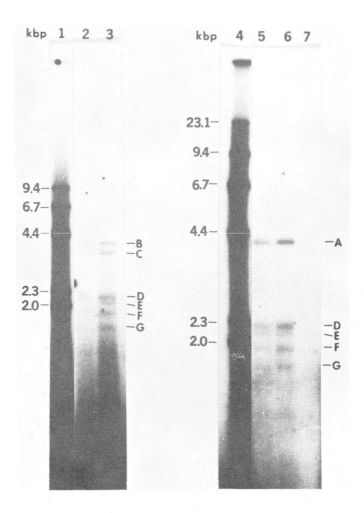

Fig. 11: The role of topoisomerase 1 in *in vitro* replication of pea ctDNA. *In vitro* replication reactions were carried out as described in the text, using pCP12–7 (lanes 2 and 3) or pCB1–12 (lanes 5–7) as template. The DNA was then digested with *Eco*RI and *Pst*I and the products analysed in a 1% native agarose gel. The resulting autoradiogram showed that fragments of the expected sizes were labelled (fragments identified at the right, see also Fig. 5). Lanes 1 and 4, phage lambda DNA digested with *Hind*III and 5' end-labelled with [^{32}P]-ATP, with sizes as shown at the left. Lanes 2 and 5, reactions at 70 mM KCl with the partially purified DNA polymerase fraction only. Lanes 3 and 6, reactions at 70 mM KCl with 10 μl of purified topoisomerase added to the partially purified DNA polymerase fraction. Lane 7, the reaction carried out with topoisomerase I addition, but at 125 mM KCl, which is inhibitory for topoisomerase I activity, but permissive for DNA polymerase activity on a nonspecific template (see Table IV).

topoisomerase I enzymes are known to have different structure and mechanism of action (23, 24), and this data suggests that they are not interchangeable in their role in replication. It was also found that if the reactions were carried out at 125 mM KCl, which was shown to inhibit topoisomerase I activity, no stimulation of activity was obtained upon addition of topoisomerase I. These results suggest that the pea chloroplast topoisomerase I is required for optimal replication activity *in vitro*.

The DNA synthesised *in vitro* in the presence and absence of topoisomerase I was analysed by gel electrophoresis (Fig. 11). A 4–5-fold increase in product was obtained when 20 units of pea chloroplast topoisomerase I was added to the replication system lacking topoisomerase I activity on the pCP12–7 template at 70 mM KCl (Fig. 11, lanes 2 and 3). When pCB1–12 was used as template under the same conditions, about a 2-fold increase in product was obtained upon addition of topoisomerase I (Fig. 11, lanes 5 and 6). Each of the expected *Pst*I + *Eco*RI restriction fragments was labelled, showing that the entire template was replicated. When the reaction was carried out at 120 mM KCl with the pCB1–12 template, no replication could be detected on the autoradiogram (Fig. 11, lane 7).

These results clearly show a role for topoisomerase I in the replication of pea ctDNA *in vitro*. The phosphocellulose replication fraction used for the replication experiments was analysed for the presence of topoisomerase II. This enzyme has been previously identified in pea chloroplasts (18), and like all topoisomerase II enzymes, is inhibited by novobiocin. A low level of topoisomerase II activity was found in our replication fraction, as assayed by the ability to supercoil relaxed circular molecules in the presence of ATP. Additional replication reactions with the pCP12–7 and pCB1–12 templates were carried out in the presence of novobiocin, with no effect on the replication activity of these templates (Table IV). These data indirectly rule out the role of topoisomerase II in the replication of ctDNA. However, the final proof for the role of topoisomerase II in replication has to be obtained using purified topoisomerase II coupled with *in vivo* experiments.

Acknowledgements

This work was supported by a grant from the N. I. H. (GM 33725). B. L. N. was the recipient of a postdoctoral fellowship from the Campus Biotechnology Training Program.

References

1. Kolodner, R. D. and Tewari, K. K. (1975) *Biochim. Biophys.* Acta 402, 372–390

2. Shinozaki, K., Ohme, M., Tanaka, M., Wakasugi, T., Hayashida, N., Matsubayashi, T., Zaita, N., Chunwongse, J., Obokata, J., Yamaguchi-Shinozaki, K., Ohto, C., Torazawa, K., Meng, B. Y., Sugita, M., Deno, H., Kamogashira, T., Yamada, K., Yamada, K., Kusuda, J., Takaiwa, F., Kato, A., Tohdoh, N., Shimada, H. and Sugiura, M. (1986) *EMBO J.* 5, 2043-2049

3. Ohyama, K., Fukuzawa, H., Kohchi, T., Shirai, H., Sano, T., Sano, S., Umesono, K., Shiki, Y., Takeuchi, M., Chang, Z., Aota, S., Inokuchi, H. and Ozeki, H. (1986) *Nature* (London) 322, 572-574

4. Chu, N. M., Shapiro, D. R., Oishi, K. K. and Tewari, K. K. (1985) *Plant Mol. Biol.* 4, 65-79

5. Crouse, E. J., Schmitt, J. M. and Bohnert, H. -J. (1985) *Plant Mol. Biol. Reporter* 3, 43-89

6. Oishi, K. K., Shapiro, D. R. and Tewari, K. K. (1984) *Mol. Cell. Biol.* 4, 2556-2563

7. Shapiro, D. R. and Tewari, K. K. (1986) *Plant Mol. Biol.* 6, 1-12

8. Kolodner, R. D. and Tewari, K. K. (1975) *Nature* (London) 256, 708-711

9. Kolodner, R. and Tewari, K. K. (1975) *J. Biol. Chem.* 250, 8840-8847

10. Tewari, K. K., Kolodner, R. D. and Dobkin, W. (1976). In: *Genetics and Biogenesis of Chloroplasts and Mitochondria* (Bucher, T., ed.) pp. 379-386, Elsevier/North-Holland Biomedical Press, Amsterdam

11. Koller, B. and Delius, H. (1982) *EMBO J.* 1, 995-998

12. Ravel-Chapuis, P., Heizmann, P. and Nogon, V. (1982) *Nature* (London) 300, 78-81

13. Waddell, J., Wang, X.-M. and Wu, M. (1984) *Nuc. Acids Res.* 12, 3843-3856

14. Wang, X.-M., Chang, C. H., Waddell, J. and Wu, M. (1984) *Nuc. Acids Res.* 12, 3857-3872

15. Wu, M., Lou, J. K., Chang, D. Y., Chang, C. H. and Nie. Z. Q. (1986) *Proc. Nat. Acad. Sci. USA* 83, 6761-6765

16. Gold, B., Carrillo, N., Tewari, K. K. and Bogorad, L. (1987) *Proc. Nat. Acad. Sci. USA* 84, 194-198

17. McKown, R. L. and Tewari, K. K. (1984) *Proc. Nat. Acad. Sci. USA* 81, 2354-2358

18. Lam, E. and Chua, N. H. (1987) *Plant Mol. Biol.* 8, 415-424

19. Seidlecki, J., Zimmermann, W. and Weissbach, A. (1983) *Nuc. Acids Res.* 11, 1523-1536

20. Kolodner, R. D. and Tewari, K. K. (1975) *J. Biol. Chem.* 250, 4888-4895.

21. Palmer, J. D. and Thompson, W. F. (1981) *Proc. Nat. Acad. Sci. USA* 78, 5533-5537

22. Lehmbeck, J., Rasmussen, O. F., Bookjans, G. B., Jepsen, B. R., Stumman, B. M. and Henningsen, K. W. (1986) *Plant Mol. Biol.* 7, 3-10

23. Kaguni, J. M. and Kornberg, A. (1984) *Cell* 38, 4888-4895

24. Wang, J. C. (1985) *Ann. Rev. Biochem.* 54, 665-697

Molecular Genetics of Herbicide Resistance in Cyanobacteria

JUDY BRUSSLAN AND ROBERT HASELKORN

Department of Molecular Genetics and Cell Biology,
University of Chicago,
Chicago, IL 60637, USA

Summary

Cyanobacteria are an excellent model for studies of herbicide resistance. Mutants resistant to classical and non-classical herbicides have been isolated, and in some cases the amino acid alteration(s) are known. Mutations in plants, algae, photosynthetic bacteria, and cyanobacteria are compared. Data concerning the question of dominance or recessiveness of herbicide resistance in cyanobacteria are also discussed.

Herbicides have been developed during the last century to inhibit processes and pathways which are unique to plants: photosynthesis, chlorophyll biosynthesis and essential amino acid biosynthesis (Brian, 1976). Although not designed for this purpose, these herbicides are also lethal to cyanobacteria which share many of the plant's unique pathways. Indeed these pathways probably originated in ancient cyanobacteria which were the ancestors of today's chloroplast.

The herbicides which inhibit photosynthesis have been studied extensively. These include the classical herbicides: triazines, ureas, triazinones, and uracils; and the non-classical herbicides which include the phenol-types (Trebst, 1987). The classical herbicides block electron transfer between the plastoquinones Q_A and Q_B by competing with Q_B for the Q_B-binding site. Binding competition was demonstrated directly between atrazine and 6-azido-Q_0C_{10}, a synthetic quinone which mimics Q_B (Vermaas et al., 1983). Q_B, the second stable electron acceptor of photosystem II, receives two electrons from Q_A before it leaves the Q_B-binding site as Q_BH_2. When Q_B is only partially reduced, as Q_B^-, its affinity for the Q_B-binding site is high and herbicides do not compete effectively. A binary oscillation of herbicide binding affinity is observed,

---

Abbreviations: DCPIP—dichlorophenolindophenol, LD$_{50}$—lethal dose killing 50%

dependent on the number of single flashes given to intact chloroplasts. After an odd number of flashes, most of the plastoquinone is in the semiquinone form, tightly bound to the Q_B-binding site, and thus the number of herbicide binding sites is decreased (Laasch et al., 1984). The Q_B-binding site is believed to be located on the D1 protein. The D1 protein forms part of the core of photosystem II which is involved in charge separation. This was demonstrated by isolation of a particle containing the D1 protein, the D2 protein, cytb559, 5 chlorophyll a, 2 pheophytin a, and one B-carotene, which is capable of pheophytin photoreduction (Nanba and Satoh, 1987). It has been postulated, based on primary structure as well as functional analogy, that the D1 and D2 proteins are homologous to the L and M proteins, respectively, of the *Rhodopseudomonas viridis* reaction centre (Trebst and Draber, 1986). X-ray diffraction data from crystals of reaction centres from photosynthetic bacteria indicate that L and M bind 4 bacteriochlorophylls, 2 bacteriopheophytins, 2 ubiquinones, and one non-heme iron (Deisenhofer et al., 1985, Michel et al., 1986a, Chang et al., 1986, Allen et al., 1987). L and M both cross the thylakoid membrane five times, and the Q_B-binding site is between the membrane-spanning helices IV and V of both proteins. Homologous helices are probably located on the D1 protein (see below).

The use of triazine herbicides has inevitably led to the selection of herbicide-resistant weeds (Gressel et al., 1982). Mutants resistant to atrazine are also cross resistant to the urea herbicide diuron, the triazinone herbicide metribuzin, and the uracil herbicide bromacil. The D1 protein was implicated as the target of these herbicides due to the labelling of D1 by azido derivatives of atrazine, urea, and triazinone (Trebst, 1986). Also, herbicide resistance is inherited maternally (Galloway and Mets, 1983) which is consistent with D1 being encoded by a plastid gene. Indeed, the cloning and sequencing of a *psbA* gene from the chloroplast DNA of a herbicide-resistant mutant of *Amaranthus hybridus* revealed a single basepair change which produced a D1 protein mutated to gly at ser264 (Hirschberg and McIntosh, 1983).

The photosynthetic reaction centres of cyanobacteria are homologous to those of higher plants (Bryant, 1986). In fact, the *psbAI* gene product, D1, from *Synechococcus* 7942 has 85% similarity to the *psbA* gene product of *Pisum sativum* (Golden and Haselkorn, 1985, Oishi et al., 1984). Photosynthetic herbicide-resistant mutants of cyanobacteria have been isolated (Golden and Sherman, 1984, Astier et al., 1986, Buzby et al., 1987, Gingrich, 1987, Hirschberg et al., 1987a), and many of these mutants contain the same amino acid changes in the D1 protein that were seen in herbicide-resistant algae and higher plants (see Table 1). In some herbicide-resistant mutants, electron transfer from Q_A to Q_B is slowed (Astier et al., 1986, Robinson et al., 1987), as in higher plants and algae

TABLE 1

Cross Resistance in Herbicide-Resistant Mutants

| Mutation: | Level of Resistance Relative to Wild Type | | | | | | | |
	F211→S	V219→I	A251→V	F255→Y	G256→E	S264→A	S264→G	L275→F
Diuron		17[2]	5[2]	0.5[2] 2[3]	10[2]	100[2] 150[3]	2[4]	5[2]
Atrazine	10[1]	2	25	15 47	80	10 18	1000	1
Bromacil		1	1	70		300	20	5
Metribuzin				0.8		5000	300	
Ioxynil				0.5		0.5	0.6	

1. *Synechococcus 7002*
2. *Chlamydomonas rheinhardtii*
3. *Synechococcus 7942*
4. *Amaranthus hybridus*

References are provided for herbicide-resistant strains in which resistance relative to wild type has been measured. In *Synechococcus 7002* there is also a V219 → I change, and in *Synechococcus 7942* there is a F255 → L/S 264 → A double mutation (J. Hirschberg, personal communication) Mutants of *Synechocystis 6714* have also been isolated (C. Astier, personal communication) for which relative resistance values are provided in the text.

(Arntzen, *et al.*, 1982, Erickson *et al.*, 1985). For these reasons, cyanobacteria are a legitimate model for studies of resistance to photosynthetic herbicides.

Cyanobacteria offer many advantages over higher plants and algae due to their fast growth, simple genome organization (Doolittle, 1979), ability to be transformed with DNA, and their efficient homologous recombination system (Tandeau de Marsac and Houmard, 1987). A large number of cells can be mutagenised and herbicide-resistant mutants can be selected. If a resistant strain contains an alteration in the DNA then total chromosomal DNA from this strain can be used to transform a herbicide-sensitive strain to herbicide resistance. In this way, isogenic strains differing solely in the nucleotide(s) responsible for the herbicide resistance phenotype can be constructed. If one is studying a herbicide which affects D1, then the *psbA* gene(s) from the resistant transformant can be cloned and sequenced in order to determine which amino acid(s) have been changed. The *psbA* gene from the herbicide-resistant strain can then be used to transform sensitive cells to herbicide resistance, unequivocally demonstrating that alteration of D1 alone results in herbicide resistance (Golden and Haselkorn, 1985). Cyanobacterial thylakoids can be isolated and the Hill reaction, measuring electron transfer to DCPIP, can be used to determine the LD$_{50}$ of many types of herbicides (Astier *et al.*, 1986, Hirschberg *et al.*, 1987a).

Herbicide-resistant mutants of plants, algae, and cyanobacteria, their relative resistance to herbicides; and changes in amino acids are listed in Table 1 and Figure 1. It appears that mutations to herbicide resistance are repeatedly found at certain amino acids (D1val219, D1phe255, and D1ser264). Two of these amino acids, D1phe255 and D1ser264, have terbutryn-resistant homologues in the L subunit of *R. sphaeroides* and/or *R. viridis* (Paddock *et al.*, 1987, Sinning and Michel, 1986) at positions Lphe215 and Lser223. In these cases the amino acid changes are different: Lphe216→ser instead of D1phe255→tyr and Lser223→pro instead of D1ser264→ala gly. A Lser223→ ala has been isolated, but is accompanied by a change at Larg217→his. In *R. sphaeroides* there is also a change at Ltyr222→gly and Lile229→met.

A decrease in the rate of electron transport from Q$_A$ to Q$_B$ has been observed in D1ser264 mutants of *Amaranthus hybridus* (Arntzen, *et al.*, 1982), *Chlamydomonas rheinhardtii* (Erickson, *et al.*, 1985), *Synechococcus* 7942 (Robinson, *et al.*, 1987), but not in the grass *Phalaris paradoxa* (Hirschberg *et al.*, 1987b). Electron transfer is also altered in

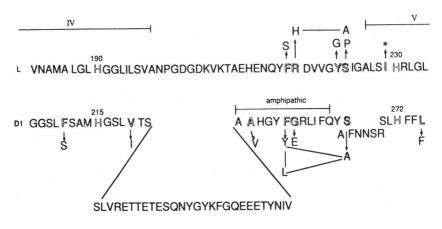

Fig. 1: Comparison of the amino acid sequence of the L subunit of *R. viridis* (Michel *et al.*, 1986b) and the D1 protein encoded by *psbAI* of *Synechococcus* 7942 (Golden and Haselkorn, 1985). The histidines which bind the non-heme iron are outlined, and the sequences are aligned as in Trebst, 1987. Helices IV and V, as well as a putative amphipathic helix in the D1 protein are indicated. Amino acids in which herbicide-resistant mutations have been localised are shadowed, and the amino acid substitution is noted above or below L or D1, respectively. Bars connecting 2 amino acid substitutions indicate double mutations. D1ser264 → ala is found as a single or as a double mutation, but D1phe255 → leu is found only as a double mutation (J. Hirschberg, personal communication). L subunit ser223→ pro is found as a single mutation (Paddock, *et al.*, 1987), but ser 223→ ala has been isolated only as a double mutation (Sinning and Michel, 1987). The asterisk above ile229 of the L subunit indicates the site of saturation mutagenesis in *R. capsulatus* (Bylina and Youvan, 1987).

D1gly256→glu, but not in D1val219→ile, D1ala251→val, D1phe255→tyr or D1leu275→phe of *Chlamydomonas rheinhardtii* (Erickson, *et al.*, 1985, Rochaix and Erickson, 1987). In *R. sphaeroides*, mutants at Ltyr222 and Lser223 have altered electron transport, but Lile229 has normal Q_A to Q_B electron transfer (Paddock *et al.*, 1987).

The different patterns of cross resistance indicate that D1ser264 is important in the binding of many herbicides, interacting with diuron and atrazine, as well as triazinones and uracils. D1phe255 interacts with atrazine, and D1val219 interacts with diuron. The D1ala251 is involved primarily in diuron binding while the D1leu275 mutation is involved in diuron and bromacil binding, although alteration of this amino acid only changes resistance five-fold. The change at D1phe211 affects binding to atrazine. In all of the mutants in which it has been tested, sensitivity to the non-classical phenolic herbicide, ioxynil, is enhanced.

Mutants with different patterns of cross resistance have been isolated in *Synechocystis* 6714 (C. Astier, personal communication). DCMU II_B (Astier, *et al.*, 1986) is 300X resistant to diuron, as compared to wild type, but only 5X resistant to atrazine. Az-IV is 120X resistant to diuron and 30X resistant to atrazine. These patterns resemble ser264→ala mutants except that electron transport is not altered. Two mutants of *Synechocystis* 6714, resistant to ioxynil, but sensitive to diuron and atrazine, have also been isolated. Iox-l and Iox-ll are both 5X resistant to ioxynil, but only Iox-l is cross resistant to bromoxynil, another phenolic herbicide. Electron transport is normal in these strains, also. At present, *psbA* and *psbD*, which encode D1 and D2 respectively, are being cloned from the four mutant strains as well as from wild type. The amino acid alterations responsible for these new patterns of resistance should be known soon.

Double mutants have been isolated in *Synechococcus* 7942 (J. Hirschberg, personal communication). Analysis of each mutation alone or in combination indicates that some herbicide resistance mutations are additive. For example, compared to wild type, D1ser264→ala is 18X more resistant to atrazine while D1phe255→tyr is 47X more resistant to atrazine. The double mutant is 350X more resistant. For other herbicides this pattern is not observed. For example, resistance to monuron is increased 570X in the D1264 mutant and 1.5X in the D1255 mutant, yet only 250X in the double mutant.

The positions of these substitutions on D1 encoded by *psbAl* from *Synechococcus* 7942 and on the L subunit of *R. viridis* are indicated in Figure 1. One striking difference between L and D1 is in the region between helices IV and V, the Q_B-binding pocket. This region is highly conserved among plants, algae, and cyanobacteria, and thus must serve a specific function. The D1phe211 mutation is one turn of helix IV (four amino acids) from D1his215 while the D1leu275 mutation is one turn of helix V (three amino acids) from D1his272. These amino acids must face into the Q_B-binding pocket.

Bylina and Youvan, using *R. capsulatus*, have performed saturation mutagenesis on ile229 of the L subunit, the amino acid residue next to Lhis230, which is bound to the non-heme iron. It was found that leu and met at position 229 confer atrazine resistance, while val and ala substitutions are herbicide-sensitive. Only these four substitutions are fully competent photosynthetically. These results suggest that hydrophobic residues of moderate size function best at this position. Homologous mutations in algae and cyanobacteria have not yet been isolated, but they could be constructed using site-specific mutagenesis. D1Leu271, adjacent to D1his272, which binds the non-heme iron, would be an excellent candidate for mutagenesis. One of the three glutamic acids at positions D1242-244 could be involved in binding the non-heme iron (Trebst and Draber, 1986), and thus these amino acids ought to be examined as well.

One major question that can be answered by cyanobacterial studies is whether resistance is dominant or recessive to herbicide-sensitivity. In plants, algae, and cyanobacteria grown in the light, the D1 protein turns over (Edelman *et al.*, 1984, P. Goloubinoff, personal communication). This turnover is inhibited by diuron in wild type strains, but is unaffected by diuron in resistant strains. It has been believed that a chloroplast or cell containing both diuron-resistant and diuron-sensitive D1 proteins would eventually contain 100% non-functional diuron-sensitive D1 proteins in its reaction centres when the cell is treated with diuron. This is because diuron-sensitive D1 proteins cannot turn over in the presence of diuron, and should become 'stuck' in the membrane. Therefore diuron sensitivity is expected to be dominant to diuron resistance.

Two different approaches to the question of dominance, both using *Synechococcus* 7942, have been pursued. *Synechococcus* 7942 contains three functional copies of the *psbA* gene (Golden *et al.*, 1986). Three strains, each containing two genes with a wild type D1ser264 and one gene with a D1ser264→ala mutation, in each of the three *psbA* gene positions, were constructed (Brusslan and Haselkorn, 1987). The phenotype of these three strains is resistance to diuron, independent of light intensity. Thus the location of the mutant gene in the gene family does not affect cell phenotype, even if the resistance allele is in *psbAlll*, the least abundantly transcribed gene of the family. Fluorescence titration, measuring variable fluorescence one μsec after a saturating flash as a function of diuron concentration, indicates that there are two populations of functional reaction centres in a *psbAlll* diuron-resistant strain of *Synechococcus* 7942. At a low concentration of diuron, 6nM, approximately one-third of the reaction centres fluoresce with fast-rise kinetics, demonstrating that Q_A to Q_B transfer has been blocked. (This is the diuron concentration required to block transfer in sensitive wild type reaction centres.) The other two-thirds of the reaction centres do not fluoresce until a higher concentration of diuron, 1800nM, is reached (Robinson *et al.*, 1987). If there are two populations of D1 proteins in the cell, and the phenotype of the cell is herbicide resistance, then it follows that resistance to diuron is dominant to sensitivity. The implications of this result for the interpretation of the turnover data remain to be determined.

On the other hand, a strain which contains a duplication of the *psbAI* gene has been constructed (Pecker *et al.*, 1987). One allele of the *psbAI* gene is herbicide-resistant while the other allele is herbicide-sensitive. The phenotype of this strain is herbicide sensitivity, independent of light intensity; yet, *in vitro* Hill reactions measured in this strain indicate an intermediate value of herbicide resistance, and DNA isolated from this strain can be used to transform wild type cells to diuron resistance. The duplication strain has four *psbA* genes, of which only one is a diuron-resistant allele, yet the herbicide-resistant strains isolated in this lab contain one diuron-resistant allele (*psbAI*) and two sensitive alleles (*psbAII* and *psbAIII*) (Hirschberg *et al.*, 1987a). The resistant phenotype of the latter strains is consistent with the results described above (Brusslan and Haselkorn, 1987). This discrepancy may be caused by a position effect resulting from duplication or by the presence of four *psbA* genes, of which three are herbicide-sensitive. It will be interesting to see the results of a fluorescence analysis of the *psbAI* duplication strain similar to the one done on the *psbAIII* diuron-resistant strain mentioned above. The phenotype of a psbAII or a *psbAIII* duplication may provide further insight into these conflicting results.

Cyanobacteria can also be used to study other herbicides. Gabaculin prevents chlorophyll biosynthesis by inhibiting glutamyl semialdehyde transaminase, an enzyme involved in an early step of chlorophyll biosynthesis (Kannangara and Schouboe, 1985). A mutant of *Synechococcus* 7942 has been isolated which is resistant to a concentration of gabaculin that is 10-fold higher than is required to kill wild type cells (David Paterson, personal communication). The mutation is due to an alteration of the DNA as demonstrated by transformation of a gabaculin-sensitive strain to gabaculin resistance by chromosomal DNA isolated from the resistant strain. An analysis of the gene conferring herbicide resistance should indicate whether the transaminase itself is mutated, whether there is a mutation which results in increased expression of the transaminase as occurs in *Chlamydomonas rheinhardtii* (Kahn and Kannangara, 1987), or whether a different gene confers gabaculin resistance.

References

1. Allen, J. P., Feher, G., Yeates, O., Komiya, H., and D. C. Rees (1987) Structure of the photosynthetic reaction centre from *Rhodobacter sphaeroides* R-26: the cofactors, *Proc. Natl Acad Sci. USA* 84: 5730–5734
2. Arntzen, C. J., Pfister, K. and K. E. Steinback (1982) the mechanism of chloroplast triazine resistance: alterations in the site of herbicide action. In: LeBaron, H. M. and Gressel, J. (eds.) *Herbicide Resistance in Plants*, pp. 185–214, New York: John Wiley and Sons

3. Astier, C., Meyer, I., Vernotte, C. and A. L. Etienne (1986) Photosystem II electron transfer in highly herbicide resistant mutants of *Synechocystis* 6714. *FEBS* Lett 207: 234–238

4. Brian, R. C. (1976) The history and classification of herbicides. In: L. J. Audus, (ed), *Herbicides: Physiology, Biochemistry, Ecology*, pp. 1–54, London: Academic Press

5. Brusslan, J. and R. Haselkorn (1987) Herbicide resistance in the *psbA* multigene family of *Synechococcus* PCC 7942. In: von Wettstein, D. and Chua, N. H. (eds.) *Proceedings of NATO ASI Plant Molecular Biology*, New York: Plenum Pub., in press

6. Bryant, D. (1986) The cyanobacterial photosynthetic apparatus: comparisons to those of higher plants and photosynthetic bacteria. In: Platt, T. and Li, W. K. W. (eds.) Photosynthetic Picoplankton, *Canad Bull Fisheries Aquatic Sci.* 214: 423–500

7. Buzby, J. S., Mumma, R. O., Bryant, D. A., Gingrich, J., Hamilton, R. H., Porter, R. D., Mullin, C. D. and S. E. Stevens, Jr. (1987) Genes with mutations causing herbicide resistance from the cyanobacterium PCC 7002. In: Biggins, J. (ed.) *Progress in Photosyn. Res.*, vol IV, pp. 757–760, Dordrecht: Martinus Nijhoff Pub

8. Bylina, E. J. and D. C. Youvan (1987) Genetic engineering of herbicide resistance: saturation mutagenesis of isoleucine 229 of the reaction centre L subunit. *Z Naturforsch* 42c: 751–754

9. Chang, C. H., Tiede, D., Tang, J., Smith, U., Norris, J. R. and M. Schiffer (1986) Structure of *Rhodopseudomonas sphaeroides* R-26 reaction centre, *FEBS* Lett 205: 82–86

10. Deisenhofer, J., Epp. O., Miki, K., Huber, R., and H. Michel (1985) Structure of the protein subunits in the photosynthetic reaction centre of *Rhodopseudomonas viridis* at 3A resolution. *Nature* 318: 619–624

11. Doolittle, W. F. (1979) The cyanobacterial genome, its expression, and the control of that expression. *Adv. Microbiol. Physiol.* 20: 1–102

12. Edelman, M., Mattoo, A. K. and J. B. Marder (1984) Three hats of the rapidly-metabolised 32 kilodalten protein of the thylakoids. In: Ellis, R. J. (ed.) *Chloroplast Biogenesis*, pp. 283–302, Cambridge: Cambridge University Press

13. Erickson, J. M., Rahire, M., Rochaix, J-D. and L. Mets (1985) Herbicide resistance and cross-resistance: changes at three distinct sites in the herbicide-binding protein. *Science* 228: 204–207

14. Galloway, R. E. and L. J. Mets (1983) Atrazine, bromacil, and diuron resistance in *Chlamydomonas. Plant Physiol.* 74: 469–474

15. Gingrich, J. (1987) this volume

16. Golden, S. S. and L. A. Sherman (1984) Biochemical and biophysical characterization of herbicide-resistant mutants of the unicellular cyanobacterium, *Anacystis nidulans* R2. *Biochim. Biophys.* Acta 764: 239–246

17. Golden, S. S. and R. Haselkorn (1985) Mutation to herbicide resistance maps within the *psbA* gene of Anacystis *nidulans* R2. *Science* 229: 1104–1107

18. Golden, S. S., Brusslan, J. and R. Haselkorn (1986) Expression of a family of *psbA* genes encoding a photosystem II polypeptide in the cyanobacterium *Anacystis nidulans* R2. *EMBO J* 5: 2789–2798

19. Gressel, J., Ammon, H. U., Fogelfors, H., Gasquez, J., Kay, Q. O. N. and H. Kees (1982) Discovery and distribution of herbicide-resistant weeds outside North America. In: LeBaron, H. M. and Gressel, J. (eds.) *Herbicide Resistance in Plants* pp. 33–55, New York: John Wiley and Sons

20. Hirschberg, J. and L. McIntosh (1983) Molecular basis of herbicide resistance in *Amaranthus hybridus. Science* 222: 1346–1348

21. Hirschberg, J., Ohad, N., Pecker, I. and A. Rahat (1987a), Isolation and characterization of herbicide-resistant mutants in the cyanobacterium *Synechococcus* R2. *Z Naturforsch* 42c: 758–761

22. Hirschberg, J., Yehuda, A. B., Pecker, I. and N. Ohad (1987b) Mutations resistant to photosystem II herbicides In: von Wettstein, D. and Chua, N. H., (eds.) Proceedings of NATO ASI Plant Molecular Biology, New York: Plenum Pub., in press

23. Johanningmeier, U., Bodner, U. and G. F. Wildner (1987) A new mutation in the gene coding for the herbicide-binding protein in *Chlamydomonas*. *FEBS* Lett 211: 221–224

24. Kahn, A. and C. Gamini Kannangara (1987) Gabaculine-resistant mutants of *Chlamydomonas rheinhardtii* with elevated glutamate 1-semialdehyde aminotransferase activity. *Carlsberg Res. Commun.* 52: 73–81.

25. Kannangara, C. Gamini and A. Schouboe (1985) Biosynthesis of $\triangle$-aminolevulinate in greening barley leaves. VII Glutamate 1-semialdehyde accumulation in gabaculine-treated leaves. *Carlsberg Res. Commun.* 50: 179–191

26. Laasch, H., Schreiber, U. and W. Urbach (1984) Binding of radioactively labelled DCMU in dependence of the redox state of the photosystem II acceptor side. In: Sybesma, C. (ed.) *Adv. Photosyn. Res.*, vol IV, pp 25–28, The Hague: Martinus Nijhoff Pub

27. Michel, H, Epp, O. and J. Deisenhofer (1986a) Pigment-protein interactions in the photosynthetic reaction centre from *Rhodopseudomonas viridis*. *EMBO J* 5:2445-2451

28. Michel, H., Weyer, K. A, Greenberg, H., Dunger, I., Oesterhelt, D. and F. Lottspeich (1986b) The 'light' and 'medium' subunits of the photosynthetic reaction centre from *Rhodopseudomonas viridis*: isolation of the genes, nucleotide and amino acid sequence. *EMBO J* 5: 1149/1158

29. Nanba, O. and K. Satoh (1987) Isolation of a photosystem II reaction centre consisting of D-1 and D-2 polypeptides and cytochrome b-559, *Proc. Natl Acad. Sci. USA* 84: 109–112

30. Oishi, K. K., Shapiro, D. R. and K. K. Tewari (1984) Sequence organisation of a pea chloroplast DNA gene coding for a 34,500-dalton protein. *Mol. Cell Bio.* 4: 2556–2563

31. Paddock, M. L., Williams, J. C., Rongey, S. H., Abresch, E. C., Feher, G. and M. Y. Okamura (1987) Characterisation of three herbicide-resistant mutants of *Rhodopseudomonas sphaeroides* 2.4.1: structure-function relationship. In: Biggins, J. (ed.) *Prog. Photosyn. Res.*, vol III, pp 775–778, Dordrecht: Martinus Nijhoff Pub

32. Pecker, I., Ohad, N. and J. Hirsehberg (1987) The chloroplast-encoded type of herbicide resistance is a recessive trait in cyanobacteria. In: Biggins, J. (ed.) *Prog. Photosyn. Res.*, vol III, pp 811–814, Dordrecht: Martinus Nijhoff Pub

33. Robinson, H., Golden, S. S. Brusslan, J. and R. Haselkorn (1987) Functioning of photosystem II in mutant strains of the cyanobacterium *Anacystis nidulans* R2 In: Biggins, J. (ed.) *Prog. Photosyn. Res.* vol IV, pp 825–828, Dordrecht: Martinus Nijhoff Pub

34. Rochaix, J–D. and J. M. Erickson (1987) Genetic and molecular analysis of photosystem II. *Trends in Biochem.*, in press

35. Sinning, I, and H. Michel (1986) Sequence analysis of mutants from *Rhodopseudomonas viridis* resistant to the herbicide terbutryn. *Z. Naturforsch* 42c: 751–754

36. Tandeau de Marsac, N. and J. Houmard (1987) Advances in cyanobacterial moleculer genetics In: Fay, P and Van Baalen, C. (eds.) *The Cyanobacteria: A Comprehensive Review*, Elsevier Scientific Pub. Co., in press

37. Trebst, A. and W. Draber (1986) Inhibitors of photosystem II and the topology of the herbicide and Q_B^- binding polypeptide in the thylakoid membrane. *Photosyn. Res.* 10:381–392

38. Trebst, A. (1987) The three-dimensional structure of the herbicide binding niche on the reaction centre polypeptides of photosystem II. *Z. Naturforsch* 42c: 742–750.

39. Vermaas, W. F. J., Arntzen, C. J., Gu, L-Q. and C-A. Yu (1983) Interactions of herbicides and azid quinones at a photosystem II binding site in the thylakoid membrane. *Biochim. Biophys.* Acta 723: 266–275

Molecular Analysis of a Gene (*irp*A) Encoding a Function Essential for Iron-limited Growth in the Cyanobacterium *Anacystis nidulans* R2

GEORGE S. BULLERJAHN, K. J. REDDY,
HAROLD C. RIETHMAN AND LOUIS A. SHERMAN

Division of Biological Sciences,
University of Missouri,
Columbia, Missouri 65211, USA

Summary

We describe the cloning, sequencing and mutagenesis of an *Anacystis nidulans* R2 gene whose function is required for growth in iron-limited medium. The gene, designated *irp*A, was cloned from a λgt11 phage expression library: *irp*A directs the synthesis of a 36 kDa membrane polypeptide which accumulates only in iron-deficient cells. DNA sequence analysis revealed an open reading frame capable of encoding a polypeptide of 38.6 kDa, and hydropathy plots suggested that the hodrophobic amino terminus of the *Irp*A protein could represent a signal sequence. A site-directed mutant of *A. nidulans* R2 lacking *irp*A function was constructed by Tn5 insertion, and the mutant, designated *A. nidulans* K7, was defective in growth in low-iron medium. Lastly, the DNA sequence of the region immediately upstream from the *irp*A bears close similarly to the consensus operator sequence conferring iron regulation in *Escherichia coli*.

Introduction

We have become increasingly interested in the global responses of cyanobacteria to iron starvation. The cyanobacteria, aerobic bacteria capable of oxygenic photosynthesis, are known to secrete iron-solubilising siderophores during iron stress. Adaptation to iron stress is also accompanied by major changes in the accumulation of pigment-proteins in the thylakoid, and the appearance of stress-specific proteins in the cytoplasmic membrane (1–3). These changes are reflected in the gross morphologial differences between normally-grown and iron-limited cyanobacteria (4). Recent work from our laboratory has focussed on the identification of iron-regulated genes in the cyanobacterium *Anacystis nidulans* R2. Many of these genes encode proteins which are believed to be important in thylakoid and cytoplasmic membrane organisation during iron stress. We have recently cloned an iron-regulated gene, *irp*A (for-iron regulated protein) from a λgt11 expression library. The *irp*A gene was cloned by using an antibody to a 36 kD membrane protein which was found in cytoplasmic membrane fractions of iron-limited *A. nidulans* cells (3). DNA sequence analysis of *irp*A revealed an open reading frame capable of encoding a polypeptide of 38.6 kD. In this report, we discuss the cloning of *irp*A and we describe the construction of a site-directed mutant o *Anacystis nidulans* R2 generated by insertion of Tn5 into the *irp*A gene. This mutant is defective in iron-limited growth; furthermore, sequence analysis of the region upstream from the proposed transcription initiation site revealed a sequence structurally similar to the *Escherichia coli* operator recognized by the Fur repressor (5).

Materials and Methods

Strains, plasmids, phage and culture conditions. The plasmids, and the *E. coli, A. nidulans* R2 and lambda phage strains employed in this study are listed in Table 1. *E. coli* strains were grown on Luria-Bertani (LB) plates (6); *A. nidulans* was grown in BG-11 medium (7) in continuous light (0.5 mW/cm^2). Low-iron BG-11 was prepared as previously described (2).

Construction of the λgt11 genomic and λEMBL-3 libraries. The *A. nidulans* R2 λgt11 expression library was constructed essentially according to Young and Davis (8) with the following modification. Excess *Eco*RI linkers were separated by centrifugation on 10 to 40% linear sucrose gradients for 24 h at 25,000 rpm. Gradient fractions containing 0.5 to 7 kb fragments were pooled and dialysed against 10 mM TrisHCl, pH 7.5, 1 mM ethylenediaminetetraacetate (TE) buffer. DNA was extracted twice with phenol/chloroform and ethanol precipitated. The purified, size-selected fragments were ligated to dephosphorylated λgt11 arms using T4 DNA ligase, and the ligated DNA was packaged using Promega Biotech *in vitro* packaging extracts. A λEMBL-3 (9) library was similarly constructed by packaging ligated genomic fragments of ¯10–20 kb.

Screening of the λgt11 expression library. Screening of the λgt11 expression library was performed according to Young and Davis (8) except the Bovine Serum Albumin (BSA) was used for blocking instead of 20% fetal calf serum. Nitrocellulose filters were blocked in 5% BSA/Tween-20 for 1 h and incubated in CPVI-4 primary antibody (1:10,000) for 1 h. Filters were washed six times for 10 min each in 10 mM Tris, pH 7.5, 150mM NaCl containing 0.05% Tween 20 (TBS/Tween), after incubation in primary and secondary antibody. For the primary and secondary screening of the library,[125] I-labelled anti-rabbit immunoglobulin G (IgG) and horseradish peroxidase-conjugated goat anti-rabbit IgG were used, respectively.

TABLE 1

Bacterial strains, plasmids and phages

Strain/Phage/Plasmid	Relevant Characteristics	Source of Reference
E. coli		
Y1089	*lacU ProA*+*lon araD rpsL hsd (r*−*m*+*) str A supF trpC* pMC9(Ap^r)	8
Y1090	*lacU proA*+ +*lon araD strA hflA hsd*(r−m+) pMC9(Ap+)	8
BD1388		
JM83	*his ara leu thr trpA sup*+	11
	ara △ *(lac, proA, B) rspL 080 lacZ*	12
Anacystis nidulans		
R2	wild type	This laboratory
R2K7	R2 *irp*A::Tn5	This study
Bacteriophage		
λgtll	λ*lac*5 cI857 *nin*5 S100	8
λEMBL-3		9
λgtAN26	derivative of λgt11 with a 1.7 kb immunopositive *irp*A insert	3
λgtAn103	derivative of λgt11 with a 1.3 kb *irp*A insert	This study
λgtAn104	derivative of λgt11 with a 2.6 kb *irp*A insert	This study
λgtAN105	derivative of λgt11 with a 3.1 kb *irb*A insert	This study
λgtAN104.17	derivative of λgtAn 104 carrying *irp*A::Tn5 yielded by recombination between pRB96.37 and gt AN104	This study
λEM127	derivative of λEMBL-3 with a 10.8 kb kb *irp*A insert λEM130	This study
	derivative of λEMBL-3 with an 11.0 kb *irp*A insert	This study
λ467	λb221 *rex::*Tn5 cI857, *Oam Pam*	10
Plasmids		
pUC8	*Ap^r*	
pRB96	derivative of pUC8 containing a 700bp *Eco*RI-*Ava*l *irp*A internal fragment; expresses a 29 kDa fusion protein in *E. coli*	*This study*
pRB96.37	derivative of pRB96 bearing Tn5 inserted into the 700 bp*irp*A sequence	This study
pRB104.17	derivative of pUC8 bearing the *Eco*RI *irp*A::Tn5 insert from λgtAN104.17	This study

Growth experiments. *A. nidulans* R2 and K7 were diluted 1:10,000 from log-phase BG-11 into low-iron BG-11 medium. After the cultures had greened (~2 × 10⁶ cells/ml), the cultures were again diluted 1:1000 in low-iron BG-11. After the primary subculture from BG-11, the R2 cultures greened in 5 days, whereas the K7 cultures grew more slowly, greening after 7–8 days. The secondary subcultures into low-iron BG-11 were monitored for growth at 750 nm. For the iron chelation experiments log-phase cultures growing in BG-11 medium were treated with 2, 2′-bipyridyl at final concentrations of 100 μM and 300 μM.

Construction of the irpA::Tn5 insertions. The original *irp*A immunopositive phage clone, λgtAN26, contains a 1.7 kilobase (kb) *Eco*RI-linker adapted insert expressing a *lacZ-irpA* fusion protein in *E.coli* Y1089. We subcloned a 700 base pair (bp) *Eco*RI/*Ava*I fragment of λgtAN26 into the plasmid pUC8 (12); this construct, pRB96, expressed a 29 kD immuno-positive polypeptide in *E. coli* JM83 (3). Based on the DNA sequece of *irp*A, the *Eco*RI/*Ava*I fragment was an internal fragment of the gene, and thus a suitable target for mutagenesis with the kanamycin resistance transposon, Tn5. This was performed essentially as described by deBruijn and Lupski (11) by transforming pRB96 into a *sup*⁺ strain, *E. coli* BD 1388 (11), and infecting this strain with λ 467 rex::Tn5 at a multiplicity of infection (m.o.i.) of 2. After plating on LB medium that contained 20 μg/ml kanamycin, Km^r colonies arose at a frequency of 2×10^{-4} per recipient cell. Total plasmid DNA from the Km^r population was isolated and transformed into *E. coli* JM83. Km^r Ap^r transformants arose at a frequency of 2×10^{-7} per recipient, and 89 colonies were retained forDNA blot analysis to determine the site of Tn5 in each plasmid clone. From this population of Tn⁵ insertions, one plasmid, pRB96.37, was chosen for further study. The site of the Tn5 insertion was 100 bp upstream from the *Ava*I site, which in turn is 220 bp from the *irp*A termination codon.

The short region of sequence identity shared by pRB96.37 and the *A. nidulans* chromosome may have been insufficient to promote site-directed gene replacement following transformation. In order to generate a construct suitable for transformation into R2, an *irp*A::Tn5 insertion bearing large (~ 1 kb) flanking DNA sequences was required to promote recombination. Our strategy to retrieve such flanking sequences involved recombination between pRB96.37 and a λgt11-derived clone. λgtAN104, a phage bearing a 2.4 kb *Eco*RI-adapted fragment of *A. nidulans* DNA covering the *irp*A region. To accomplish this, we transformed pRB96.37 into *E. coli* Y1090, the host for λgt11 growth (8). A culture of *E. coli* Y1090 (pRB96.37) was infected with λgtAN104 at an m.o.i. of 2, and the resulting transducing lysate was then tested for the ability to form Km^r lysogens on *E. coli* Y1089. After plating, Km^r Y1089 derivatives arose at 4.8×10^{-7} per recipient. Twenty-four Y1090 lysogens were picked and screened by DNA blot analysis, and 19 of these yielded the same restriction pattern as the original *irp*A::TN5 insertion into pRB96 (3). One of these lysogens was induced, yielding phage λgtAN104.17. Since Tn5 lacks an *Eco*RI site, the 8.0 kb *Eco*RI insert containing the *irp*A::Tn5 insertion was subcloned into the *Eco*RI site of pUC8 (12). Two µg of this plasmid, pRB104.17, was used to transform *A. nidulans* R2 to Km^r. Km^r R2 colonies arose at a frequency of 2.5×10^{-7}/recipient, and these isolates were checked for altered growth in low-iron BG-11 medium. One clone, *A. nidulans* K7, was retained for further characterisation.

Results

Identification and restriction mapping of the *irp*A gene region. Restriction mapping of the *irp*A gene region in the chromosome was approached in two ways: (1) Southern analysis of the chromosomal DNA using inserts from λgtAN26 and λgtAN103 as probes; and (2) using larger *Sal*/I fragments from the λEMBL-3 recombinants (λEM127 and λEM130) as probes. Southern blots of the chromosomal DNA using different DNA probes are shown in Figure 1. The 1.7 kb *Eco*RI fragment from λgtAN26 hybridised to a unique band in many enzyme digestions. The probe identified two 1.7 and 1.5 kb *Pst*I bands, a 6.0kb *Sal*I band and 12.5 kb *Sst*I fragment. When the 1.3 kb *Eco*RI fragment from λAN103 was used as a probe, the Southern revealed a single band in *Pst*I. *Acc*I and *Sal*I digestions. Enzymes such as *Bgl*I and *Pvu*II showed hybridization to two bands (Fig 1).

To retrieve large fragments of the *A. nidulans* R2 chromosome containing the *irp*A gene, the λEMBL-3 library was screened by probing with the 3.1 kb DNA fragment covering *irp*A from λgtAN105. The screening of 8 plates, containing about 2000 plaques each, generated about 20 positives, 6 of which were characterised further. DNAs from four of these clones were digested with *Sal*I to release chromosomal fragments; all four clones retained chromosomal DNA ranging in size from 10 to 14 kb. Three DNA probes from two λEMBL-3 recombinants were used to map the entire *irp*A region. The 6.0 and 4.5 kb *Sal*I fragments from λEM130 and the 3.5 kb *Sal*I fragment from λEM127 were used individually and in combination to probe chromosomal DNA digested with *Sal*I. *Bam*HI, and *Hind*III (Fig IC). The Southern blots allowed us to map a 20 kb region on the *A. nidulans* R2 chromosome.

DNA sequence analysis of the *irp*A gene and flanking regions. M13 clones carrying deletions on either side of the insert were sequenced using a universal primer; as a result, both strands of the gene were completely sequenced. The DNA sequence (Fig. 2) shows an 1068 bp open reading frame with a 300 bp 5'-and 200bp 3'-sequence. This open reading frame is capable of coding for a protein of 356 amino acids with a predicted molecular weight of 38,584 daltons. The 5'-terminus of the *irp*A gene

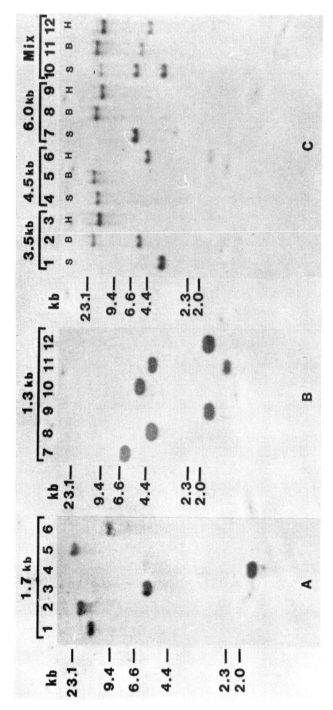

Fig. 1: Hybridisation of DNA probes to A. nidulans R2 chromosomal DNA digested with various restriction enzymes. Panel A was probed with the 1.7 kb *EcoRI* fragment from λgtAN26 and panel B was probed with the 1.3 kb *EcoRI* fragment from λgtAN103. Restriction enzyme digestions of the chromosomal DNA was as follows: Lane 1, *HindIII*; lane 2, *BamHI*; land 3, *SalI*; lane 4, *PatI*; lane 5, *KpnI*; lane 6, *SstI*; lane 7, *SalI*; lane 8, *AccI*; lane 9, *PstI*; lane 10, *BglI*; lane 11, *PvuII* and lane 12, *Aval*. In panel C, *SalI* fragments from λEMBL-3 recombinants were used as probes. The *A. nidulans* R2 chromosomal DNA was digested with *SalI* (lanes 1, 4, 7 and 10), *BamHI* (Lanes 2, 5, 8 and 11), and *HindIII* (lanes 3, 6, 9 and 12). The 6 kb and 3.5 kb *SalI* fragments were from two λEMBL-3 recombinants, λEM120 and λ130. The DNA probes used in hybridisation are indicated on the top of the lanes (S, *SalI*; B, *BamHI* and H, *HindIII*).

```
                    -300                                              -250
                    CAAAGGCGATCGTGGCTGGTGCTAGCCCCAACCAAAGTCCTGTTCCCATCCAAAGGGCTGCAG
       pKJ49
        ↓                                    -200
       GTACGTGCATACTCACTCCTCTTTCTAAAAATGATTATTATTCTCATTTTTAAGAGTGTCAATTGACTACTTTTCTTCC

       -150                                              -100
       GTAACTCATGGCCTAGCTCTGTTTCTGACGTTGCTATGGCCAAGCTTGCGAGACACCGATTCACCAGGGAAGTAGTGGA
                                    pKJ34
                              -50    ↓
       GCCTACGTGGAGTGTGATGCCACACTTCCTTGCCCGCGAGTGATCGCTTTAGTGAGAGGTTGTTCGGTCGTAGTGGCTG
```

```
                                                             50
GTG ATC GTG ACA GGC TCT CAG GTC AGG CAA GGT TTA AAC ACT TGG TTT GTG CTC CCG CTG
Val Ile Val Thr Gly Ser Gln Val Arg Gln Gly Leu Asn Thr Trp Phe Val Leu Pro Leu    20
pKJ42                          pKJ33
  ↓                               ↓                     100
CGT AGG ACT GCG ATC GGC CTG GGC TGC GCC GGA GTT GCA ACG CTC TTC TCT GCC TGT GGT
Arg Arg Thr Ala Ile Gly Leu Gly Cys Ala Gly Val Ala Thr Leu Phe Ser Ala Cys Gly    40

                              150
CAA ACC CAG GCA TTG ATT ACC AAT CAG ACC ATT CAA GGA TTT GTC GAT CAG GTT GTC GTT
Gln Thr Gln Ala Leu Ile Thr Asn Gln Thr Ile Gln Gly Phe Val Asp Gln Val Val Val    60

              200
CCT AGC TAT GTC AGC GTT GCT GCT GGC GCA ACT CAG CTG GAA CAA GCC CTC CAA ACC TAT
Pro Ser Tyr Val Ser Val Ala Ala Gly Ala Thr Gln Leu Glu Gln Ala Leu Gln Thr Tyr    80

      250                                                             300
CAG CAG GCA CCG ACT GCT GCC AAT TTG GAG GCG GCT CGA CAA GCC TGG CGG GTC GCC CGC
Gln Gln Ala Pro Thr Ala Ala Asn Leu Glu Ala Ala Arg Gln Ala Trp Arg Val Ala Arg   100

                                                350
GAT CGC TGG GAG CAG ACT GAA TGT TTT GCT TTT GGG CCA GCG GAT AGC GAA GGG TTT GAT
Asp Arg Trp Glu Gln Thr Glu Cys Phe Ala Phe Gly Pro Ala Asp Ser Glu Gly Phe Asp   120

                                        400
GGG GCA ATG GAC ACC TGG CCT ATC GAT CGC CAA GGC TTG AAA ACT GCC GCA GCT CAG CCA
Gly Ala Met Asp Thr Trp Pro Ile Asp Arg Gln Gly Leu Lys Thr Ala Ala Ala Gln Pro   140
pKJ43
  ↓                         450
GTG GAG CAA CGG GAA GAT AGC CGT AAG GGC TTC CAC GCG ATC GAG GAG TTG TTG TTT GCC
Val Glu Gln Arg Glu Asp Ser Arg Lys Gly Phe His Ala Ile Glu Glu Leu Leu Phe Ala   160

              500
GCA ACG GAA CCG ACG CTG AGC GAT CGC CAG CAT CTT GTG ATC TTG GCG ACG GAC CTT ACC
Ala Thr Glu Pro Thr Leu Ser Asp Arg Gln His Leu Val Ile Leu Ala Thr Asp Leu Thr   180

      550                                                     600
AAG CAA GCA CAG GGG TTG GTC ACC CGT TGG CAA CAA GCG AGT GAT CAG CCT GCC TAT CGC
Lys Gln Ala Gln Gly Leu Val Thr Arg Trp Gln Gln Ala Ser Asp Gln Pro Ala Tyr Arg   200

                                              650
TCA GTT TTG CTC AGC GCT GGC TCG ACA GAT TCG GCC TAT CCC ACC CTG AAT GCT GCG GGA
Ser Val Leu Leu Ser Ala Gly Ser Thr Asp Ser Ala Tyr Pro Thr Leu Asn Ala Ala Gly   220

                                      700
ACC GAG ATT GTT CAA GGC CTG GTT GAT AGC CTC TCA GAG GTC GCC AGC GAA AAG ATC GGC
Thr Glu Ile Val Gln Gly Leu Val Asp Ser Leu Ser Glu Val Ala Ser Glu Lys Ile Gly   240

                              750
GGG CCA CTC GAG ACT CAA GAA CCC GAT CGC TTT GAA AGT TTT GTT AGC CGC AAT ACT CTG
Gly Pro Leu Glu Thr Gln Glu Pro Asp Arg Phe Glu Ser Phe Val Ser Arg Asn Thr Leu   260

                      800
TCT GAC CTG CGC AAC AAC TGG ACT GGC GCT TGG AAT GTC TAT CGC GGT CAG CGG TCT GAT
Ser Asp Leu Arg Asn Asn Trp Thr Gly Ala Trp Asn Val Tyr Arg Gly Gln Arg Ser Asp   280

          850                                                     900
GGG GTC GCG GCA GGA AGT CTG CAA CAG CGT TTA CAG CAA CAA CAT CCA GTG ATC GCT CAG
Gly Val Ala Ala Gly Ser Leu Gln Gln Arg Leu Gln Gln Gln His Pro Val Ile Ala Gln   300

                                              950
CAA CTC GAT CAG CAA TTT GCA ACT GCC CGC CAA GCC CTT TGG GCT ATT CCT GAA CCG ATT
Gln Leu Asp Gln Gln Phe Ala Thr Ala Arg Gln Ala Leu Trp Ala Ile Pro Glu Pro Ile   320

                                      1000
GAA ACC AAC CTT GCC AGC CCA AGA GGC AAA GTG GCT GTC CTC ACG GCT CAA ACT GCG ATC
Glu Thr Asn Leu Ala Ser Pro Arg Gly Lys Val Ala Val Leu Thr Ala Gln Thr Ala Ile   340

                              1050
GCA GCA GTC AGC GAC ACC CTA GAG CGT CAA GTT CTC CCG CTG GTT CAG TAG CTGATGACTCT
Ala Ala Val Ser Asp Thr Leu Glu Arg Gln Val Leu Pro Leu Val Gln  *

              1100                                              1150
CCCAGGGATTCCCCAACGACGCTGGATCAAACGCTCTCGACTGAGCTGGGCTGGACTCGGGTTAATTCTGAGCCTCGTT

                              1200
CTGGGCTGGTGGTGGATGGCGCCAGAATCAGTTGCAGCA
```

Fig. 2: The nucleotide sequence of the *irp*A gene, and the derived primary structure of the IrpA protein. The dyad symmetry found in the upstream DNA sequence is indicated by paired arrows above the sequence.

contains a 9 bp inverted repeat about 200 bp upstream from the predicted start codon and the inverted repeats are separated by a 9 bp AT-rich segment. The stop codon TAG was immediately followed by two more stop codons in another reading frame.

Analysis of the derived amino acid sequence. The derived amino acid sequence of the *irp*A gene was analysed with the aid of the Dnastar protein analysis programmes. The protein contained 142 hydrophobic and 107 polar amino acids, yielding an approximate derived isoelectric point of 4.87. Several repeats of 3 or 4 amino acids were found within the sequence: Val-Ala-Ala-Gly and Gln-Gly-Leu-Val were each repeated twice. Similarly, Pro-Thr-Leu was also repeated twice. The hydropathic index of the protein revealed some interesting features, as the amino terminus through residue 44 had characteristics typical of a signal sequence. In *A. nidulans* R2, one of the thylakoid lumenal proteins involved in water oxidation has been shown to possess a signal sequence of 28 amino acids (13). When the presumptive signal sequence was excluded from the IrpA protein, the remaining sequence showed bilateral symmetry in the hydropathy plot. As shown in Figure 3, the two ends of the protein were hydrophobic and amino acid sequences from 75 to 145 and 240 to 310 were hydrophilic in nature.

Characterisation of the Tn5 insertion into A. nidulans K7. DNA was prepared from *A. nidulans* K7, digested with restriction endonucleases and

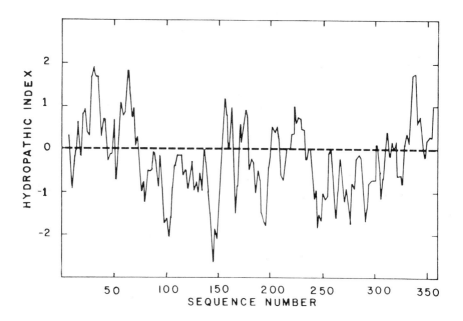

Fig. 3: The hydropathy profile of the IrpA protein according to Kyte and Doolittle (16), using a window of 11 amino acid residues.

transferred to nitrocellulose. Probing these blots with the insert of phage λgtAN103, a 1.3 kb fragment containing the *irp*A gene, revealed that Tn*5* was inserted into the correct site in the *A. nidulans* genome (Figure 4). Since *irp*A resides on a 24 kb *Eco*RI fragment, and that Tn*5* lacks *Eco*RI sites (14); the Southern blot of K7 DNA revealed that *irp*A was carried on a larger *Eco*RI fragment than in the wild type strain (Figure 4). Furth-ermore, knowing both the position of *Ava*I sites 500 bp from the ends of Tn*5* and the restriction map of the *irp*A region (14), *Pst*I/*Ava*I double digests confirm the position of Tn*5* 100 bp upstream of the *Ava*I site in *irp*A. Insertion of Tn*5* yields two fragments of 1300 and 600 bp which are absent in the wild type strain (Figure 4). From these studies it is evident that the *irp*A::Tn*5* insertion constructed in *E. coli* has replaced the *irp*A gene by homologous recombination, following transformation into *A. nidulans* R2.

Growth characteristics of K7. *A. nidulans* K7 was chosen from the Km^r

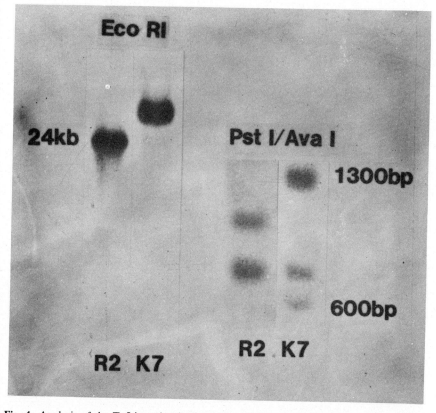

Fig. 4: Analysis of the Tn*5* insertion in *A. nidulans* K7 (*irp*A::Tn*5*). Left panel, Southern blots of *Eco*RI digests of *A. nidulans* R2 and K7 DNA were probed with the *Eco*RI fragment of λgtAN103; right panel, λgtAN103 was used to probe *Pst*I/*Ava*I double digests of R2 and K7 DNA.

population of *A. nidulans* R2 transformants because *A. nidulans* K7 grew very poorly in iron-depleted BG-11 medium. Whereas R2 can be continuously subcultured in low-iron BG-11, K7 cannot grow after two

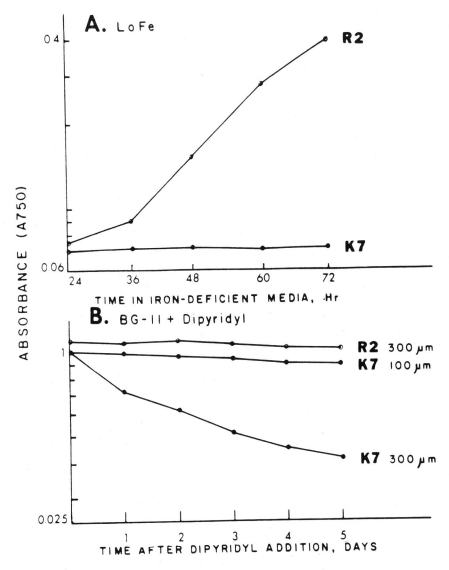

Fig. 5: *Panel A*—Growth characteristics of *A. nidulans* R2 and the *irp*A::Tn5 derivative, K7. *A. nidulans* R2 and K7 cells grown in iron-containing medium were diluted 1:1000 into low-iron BG-11 (2), subcultured once more at a 1:1000 dilution and monitored for growth by light scattering at 750 nm. *Panel B*—Sensitivity of *A. nidulans* R2 and K7 to 2.2′-bipyridyl. Log phase cells grown in iron-containing medium were treated with bipyridyl at the indicated concentrations. Note that bipyridyl inhibits growth of the wild-type cells; however, only the K7 culture is killed at the 300 μM concentration.

subcultures in this medium (Figure 5A). Furthermore, BG-11-grown K7 exhibits increased sensitivity to the iron-chelating agent 2.2′-bipyridyl (Fig. 5B). Addition of bipyridyl to a final concentration of 300 μM rapidly kills log phase K7 cultures but the wild-type strain survives this treatment. Bipyridyl, however, affects growth of the wild-type strain; presumably this is due to secondary toxic effects of the agent (Figure 5B).

Bipyridyl-treated R2 and K7 were also examined spectroscopically to determine whether these cultures exhibited the spectral alterations diagnostic for iron limitation. Previous work has shown that visible absorption spectra of iron-stressed *A. nidulans* yield a chlorophyll absorption peak at 672 nm, 8 nanometers blue-shifted relative to that of normally-grown cells. Furthermore, the 77K fluorescence emission spectrum of low-iron cells is composed of a single major peak at 686 nm (2). Whereas *A. nidulans* R2 could be treated with 300 μM bipyridyl without exhibiting these iron-stress-specific spectral characteristics, the K7 strain yielded spectra that were indicative of iron starvation after addition of 100 μM bipyridyl (data not shown).

Discussion

We have described the construction of a Tn5-induced site-directed mutant of *A. nidulans* R2 which inactivates the *irp*A gene. The construction strategy of this mutant has proven to be an efficient means for isolation of Tn5-mutagenised clones suitable for transformation into *A. nidulans* R2. This study has shown that the *irp*A region encodes a function of functions necessary for prolonged growth of *A. nidulans* R2 under iron-limiting conditions. Furthermore, the fact that K7 can yield the spectral changes associated with iron stress after bipyridyl treatment suggests that the K7 mutant is not altered in the ability to reorganise the thylakoid membrane as a consequence of low-iron growth. Therefore, it is likely that *irp*A encodes a function involved in iron metabolism and not in iron-stress-mediated thylakoid biogenesis.

A feature of the *irp*A gene which has attracted our attention is the structural similarity of the 5′ upstream region of *irp*A to the iron-regulated operator of *E. coli*(5, 15); this is the sequence which is recognised by the Fur repressor (Figure 6). Thus region of similarity may represent a conserved sequence involved in iron regulation in bacteria. We will soon begin functional studies to examine whether this sequence confers iron-regulated negative control on a reporter gene. If so, this will be the first report of an operator sequence in a cyanobacterium.

The antibody preparation that led to the cloning of the *irp*A gene has also allowed us to obtain a great deal more information about the response of *Anacystis nidulans* R2 to iron-deficiency. The antibody was originally generated against acrylamide gel-purified chlorophyll-protein complex

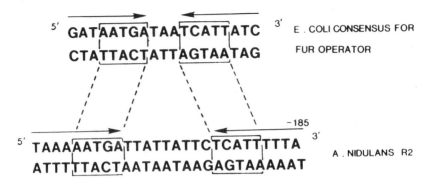

Fig. 6: Sequence comparison between the iron-regulated *fur* operator of *E. coli* and the 5' upstream region of the *A. nidulans* R2 *irp*A gene.

CPVI-4; this complex is the major chlorophyll binding complex in iron-deficient cells (1). This antibody preparation was eventually determined to cross-react against three iron inducible proteins at 36, 35 and 34 kDa. The *irp*A gene product corresponds to the 36 kDa protein and is the most strongly inducible upon growth in iron-deficient conditions. Using a variety of techniques, we have determined that *irp*A is essentially absent in normally grown cells, whereas the other two antigens are present in low quantities under normal conditions. This antibody preparation has enabled us to purify the individual proteins and to learn a great deal more about the metabolic changes that occur in cyanobacteria under iron stress. The other proteins that cross-react with this antibody include the apoprotein of the chlorophyll-protein complex CPVI-4 and a 35 kDa protein which appears to be membrane-associated.

The 35 KDa protein (GP35) is one of the most abundant membrane-associated proteins in iron-stressed cells. GP35 has been purified to homogeneity and future analysis of its structure should prove to be very interesting. The protein is very basic, extrinsic, and ConA-reactive; ConA-reactivity can indicate that this protein is glycosylated. We have also determined that four of the phycobilisome linker polypeptides are ConA-reactive and definitely contain carbohydrate (17). Therefore, there may be some relationship between this protein and the phycobilisome linkers. Interestingly, GP35 is produced is large quantity at about the same time that glycogen granules appear within the cell during iron-stress. Therefore, this glycoprotein may be somehow involved in the steps that lead to glycogen granule formation. We have now obtained a 25 residue N-terminal amino acid sequence of this protein, and we should be able to study its structural properties in more detail.

The chlorophyll-protein complex, CPVI-4, has been biochemically purified, using anion exchange chromatography of maltoside-solubilised low-iron thylakoids (Riethman and Sherman, unpublished results). The complex consists of polypeptides migrating at apparent molecular masses of 36, 34, and 12 kDa, and both the 36 and the 34 kDa proteins are involved in chlorophyll binding. It is the 34 kDa protein that cross-reacts with the CPVI-4 antibody; on the other hand, the 36 kDa protein is not cross-reactive. This protein has some interesting physical properties, and does not easily transfer from gels to nitrocellulose; thus, we originally did not detect this protein. Now that the complex can be purified, we will isolate more of the 36 kDa protein and prepare antibdoy against the protein. We will thus be in a position to clone and sequence both of these chlorophyll-binding proteins in the near future.

The functional and structural results obtained to date indicate that CPVI-4 functions as a poor antenna complex in membranes of iron-deficient cells. However, analyses of membranes during reconstitution to normal conditions indicate a potentially more important role for CPVI-4. It is possible that newly synthesised apoproteins of the larger PSI complexes and of PSII interact physically with CPVI-4 in the membrane. Since CPVI-4 binds chlorophyll less stably than the other chlorophyll proteins, the net result of such interactions is a transfer of chlorophyll from CPVI-4 to newly synthesised apoproteins. Therefore, it is possible that his chlorophyll-protein complex is involved in the normal assembly of chlorophyll-protein complexes. The complex is present in very low amounts in normal cells, but becomes the predominant chlorophyll species in iron-deficient membranes. It will be of great interest to determine if this chlorophyll complex is involved either actively or passively with membrane development during recovery from iron stress.

Therefore, we have now developed a number of molecular tools that should help us to analyse metabolic events that take place during growth of cyanobacteria under iron stress and subsequent reconstitution to normal cells. These studies will be greatly aided by the mutant. *A. nidulans* K7 which is unable to grow under iron-deficient conditions. This mutant should allow us to synchronise conditions very carefully, and to precisely study reconstitution to normal conditions. It will also be very valuable for analysis of iron storage and acquisition in the cyanobacteria. Since the nucleotide sequence of *irp*A indicates that iron regulation in cyanobacteria may be very similar to that of *E. coli*, we will be able to rely heavily on the great body of knowledge accumulated in *E. coli* as we continue our analysis in the cyanobacteria.

References

1. Pakrasi, H. B., Riethman, H. C. and Sherman, L. A. (1985). *Proc. Natl Acad. Sci. USA*. 82, 6903–6907
2. Pakrasi, H. B., Goldenberg, A. and Sherman, L. A. (1985), *Plant Physiol.* 79, 290–295
3. Sherman, L. A., Reddy, K. J., Riethman, H. C. and Bullerjahn, G. S. (1987). In: *Progress in Photosynthesis Research*, Vol. IV (Biggins, J., ed.) pp. 773–776, Martinus Nijhoff Publishers, Dordrecht

4. Sherman, D. M. and Sherman. L. A. (1983). *J. Bacteriol.* 156, 393–401
5. DeLorenzo, V., Wee, S., Herrero, M. and Nielands, J. B. (1987). *J. Bacteriol.* 169, 2624–2630.
6. Davis, R. W., Botstein, D. and Roth J. R. (1980). *Advanced Bacterial Genetics.* Cold Spring Harbor Laboratory, New York
7. Allen, M. M. (1968). *J. Phycol* 4, 1–4
8. Young, R. A., Bloom, B. R., Grosskinsky, C. M., Ivanyi, J., Thomas, D. and Davis, R. W. (1985). *Proc. Natl Acad. Sci. USA.* 82, 2583–2587
9. Frischauf, A., Lehrach, H., Poustka, A. and Murray, N. (1983). *J. Mol. Biol.* 170, 827–842.
10. deBruijn, F. J. and Lupski, J. R. (1984). *Gene* 27, 131–149
11. Warner, H. R., Thompson, R. B., Mozer, T. J. and Duncan, B. K. (1979). *J. Biol. Chem.* 254, 7534–7539
12. Vieira, J. and Messing, J. (1982). *Gene* 19, 259–268
13. Kuwabara, T., Reddy, K. J. and Sherman, L. A., (1987). *Proc. Natl Acad. Sci. USA* 84, 8230–8234
14. Jorgensen, R. A., Rothstein, S. J. and Reznikoff, W. S. (2979). *Mol. Gen. Genet.* 177, 65–72
15. Bagg, A. and Neilands, J. B. (1987). *Microbiol. Rev.* 51, 509–518
16. Kyte, J. and Doolittle, R. F. (1982). *J. Mol. Biol.* 157, 105–132
17. Riethman, H. C., Mawhinney, T. P. and Sherman, L. A. (1987). *FEBS Lett.* 215, 209–214

Primary Structure and Expression in *Escherichia coli* of the Mn-stabilising Protein Involved in Photosystem II Water Oxidation

TOMOHIKO KUWABARA

Department of Chemistry,
Faculty of Science,
Toho University, Funabashi 274, Japan

Summary

Comparison of amino acid sequences of Mn-stabilising protein (MSP) of *Anacystis nidulans* R2 and spinach reveals 5 clusters of conserved residues. Domains essential for the functioning of MSP are likely to be situated in these clusters. The two cysteine residues of MSP are conserved. They are likely to form a disulfide bond which plays an important role in maintaining the tertiary structure of the protein.

The precursor of the *Anacystis* MSP possesses an N-terminal signal peptide composed of 28 amino acid residues, the structure for translocation across the thylakoid membrane. The following has been found upon expression of the MSP gene in *Escherichia coli*. (I) The expression gives a polypeptide of 30 kDa, which is indistinguishable in size from the authentic mature MSP although the gene directs the precursor. (II) The 30-kDa polypeptide is located at the outer surface of the cytoplasmic membrane. These findings suggest that the precursor MSP synthesised in *E. coli* is processed to the mature protein by a signal peptidase of *E. coli*.

Introduction

Photosystem II (PSII) is a multi-subunit complex which catalyses oxidation of water and reduction of plastoquinone by utilising light energy. The water-oxidising centre of the PSII complex is located at the lumenal surface of the thylakoid membrane and contains four Mn atoms. Two of the four Mn atoms are tightly bound to the membrane, whereas the other two are weakly bound, probably sandwiched between the intrinsic Mn-binding protein and the Mn-stabilising protein (MSP, the so-called extrinsic 33-kDa protein) (1). The MSP of spinach can be removed from the oxygen-evolving PSII particles with a solution of pH 9.3 (2), 2.6 M urea (3) or 1 M $CaCl_2$ (4). When the removal is performed in the absence of concentrated Cl^-, the weakly bound Mn atoms are released and the oxygen evolution activity is inactivated (3). If the Mn atoms are kept bound by concentrated Cl^- in the course of the removal (3,4), the MSP can rebind to the centre and restore the activity (3–5). The MSP is one of the components of the minimum PSII unit capable of evolving oxygen [6–8]. Recent studies on kinetics of thermoluminescense (9) and flash-induced oxygen release (10) indicate that the MSP accelerates the S-state transition from S_3 to S_4.

In this article, studies on the primary structure of MSP are summarised, and the structure and function of the MSP as well as its N-terminal presequence discussed. Expression and processing of the MSP of cyanobacterium *Anacystis nidulans* R2 in *Escherichia coli* are also described.

Primary Structure

Primary structure of mature MSP was first determined on the spinach protein by amino acid sequencing (11). It was shown that there are two types of molecules which differ only in the C-terminal amino acid; one type ends with a single Gln residue and the other with two Gln residues. The former consists of 247 amino acid residues with a molecular weight of

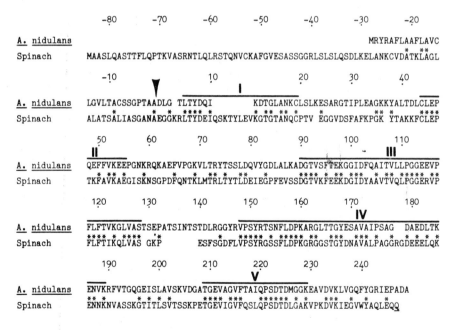

Fig. 1: Alignment of the deduced precursor MSP sequences of *A. nidulans* R2 and spinach. Conserved amino acid residues are indicated by asterisks. Tentatively assigned clusters of conserved residues are shown by horizontal bars and roman numerals. The arrowhead indicates the processing site. The numbering of residues is according to the *Anacystis* MSP. The sequences are from Kuwabara *et al.* (14) and Tyagi *et al* (13). Note that the underlined ln residue at the C-terminus of the spinach protein was not found in the cDNA clones of Tyagi *et al.* (13), but detected in the amino acid sequencing by Oh-oka *et al.* (11).

26,535; the latter has 248 residues with a molecular weight of 26,663. Watenabe *et al.* (12) sequenced a cDNA clone encoding the pea MSP, and indicated that the mature pea MSP is very similar to the spinach protein, showing 86% homology. This cDNA clone suggested the presence of the N-terminal presequence, but unfortunately, did not contain the complete coding region for it. Tyagi *et al.* (13) sequenced a full-size cDNA encoding the spinach MSP. The deduced amino acid sequence suggested that the N-terminal presequence is composed of 84 residues (Fig. 1). The deduced amino acid sequence for the mature protein agrees well with the sequence obtained by amino acid sequencing (11), and ends with a single Gln residue. Whether the heterogeneity at the C-terminus (11) comes from a post-translational modification or from a multigene family remains to be clarified. The gene coding for MSP was cloned and sequenced from the genome of cyanobacterium *Anacystis nidulans* R2 (14). The gene was shown to be present in a single copy in the cyanobacterial genome, and named *woxA* after *w*ater *ox*idation (14). The deduced amino acid sequence suggested that the translation product is a precursor consisting of 277 amino acid residues with a molecular weight of 29,306. Comparison of the sequence with that of the mature spinach MSP suggested that the N-terminal presequence of the *Anacystis* protein contains only 28 residues, and that the mature protein has 249 residues with a molecular weight of 26,462 (Fig. 1).

Conserved Amino Acid Residues in Mature MSP

In the alignment of the mature MSP sequences of *Anacystis* and spinach, there are several insertions and deletions (Fig. 1). The sequence homology is as low as 49%, reflecting the taxonomical distance between the cyano-bacterium and spinach. However, the conserved amino acid residues are not evenly distributed but clustered. In tentatively assigned five clusters (Fig. 1), the local homology values are in a range of 66–70%, suggesting that domains essential for the functioning of MSP are situated in these clusters.

It is noteworthy that the two cysteine residues in MSP are conserved, one in cluster I and the other in cluster II. These cysteine residues are likely to form a disulfide bond; it has been suggested that the spinach MSP has a disulfide bond, and that its reduction causes a significant conforma-tional change to the protein (15). Tanaka and Wada (16) have recently shown that the spinach MSP retaining the disulfide linkage can rebind to the MSP-depleted PSII particles and restore the oxygen-evolving activity, but the protein reduced with a SH-reductant cannot. This implies that the tertiary structure conferred by the disulfide linkage is essential in the functioning of MSP. The cysteine residue in cluster I and the neighbouring regions in spinach MSP have been proposed as a candidate for the site interacting with Mn atom(s), based on partial sequence homology to the putative Mn-binding site of bacterial Mn superoxide dismutases (11). Considering the presence of the disulfide linkage, the N-terminal portion of cluster II would also be important in the possible interaction. Nevertheless, specific amino acid residues responsible for the interaction with the Mn atom(s), the intrinsic protein(s) or the extrinsic 24-kDa protein (17,18) have not yet been identified. Site-directed mutagenesis study with the *woxA* gene and/or the cDNA clones seems to be one of the most promising approaches to elucidate the structure-function relationship of MSP.

Structure and Function of the N-terminal Presequence

The N–terminal presequence of MSP represents the structure which directs the protein to assemble to the PSII complex. In higher plants, the MSP is encoded by nuclear DNA and synthesised in cytoplasm as a precursor of 39 kDa (19,20). The precursor is post-translationally transported into chloroplast, and undergoes processing to reach its final destination, the lumenal surface of the thylakoid membrane. It has been shown by *in vitro* experiments that the 39-kDa precursor is processed to an intermediate size of 35 kDa by incubation with stroma, and then to the mature size by subsequent incubation with thylakoid membranes, and that the first processing is the prerequisite of the second one (12). It should be noted that the transit peptide of the spinach MSP consists of 84 amino acids (Fig. 1 and Ref. 13), which is much longer than that of nuclear-encoded stromal or thylakoidal proteins, such as the small subunit of ribulose-1,5-bisphosphate carboxylase/oxygenase and the apoproteins of the light-harvesting chlorophyll *a/b*-protein complex II (21). It is very likely that the N-terminal portion of the transit peptide of the thylakoid lumenal protein is responsible for import into chloroplast and cut off in the stroma, and the C-terminal portion is for transport across the thylakoid membrane and excised at the lumenal surface of the thylakoid membrane. The occurrence of this kind of two-step processing has been proposed also for other nuclear-encoded proteins located in the thylakoid lumen, the plastocyanin(22) and the extrinsic 24-kDa protein of PSII (23). The intermediate processing site has not yet been determined in any of the protein species.

In contrast to the transit peptide, the presequence of *Anacystis* MSP is noticeably short, composed of 28 amino acids (Fig. 1). Importantly, the structural profile of the presequence is quite similar to that of the C-terminal portion of the transit peptide; both show a hydrophobic stretch of non-charged amino acid residues, a couple of positively charged residues at the N-side of the hydrophobic stretch, an alanine residue at the -1 position of the processing site, and amino acid residues known to break ordered secondary structures (24) at the -4 and -5 positions. This similarity strongly suggests that these two polypeptide parts have the same function, namely transport of the protein across the thylakoid membrane. The absence of the N-terminal portion of the transit peptide in the cyanobacterial presequence is consistent with the identification that the N-terminal portion is the structure required for import into chloroplast; such import event does not occur in cyanobacteria.

Expression and Processing of the *Anacystis* MSP in *E. coli*

The *Anacystis* MSP can be expressed in *E. coli* JM105 transformed with a pUC18-derived plasmid, pTK2, which has the *XbaI-Hind*III *Anacystis* DNA fragment containing the *woxA* gene (14). In pTK2, the direction of the transcription of *woxA* is the same as that of *lacZ*. The following is a fundamental profile of the expression. The cells harbouring pTK2 produce two immuno-stainable polypeptides, a heavily stained one at 30 kDa and a faintly stained one at 31 kDa (Fig. 2), in the presence of the inducer isopropyl β-D-thiogalactopyranoside (IPTG). On the other hand, the cells with pTK1 (14), the pUC19-derived plasmid carrying the *Anacystis* DNA

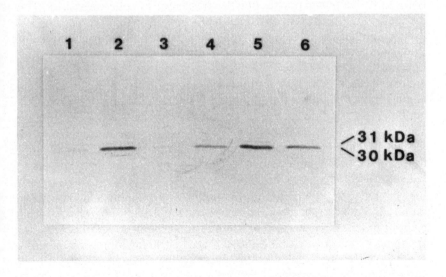

Fig. 2: Immunoblot of polypeptides of *E. coli* JM105 harbouring pTK1 or pTK2, and of *Anacystis* thylakoid membranes. *E. coli* cells were grown at 37°C in M9CA medium (25) containing 10 μg ml⁻¹ thiamine, 40 μg ml⁻¹ ampicillin and 25 μg ml⁻¹ streptomycin in the presence or absence of 1mM IPTG. The cells in the early stationary state were subjected to SDS-PAGE and Western-blotted using antiserum raised against spinach MSP. Lane 1, cells harbouring pTK1 grown in the presence of IPTG, (pTK1/+IPTG); lane 2, (pTK2/+IPTG); lane 3, (pTK1/−IPTG); lane 4, (pTK2/−IPTG); lane 5, (pTK2/+IPTG) plus *Anacystis* thylakoid membranes; lane 6, *Anacystis* thylakoid membranes.

fragment, gives only a faintly stained polypeptide at 30 kDa. The MSP-related polypeptides are produced from pTK2 even in the absence of IPTG, although the productivity is rather low. The expression level with pTK1 in the absence of IPTG is similar to that in its presence. These findings suggest that the expression of woxA in pTK2 is controlled by the lac promoter in the vector. The expression in the absence of IPTG may be explained by the high copy number of pUC18; the lac repressor molecule is produced by a single gene located on the chromosome of JM105. The production of the small amount of the 30-kDa polypeptide from pTK1 may suggest that the Anacystis DNA fragment possesses a promoter for the woxA but it is not well recognised by the E. coli RNA polymerase.

Interestingly, the 30-kDa expression product is indistinguishable in size from the authentic mature Anacystis MSP (Fig. 2). This suggests that the translation product is processed by a signal peptidase of E. coli. It is well known that processing is closely related to translocation across membrane. The following experimental results have indicated that both of the two polypeptides are located at the outer surface of the cytoplasmic mambrane. (I) When the cell materials were fractionated into periplasm, cytoplasm and membrane fraction (26), these polypeptides were found almost exclusively in the membrane fraction. (II) Trypsin treatment of cells did not digest the polypeptides, but the treatment of spheroplasts did (Fig. 3A). The integrity of the spheroplasts was proved by the fact that a 43-kDa polypeptide of E. coli, which has been shown to be localised in the cytoplasm (date not shown), is not digested by the treatment (Fig. 3B). The treatment of broken spheroplasts digested the 43-kDa polypeptide, indicating that it is not intrinsically trypsin-resistant. Failure to find degradation products of the 30- or 31-kDa polypeptide on the immunoblot implies that these polypeptides are not deeply buried in the membrane. The location of the 30-kDa polypeptide reinforces the interpretation that the translation product is processed to the mature MSP in E. coli. Whether the 31-kDa polypeptide is the translocated precursor or a processing intermediate remains to be clarified.

It has been reported that expression of a cDNA encoding preprospo-ramin A of sweet potato in E. coli gives a polypeptide with the size of prosporamin A (27), suggesting the occurrence of the processing in E. coli. It should be surprising that the N-terminal portion of the presequence of the sporamin A, which is believed to be excised in the in situ co-translational processing from preprosporamin to prosporamin by the processing enzyme in the endoplasmic reticulum (27), is quite similar to the presequence of the Anacystis MSP (Fig. 4). This similarity suggests that the processing enzymes of the thylakoid membrane, endoplasmic reticulum and E. coli have a similar substrate specificity. In contrast to the expression product of the woxA, which is bound to the outer surface of the cytoplasmic membrane, the prosporamin-A-like polypeptide produced in E. coli (27) has recently been shown to be localised in the soluble

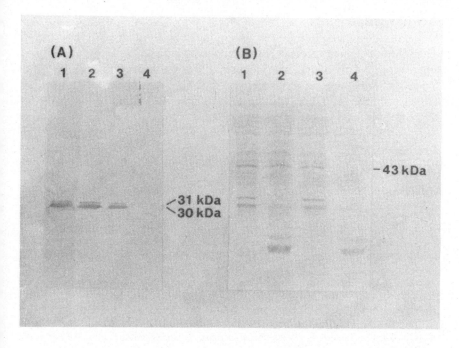

Fig. 3: Change of polypeptide profiles of *E. coli* cells, spheroplasts and broken spheroplasts by trypsin treatment. *E. coli* cells harbouring pTK2 were cultured in the presence of IPTG as stated in the legend to Fig. 1. Spheroplasts and broken spheroplasts were prepared, and trypsin treatment was performed according to Minsky *et al.* (26). (A), Immunoblot of the polypeptides of the cells and spheroplasts. Lane 1, untreated cells; lane 2, trypsin-treated cells; lane 3, untreated spheroplasts; lane 4, trypsin-treated spheroplasts. (B), CBB-stained gel showing the polypeptides of the spheroplasts and broken spheroplasts. Lane 1, untreated spheroplasts; lane 2, trypsin-treated spheroplasts; Lane 3, untreated broken spheroplasts; lane 4, trypsin-treated broken spheroplasts.

Fig. 4: Alignment of presequences of *A. nidulans* R2 MSP and sweet potato sporamin A. Arrowheads indicate the sites at which presumable processing in *E. coli* takes place. Homologous and similar amino acids are indicated by asterisks and colons, respectively. The presequence of sweet potato sporamin A is from Hattori *et al.* (27).

periplasmic fraction (K. Nakamura, personal communication). The difference in the location may come from the bulk hydrophobicity of the processed product; the *Anacystis* MSP is water-insoluble (unpublished finding) and the (pro)sporamin A is water-soluble.

Acknowledgements

The author thanks Professor K. Nakamura, Nagoya University, for helpful suggestions. This work was supported in part by Grants-in-Aid for Scientific Research on Priority Areas of the Molecular Mechanism of Photoreception (62621001) and for Cooperative Research (61304004) from the Ministry of Education, Science and Culture, Japan.

References

1. Murata, N. and Miyao, M. (1985) *Trends Biochem. Sci.* 10, 122–124
2. Kuwabara, T. and Murata, N. (1982) *Plant Cell Physiol.* 23, 533–539
3. Miyao, M. and Murata, N. (1984) *FEBS Lett.* 170, 350–354
4. Ono, T. and Inoue, Y. (1983) *FEBS Lett.* 164, 255–260
5. Kuwabara, T., Miyao M., Murata, T. and Murata, N. (1985) *Biochim. Biophys. Acta* 806, 283–289
6. Tang, X.-S. and Satoh, K. (1985) *FEBS Lett.* 179, 60–64
7. Satoh, K., Ohno, T. and Katoh, S. (1985) *FEBS Lett.* 180, 326–330
8. Ikeuchi, M., Yuasa, M. and Inoue, Y. (1985) *FEBS Lett.* 185, 316–322
9. Ono, T. and Inoue, Y. (1985) *Biochim. Biophys. Acta* 806, 331–340
10. Miyao, M., Murata, N., Lavorel, J., Maison-Peteri, B., Boussac, A. and Etienne, A.-L. (1987) *Biochim. Biophys. Acta* 890, 151–159
11. Oh-oka, H., Tanaka, S., Wada, K., Kuwabara, T. and Murata, N. (1986) *FEBS Lett.* 197, 63–66
12. Watanabe, A., Minami, E., Murase, M., Shinohara, K., Kuwabara, T. and Murata, N. (1987) in: *Progress in Photosynthesis Research*, Vol. 4 (Biggins, J., ed.) pp. 629–636, Nijhoff, Dordrecht
13. Tyagi, A., Hermans, J., Steppuhn, J., Jansson, Ch., Vater, F. and Herrmann, R.G. (1987) *Mol. Gen. Genet.* 207, 288–293
14. Kuwabara, T., Reddy K.J. and Sherman, L.A. (1987) *Proc. Natl Acad. Sci. USA* 84, 8230–8234
15. Kuwabara, T. and Murata, N. (1979) *Biochim. Biophys. Acta* 581, 228–236
16. Tanaka, S. and Wada, K. (1987) *Photosynthesis Res.*, in press
17. Murata, N., Miyao, M. and Kuwabara, T. (1983) in: *The Oxygen Evolving System of Photosynthesis* (Inoue, Y., Crofts, A.R., Govindjee, Murata, N., Renger, G., Satoh, K., eds.) pp. 213–222, Academic Press, Tokyo
18. Andersson, B., Larsson, C., Jansson, C., Ljungberg, U. and Åkerlund, H.-E. (1984) *Biochim. Biophys. Acts* 766, 21–28
19. Westhoff, P., Jansson, C., Klein-Hitpass, L., Berzborn, R., Larsson, C. and Bartlett, S.G. (1985) *Plant Mol. Biol.* 4, 137–146
20. Minami, E., Shinohara, K., Kuwabara, T. and Watanabe, A. (1986) *Arch. Biochem. Biophys.* 224, 517–527
21. Karlin-Neumann, G.A. and Tobin, E.M. (1986) *EMBO J.* 5, 9–13
22. Smeekens, S., Bauerle, C., Hageman, J., Keegstra, K. and Weisbeek, P. (1986) *Cell* 46, 365–375
23. Chia, C.P. and Arntzen, C.J. (1986) *J. Cell Biol.* 103, 725–731
24. Chou, P.Y. and Fasman, G.D. (1978) *Annu. Rev. Biochem.* 47, 251–276
25. Maniatis, T., Fritsch, E.F. and Sambrook, J. (1982) in: *Molecular Cloning: A Laboratory Manual*, pp. 440–441, Cold Spring Harbor Laboratory, NY
26. Minsky, A., Summers, R.G. and Knowles, J.R. (1986) *Proc. Natl Acad. Sci. USA*, 83, 4180–4184
27. Hattori, T., Ichihara, S. and Nakamura, K. (1987) *Eur. J. Biochem.* 166, 533–538

Further Characterisation of the Isolated PS2 Reaction Centre

J. Barber, D. J. Chapman, K. Gounaris and A. Telfer

AFRC Photosynthesis Research Group, Department of Pure and Applied Biology, Imperial College of Science & Technology, London SW7 2BB, UK

Summary

The reaction centre of photosystem two (PS2) consisting of the D1/D2/cyt b559 complex isolated from peas has been further characterised both in terms of its composition and electron transport properties. Experiments are presented which support the earlier claims (Barber *et al*. FEBS Lett. 1987, 67–83) that silicomolybdate (SiMo) can be used as a convenient artificial electron acceptor when suitable donors such as diphenylcarbazide (DPC), MnCl$_2$, NH$_2$OH or KI are present. However, when the Triton X-100 associated with the isolated complex is exchanged with beta-lauryl maltoside, a light-induced signal is observed on addition of decyl plastoquinone. This signal corresponds to the photoreduction of about 30% of the cytochrome b-559 within the complex. A similar proportion of the cytochrome is reduced in the dark by decyl plastoquinone when the reductant, sodium borohydride, is added. This result, coupled with the finding that the photoreduction of cytochrome b559 is inhibited by DCMU, suggests that the added quinone is also photoreduced possibly by the Q_A and/or Q_B sites.

By comparing the isolated PS2 reaction centre with various PS2 core preparations showing different sensitivities to herbicides (DCMU and atrazine) and having different levels of a 22 kD polypeptide (as judged by immunoblotting), it is suggested that this latter polypeptide may be functionally analogous to the H-subunit of the reaction centre of purple bacteria. Indeed, immunoblotting suggested that the 22 kD polypeptide was present to some extent in our PS2 reaction centre preparation.

Introduction

In recent years there have been many reports of methods to isolate complexes which are able to carry out photosystem two (PS2) reactions. Some preparations maintain the ability to oxidise water (1,2) while in

---

Abbreviations: DCMU, 3-(3,4-dichlorophenyl)-1, 1-dimethylurea; DPC, diphenylcarbazide; EDTA, ethylenediaminetetracetic acid; MES, 2-(N-morpholino) ethanesulphonic acid; PS2, photosystem two; Tris, tris (hydroxymethyl) aminomethane hydrochloride; DCPIP, 2,6-dichlorophenol indiphenol; SDS, sodium dodecyl sulphate.

others there are fewer polypeptides and the ability to evolve oxygen is lost (3–5). Despite these advances it was not clear until recently which proteins in the PS2 complex make up the reaction centre where primary photochemical charge separation occurs. However, the recent isolation of a complex consisting of the D1 and D2 polypeptides (*psb*A and *psb*D gene products, respectively) and the alpha-and beta-subunits of cytochrome b-559 (6,7) and its characterisation as the photosystem two (PS2) reaction centre (8–10), has fulfilled a prediction that the D1 and D2 polypeptides are structurally and functionally comparable with the L and M subunits of the reaction centre of purple bacteria (11–13). Although there was some resistance to this notion (14) there is now no doubt that the isolated D1/D2 cyt b-559 complex contains the chlorophylls which constitute the primary electron donor P680 and the pheophytin molecule which acts as the primary electron acceptor. For some reason, which is not yet clear, the isolated complex appears to have no associated plastoquinone (6,15). In the absence of a quinone acceptor, the radical pair $P680^+$ $Pheo^-$ produced by flash excitation of the isolated complex back-reacts with a halftime of 36 ns (9,10) with a small 30 μs component due to chlorophyll triplet state formation (8,10).

Nanba and Satoh (6) were able to show that when the D1/D2/cyt b559 complex is illuminated in the presence of excess dithionite the P680 $Pheo^-$ state photoaccumulated. We have confirmed this result and further shown that an optical absorption difference signal indicative of the state $P680^+$Pheo is obtained when silicomolybdate is used as an artificial electron acceptor (16). If, under these latter conditions, diphenyl-carbazide is added, the $P680^+$ signal is no longer detected and a net rate of electron flow to silicomolybdate occurs. It has been shown that PS2 electron donors, like diphenylcarbazide, donate to $P680^+$ via the intermediate, usually referred to as Z (17). Therefore our results indicate that Z is located within the D1/D2/cyt b559 complex and cannot be plastoquinone as previously suggested (18).

In this paper we characterise further the net electron transport capacity of the isolated D1/D2/cyt b559 complex and investigate the possibility that a 22 kD nuclear coded protein could be the PS2 equivalent to the H–subunit of the bacterial reaction centre.

Materials and Methods

The PS2 reaction centre was isolated from the leaves of pea seedlings (*Pisum sativum* var. Feltham First) using a modified form of the method outlined previously (7, 16) and which was based on the approach suggested by Nanba and Satoh (6) in which solubilisation of PS2-enriched membranes in Triton X-100 is followed by ion-exchange chromatography. Chloroplast thylakoid membranes were isolated according to a published procedure, except for the use of a maceration medium of 50 mM KH_2PO_4, pH 7.5 (NaOH), 0.35 M KCl, 0.5 mM EDTA and centrifugation at 5,000 xg at 4°C for 10 min., to produce a chloroplast pellet. This was resuspended in 6 mM $MgCl_2$ to rupture any intact plastids by osmotic shock and buffer then added to give a final composition of 0.2 M sucrose, 0.1 M NaCl, 50 mM Tricine pH 8.0 before further centrifugation. A PS2-enriched membrane fraction (BBY) was prepared from this pellet, essentially according to Berthold *et al.* (19), by resuspension to a concentration of 3 mg chlorophyll.ml^{-1} in a solubilisation buffer of 5 mM $MgCl_2$, 15 mM NaCl, 20 mM MES, pH 6.3 and incubating on ice in the dark for 45 min. before adding 0.5 volume 10% Triton X-100 (w/v) in solubilisation buffer. A further 30-min. incubation, on ice, in the dark and centrifugation at 40,000 xg for 30 min. gave a pellet which was resuspended to 2 mg chlorophyll.ml^{-1} in a wash and storage medium of 10% (w/v) glycerol in solubilisation buffer. The membranes were pelleted once more, resuspended to 4 mg chlorophyll.ml^{-1} and frozen in liquid nitrogen before storage at –80°C.

To isolate the reaction-centre complex, a PS2-enriched membrane sample of 200 mg chlorophyll was thawed and washed to deplete extrinsic membrane polypeptides by dilution to 0.8 mg chlorophyll.ml^{-1} in 50 mM Tris pH 9.0, incubated on ice in the dark for 10 min. and centrifuged at 40,000 xg, 4°C for 20 min. The pellets were resuspended in 50 mM Tris, pH 7.2 (200 mls) and 33 mls of 30% Triton X-100 added to give a final chlorophyll concentration of 0.8 mg.ml^{-1} and a Triton to chlorophyll ratio of 50 to 1 (w/w). A 60-min. incubation with stirring, in the dark, on ice was followed by centrifugation at 100,000 xg for 60 min. and application of the supernatant to a column (16 × 300 mm) of Fractogel TSK DEAE–650 (S) (Merck–BDH) maintained at 6°C. Extensive washing at 0.4 ml.min^{-1} was carried out with 350 ml, 30 mM NaCl in a running buffer of 0.2% Triton X-100, 50 mM Tris-Cl, pH 7.2. This resulted in the return of the absorbance (at 280 nm) of the eluant to the same level as for the running buffer itself and removed more than 98% of the chlorophyll applied. The material remaining on the column was eluted by a linear concentration gradient of 2 mM NaCl.ml^{-1}. Appropriate 2 ml fractions at about 100 mM NaCl were pooled, diluted four-fold in running buffer and loaded on a smaller column (9 × 100 mm) of the same DEAE-Fractogel. After further extensive washing with 30 mM NaCl in running buffer (about 50 ml at 0.5 ml.min^{-1}) and a linear NaCl gradient (5mMl^{-1} at 0.5 ml.min^{-1}) the complex eluted as a sharp peak (O.D.280 nm) at about 110 mM NaCl. For preparation of the reaction-centre in beta-lauryl maltoside the material resulting from the second column was diluted four-fold with running buffer without Triton and applied to a 9 × 30 mm DEAE-Fractogel column which had been equilibrated in a buffer containing 4 mM beta-lauryl maltoside, 50 mM Tris-Cl pH 7.2. Extensive washing with this buffer was carried out at 0.5 ml.min^{-1}. The complex was finally eluted with the above buffer containing 130 mM NaCl at 0.1 ml.min^{-1}. Each preparation was checked for purity by comparison with previous determinations (16) of the optical absorbance spectrum, the cytochrome b559 to chlorophyll a ratio and the polypeptide composition using Coomassie blue staining after SDS-polyacrylamide gel electrophoresis on either 12–17% gradient gels, or 10–20% gels containing 6 M urea.

Chlorophyll levels were measured according to Arnon (20) and cytochrome b559 estimated from reduced (dithionite) minus oxidised (ferricyanide) absorbance difference at 559 nm, taking an extinction coefficient of 15 mM^{-1}.

Photosystem 2 core complexes capable of oxygen evolution were isolated from BBYs obtained from pea thylakoids using the procedures of Ikeuchi et $al.$ (2) with the modification of Ikeuchi and Inoue (21) (preparation I) and of Ghanotakis and Yocum (22) (preparation II). Briefly, preparation I was obtained by solubilisation of PS2-enriched membranes (BBYs) by octyl-beta-D-glucopyranoside and centrifugation through a discontinuous sucrose gradient containing the above detergent. Preparation II was isolated by means of octyl-beta-D-glucopyranoside solubilisation in the presence of 0.5 M NaCl and subsequent centrifugation and dialysis steps as described in (22). The pellets obtained from both procedures were resuspended in a medium containing 0.4 M sucrose, 40 mM Mes (pH 6.0), 10 mM NaCl and 10 mM CaCl$_2$ (medium A). Preparation II contained considerable amounts of a polypeptide in the molecular weight region of 20–22 kD. This polypeptide has been suggested as playing a role in the binding of the extrinsic 23 kD protein (23) to the thylakoid membranes as well as being involved in the functioning of the reducing side of PS2 (22,24). In order to reduce the amount of the 22 kD protein associated with preparation II we used the procedure described by Ljunberg et $al.$ (23) with the following modifications: PS2 core complexes (preparation II)—rather than PS2 membranes—were subjected to 1 M NaCl, 0.06% Triton X-100 treatment in a buffer containing 40 mM Mes, pH 6.0. Samples were incubated for 15 min. on ice in the dark at 0.7 mg chlorophyll/ml, centrifuged at 40,000 xg for 30 min. and the resultant pellet resuspended in medium A (preparation III). Sodium dodecyl sulphate polyacrylamide gel electrophoresis (SDS-PAGE) of these various PS2 core complexes was performed according to (25) using a 10–17% acrylamide gradient and 7 M urea in the resolving gel. The buffers were according to Laemmli (26) and electrophoresis was carried out under constant voltage (100 V) for 24 hours. Western blots of polyacrylamide gels were carried out by electrophoretic transfer onto nitrocellulose sheets in a buffer containing Tris-glycine-SDS. The protein-binding capacity of the nitrocellulose was saturated by incubation overnight at 4°C in skimmed milk containing 0.3% (v/v) Tween 20. This was followed by incubation with the antibody raised against the 22 kD protein and subsequent incubation with rabbit immunoglobulin conjugated to alkaline phosphatase. Visualisation was achieved by the use of nitro blue tetrazolium and 5-bromo-4-chloro-3-indolyl phosphate. Rabbit antiserum against the 22 kD polypeptide was kindly supplied by Professor B. Andersson.

Light-induced absorbance changes were measured with a Perkin Elmer 557 dual-beam spectrophotometer with side illumination of a 1 ml, 1 cm pathlength, sample cuvette from a quartz-iodine source equipped with appropriate light guides and transmission filters (Calflex heat filter and 2 mm Schott RG660 cut-off filter). The intensity at the cuvette surface (300 $\mu E.^{-2}.s^{-1}$) was attenuated with neutral density filters, the photomultiplier shielded by a Schott BG38 filter and the cuvette temperature maintained at either 4°C or 10°C. The reduction of silicomolybdate (0.2 mg.ml^{-1}; Pfalts and Bauer, Inc. Stanford. Conn. USA) was monitored by an increase in absorbance at 600 nm in 1 ml of 50 mM Tris, pH 8.0 with 2 μg chlorophyll a.ml^{-1}. This wavelength was chosen to monitor silicomolybdate reduction to avoid the supersition of any other light-induced signal. An approximate extinction coefficient of 4.8 mM^{-1} was used after determination by recording the reduced minus oxidised difference change at 600 nm for standards dissolved in the Tris assay medium (pH 8.0). The photoreduction of 2,6-dichlorophenol-indophenol (DCPIP) was followed as an absorbance change at 560 nm minus absorbance at 520 nm using dual wavelength mode in 1 ml of 20 mM MES (pH 6.0) with 10 μg chlorophyll.

Results

Earlier work from our laboratory has shown that the isolated PS2 reaction centre complex consisting of D1/D2/cyt b559 can catalyse the net photo-reduction of silicomolybdate with diphenylcarbazide (DPC) as an electron donor (16). We have tested a range of compounds, previously shown to act as artificial electron donors to PS2 in intact and fragmented thylakoids (17,27,28) as possible alternatives to DPC. In all cases the acceptor was silicomolybdate whose light-induced reduction was monitored as an increase in absorbance at 600 nm. Some of the compounds tested gave no net electron transport activity (ascorbate and hydroquinone), while others could not be used because silicomolybdate was reduced directly without the addition of PS2 reaction centre (e.g. benzidine and p-phenylenediamine). However, MnCl$_2$, NH$_2$OH and KI all gave significant activity. A comparison of the relative rates is given in Table 1 and demonstrates that DPC was the most effective donor. For DPC, rates of about 1000 μ equivalent electrons.mg chl^{-1}.hr^{-1} were recorded with 2 μg chl.ml^{-1} in a cuvette illuminated at 300 μE.m^{-2}.s^{-1} and with an assay temperature of 4°C. Light intensity curves indicated that, with this low level of chlorophyll in the measuring cuvette, 300 μE.m^{-2}.s^{-1} of light transmitted by the Schott RG660 cut-off filter was just about saturating for the net electron transport activity. Addition of EDTA to the

TABLE 1

Light-dependent reduction of silicomolybdate by various donors

(a) *Relative rates*

Donor	−EDTA	+EDTA
DPC	100	101 ± 4
MnCl$_2$	48 ± 5	18 ± 5
NH$_2$OH	42 ± 5	20 ± 4
KI	12 ± 2	10 ± 3

(b) *Action of combined donors on relative rates*

Donors	No addition	+DPC	+MnCl$_2$	+NH$_2$OH	+KI
DPC	100	–	80 ± 19	62 ± 12	103 ± 10
MnCl$_2$	48 ± 5	80 ± 19	–	43 ± 8	40 ± 3
NH$_2$OH	42 ± 5	62 ± 12	43 ± 8	–	31 ± 5
KI	12 ± 2	103 ± 10	40 ± 3	31 ± 5	–

The absolute rate corresponding to 100% was estimated to be 870±189 μequiv. electrons mg^{-1}.hr^{-1}. The values given are the average (±S.E.) of three to five separate preparations of the PS2 reaction centre.

assay medium resulted in an inhibition of the MnCl$_2$ supported rate (see Table 1a) but there was no effect of this divalent cation chelator on the DPC and KI supported rates. It should be noted, however, that the ability of NH$_2$OH to donate electrons was inhibited by EDTA. The reason for the latter observation is as yet unclear and could be due to a requirement of divalent cations for this reaction. The effect of adding combinations of the donors is shown in Table 1b. Despite the fact that it has been claimed that some donors (e.g. MnCl$_2$ and DPC) use different sites (17) no synergism was noted and indeed in the case of DPC and NH$_2$OH their joint addition caused a significant decrease when compared with the rate measured with only DPC present.

All the above measurements were made at about 4°C because exposure to higher temperatures reduced the ability of the isolated PS2 reaction centre to catalyse the light-induced reduction of silicomolybdate. This inhibition at temperatures above 4°C was time-dependent (see Fig. 1A). For example, at 26°C, 50% loss of activity occurred in 4 min. when the preparation was incubated in the dark before assaying its activity. Although the reaction centre remained active at 4°C in the dark, this activity was progressively lost when the sample was illuminated at 4°C (see Fig. 1B).

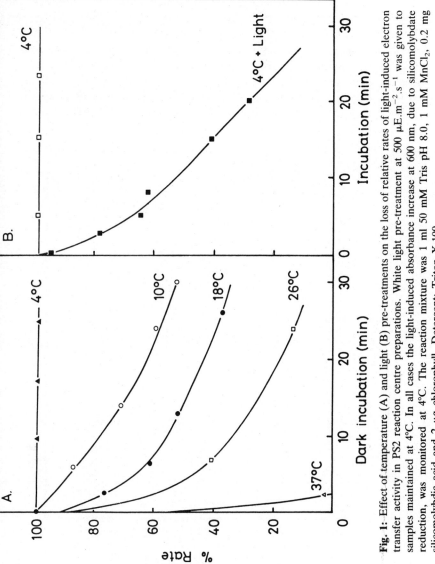

Fig. 1: Effect of temperature (A) and light (B) pre-treatments on the loss of relative rates of light-induced electron transfer activity in PS2 reaction centre preparations. White light pre-treatment at 500 $\mu E.m^{-2}.s^{-1}$ was given to samples maintained at 4°C. In all cases the light-induced absorbance increase at 600 nm, due to silicomolybdate reduction, was monitored at 4°C. The reaction mixture was 1 ml 50 mM Tris pH 8.0, 1 mM MnCl$_2$, 0.2 mg silicomolybdic acid and 2 μg chlorophyll. Detergent: Triton X-100.

The sensitivity of the light-induced net reduction rate of silicomolybdate to incubation temperatures above 4°C in the dark or to illumination at 4°C was also reflected in the inability of the PS2 reaction centre to photoaccumulate reduced pheophytin in the presence of excess dithionite (data not shown). The loss of these activities seem to be due to degradation of the isolated complex which can be monitored by a blue shift in the optical absorption and emission peaks in the red. The fact that the silicomolybdate reaction responds in this way suggests that it is a valid assay of the functional activity of the isolated D1/D2/cyt b559 complex, a conclusion already reached from proteolytic digestion studies (16). To date we have been unable to replace silicomolybdate with any other classical artificial electron acceptor, such as ferricyanide or DCPIP. However,

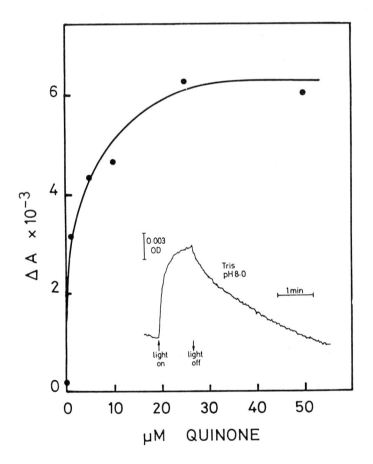

Fig. 2: Light-induced absorbancy change at 430 nm (insert) and its concentration dependence for decyl plastoquinone (DPQ). The sample contained 1ml 50 mM Tris pH 8.0, 1 mM MnCl$_2$ and 2 μg chlorophyll. The insert-time course was obtained with 50 mM DPQ. Detergent: lauryl maltoside.

under certain conditions the addition of quinones seems to catalyse an electron transfer process. We have found that, when the Triton X-100 associated with the isolated reaction centre is exchanged with lauryl maltoside in 50 mM Tris pH 8.0, a light induced increase in absorbance is consistently observed at 430 nm when decyl plastoquinone is added (see Fig. 2). Decyl plastoquinone is much more hydrophylic than the naturally occurring plastoquinone-9, as judged by reverse phase high pressure liquid chromatography. In our hands we found that the absorption increase at 430 nm is relatively small until electron donors, such as $MnCl_2$ or DPC, are

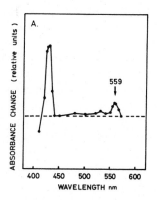 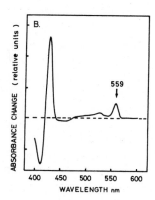

Fig. 3: Absorbance difference spectra of reaction centre preparations. (A) Light minus dark difference spectrum drawn from measurement of separate samples of a PS2 reaction centre preparation, with 1 mM $MnCl_2$ and 0.05 mM decyl plastoquinone present as in Fig. 2. (B) Reduced (dithionite) minus oxidised (ferricyanide) difference spectrum. Detergent: lauryl maltoside.

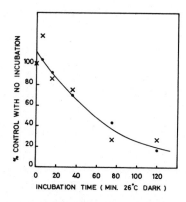

Fig. 4: Inhibition of light-dependent silicomolybdate reduction and 430 nm absorbance increase in the presence of decyl plastoquinone by incubation at 26°C in the dark. A PS2 reaction centre sample was incubated in 50 mM Tris pH 8.0 and aliquots taken for assay of silicomolybdate reduction (●) as in Fig. 1 and 430 nm absorbance increase (×) as in Fig. 2 with 0.05 mM decyl plastoquinone and 1 mM $MnCl_2$. Detergent: lauryl maltoside.

added. In the latter case the signal reaches a maximum and slowly reverses in the dark. The concentration of quinone required for maximum effect is also given in Fig.2. The light–dark difference spectrum shown in Fig. 3A suggests that the optical density change observed when donor and decyl plastoquinone are present is due to the reduction of cytochrome b559 as judged by its comparison with the reduced (dithionite) minus oxidised (ferricyanide) difference spectrum for this cytochrome measured in the isolated reaction centre (Fig. 3B). The ability of decyl plastoquinone to catalyse the light-induced reduction of cytochrome b559 was inhibited when the isolated reaction centre was preincubated at temperatures above 4°C in a way which directly parallels the inhibition of silicomolybdate reduction (see Fig. 4). It was noted, however, that in the presence of lauryl maltoside the complex was more stable than in Triton x-100 (compare Figs. 4 and 1A). Nevertheless, similar light-induced changes dependent on the presence of decyl plastoquinone could also be observed without carrying out the detergent exchange, but the signal sizes were variable.

A possible explanation for the above result is that the added quinone can in some way accept electrons from the reaction centre and pass them to cytochrome b559. This could involve interaction with the vacant Q_A or Q_B sites which, by analogy with the purple bacterial reaction centre, should reside on the D1/D2 heterodimer. Alternatively, the added quinone may perturb the conformation of the isolated complex in some

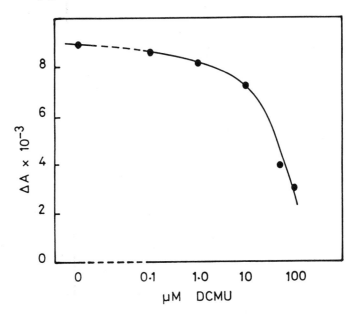

Fig. 5: Inhibition of light-induced absorbance increase of 430 nm by DCMU. Assays were as in Fig. 2 with 0.05 mM decyl plastoquoinone and addition of DCMU in 10 µl ethanol. Detergent: lauryl maltoside.

other way and allow electron transfer to the cytochrome. To check these two possibilities DCMU was added to the reaction medium. From binding studies it has already been shown that this herbicide will associate stoichiometrically with the isolated D1/D2/cyt b559 complex (M. T. Giardi, J. Marder and J. Barber, in press (1988), *Biochim, Biophys,* Acta). This binding, however, has a low affinity (Km 12.6 μM) when compared with intact thylakoid membranes (Km 34 nM). Fig. 5 shows that DCMU does significantly block the decyl plastoquinone-catalysed, light-driven reduction of cytochrome b559 and, as with electron transfer in PS2 core preparations (5,22,24), at a higher concentration than usual when compared with that needed to inhibit the PS2 activity of thylakoids or membrane fragments. The determination of the redox properties of cytochrome b559 with different reducing agents given in Fig. 6 shows that virtually all the cytochrome is present in its low potential form, being reducible by dithionite but not by hydroquinone. However, on addition of decyl plastoquinone a significant proportion (34±4%) becomes reducible by sodium borohydride. The size of this portion is consistent with the redox

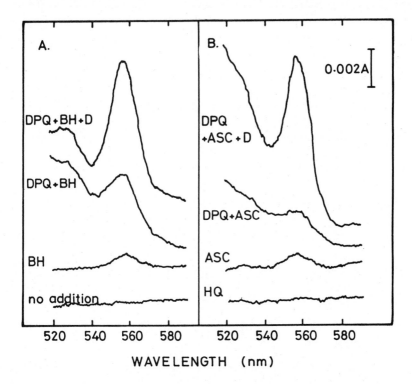

Fig. 6: Reduced, minus-oxidised difference spectra of cytochrome b559. Reaction centre samples (2 μg chlorophyll in 1 ml) were subjected to the combinations indicated, where the symbols mean: 10 mM hydroquinone (HQ), 100 mM sodium ascorbate (ASC), 2 mM sodium borohydride (BH), 10 mM sodium dithionite (D) and 0.05 mM decyl plastoquinone (DPQ).

properties of reduced quinone (E_m = +110 mV) and that of the low potential cytochrome b559 (E_m = +100 mV). No ascorbate reducible form was detected in the reaction centres even on addition of quinone (Fig. 6), a result which differs from that obtained with PS2 cores by Satoh *et al.* (29).

The low affinity of the D1/D2/cyt b559 complex for DCMU and the apparent destabilisation of the Q_A and Q_B plastoquinone binding sites could be due to the absence or depletion of a 22 kD polypeptide. Ghanotakis and Yocum (22) have suggested that it is differences in the content of this polypeptide in various PS2 preparations which gives rise to variations in the acceptor requirement and DCMU sensitivity, a conclusion also supported by Henrysson *et al.* (24). To investigate this further we have isolated PS2 cores according to the method of Ikeuchi *et al.* (2, 21) (preparation I) and Ghanotakis and Yocum (22) (preparation II). To reduce the 22 kD content of preparation II the procedure of Ljungberg *et al.* (23) was followed (with some modifications—see 'Materials and Methods') in order to obtain preparation III. Fig. 7 shows the SDS-PAGE polypeptide profiles of these preparations and compares them with those of PS2 enriched membranes (BBYs) and the PS2 reaction centre (D1/D2/cyt b559 complex). Also shown are the results of immunoblotting of these

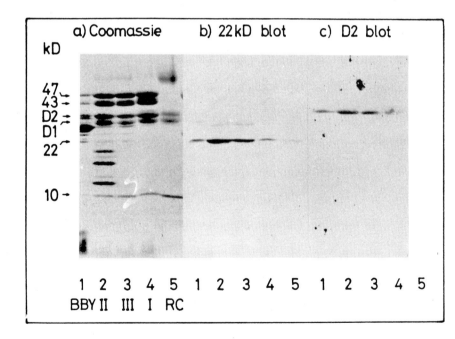

Fig. 7: Coomassie blue stained SDS-polyacrylamide gel (a) of PS2-enriched membranes (BBY) (1), preparation II (2), preparation III (3), preparation I (4) and PS2 reaction centre (5). Also shown are the Western blots using antibodies to 22 kD (b) and D2 (c) polypeptides (see 'Materials and Methods' for further details).

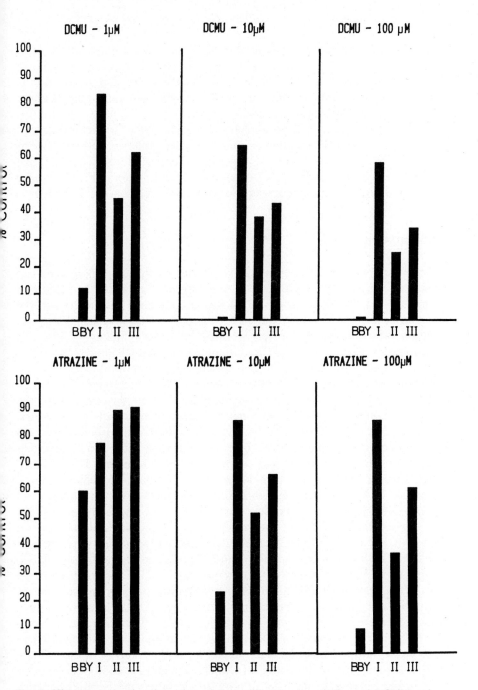

Fig. 8: Histograms showing the action of three different concentrations of DCMU and atrazine on the light-induced DPC to DCPIP reaction measured with PS2-enriched membranes (BBY) and preparations I, II and III. Each reaction was measured with 10 µg chlorophyll contained in 20 mM MES (pH 6.0), 500 µM DPC and 10 µM DCPIP. Assay temperature was 10°C.

polypeptide patterns using antibodies to D2 and 22 kD polypeptides. The blots clearly show that the 22 kD is present in PS2-enriched membranes and in preparation II but is depleted in preparation III and preparation I. Interestingly the antibody blotting did detect small amounts of the 22 kD in the D1/D2/cyt b559 preparation.

Fig 8 shows that when the diphenylcarbazide (DPC) to dichlorophenol indophenol (DCPIP) reaction is used as an assay system, the sensitivity to DCMU of the BBYs and various types of PS2 cores follows, in a qualitative way, the presence of the 22 kD protein. As can be seen in Fig. 8, similar results were also obtained when atrazine was used as the inhibitor.

Discussion

Several lines of evidence now exist which suggest that silicomolybdate can act as a legitimate secondary electron acceptor for assaying the activity of the isolated PS2 reaction centre. It has already been shown that, in the presence of this acceptor, but in the absence of an artificial donor, an optical absorption signal indicative of the photoaccumulation of P680$^+$ is observed with a concomitant quenching of chlorophyll fluorescence (16). Both signals are lost when the isolated preparation is incubated at temperatures above 4°C. They are also lost or reduced significantly when DPC is added to the suspension medium (16). In this latter case the preparation is still active and a net reduction of silicomolybdate occurs. The rate of reduction is sensitive to the condition of the isolated reaction centre complex, being inhibited by proteolytic treatments and by exposure to SDS (16). In this communication we have shown that DPC can be replaced by other PS2 electron donors, and in particular we have favoured the use of MnCl$_2$. Although this compound gives lower maximum rates than DPC, it does not catalyse a dark reduction of silicomolybdate and is more comparable with the natural Mn cluster of the water electron donor system. As with DPC, Mn catalysed, light-induced reduction of silicomolybdate is inhibited by degradation of the PS2 reaction centre complex due to light or to dark incubations at temperatures above 4°C or preillumination at 4°C. This inhibition parallels the inhibition of decyl plastoquinone catalysed, light-induced reduction of cytochrome b559, which perhaps can be considered a legitimate physiological assay. Therefore, taken as a whole, it can be concluded that the DPC or MnCl$_2$ to silicomolybdate electron transport reaction is a true measure of the activity of the isolated PS2 reaction centre consisting of the D1, D2 and cyt b559 polypeptides.

In the absence of artificial donors or acceptors, the isolated complex forms the radical pair P680$^+$Pheo$^-$ which back-reacts with a half-time of about 36 ns, although a small portion de-excites via a longer-lived triplet state (9,10). Presumably, silicomolybdate can compete for electrons from the reduced pheophytin. In so doing, some of the P680$^+$ can be reduced by the added donor. In the absence of silicomolybdate, DPC or $MnCl_2$ are unable to compete with the back-reaction and only excess dithionite is able to reduce $P680^+$ so as to allow the photoaccumulation of reduced pheophytin (6,16). Thus the net rate of electron flow seems to be partially governed by the ability of silicomolybdate to compete with the back-reaction. However, limitation must also be imposed on the donor side and, indeed, some compounds are less efficient than others. Presumably silicomolybdate receives its electrons from reduced pheophytin, possibly via a modified Q_A site, since the process is not inhibited by DCMU (16). It is usually assumed that artificial PS2 electron donors such as DPC donate to $P680^+$ via Z (17). If this is true, then, as stated in the introduction, Z must be located within the D1/D2/cyt b559 complex, in line with the recent suggestion that it is a tyrosine residue on the D1 polypeptide (31 and W. Vermaas, personal communication). The fact that decyl plastoquinone can catalyse the photoreduction of cytochrome b559 within the PS2 reaction centre is an interesting and important observation. The reaction is inhibited by DCMU, suggesting that the electron pathway involves a herbicide-binding site. Normally this site is restricted to the Q_B binding pocket on the D1 polypeptide (32) but, in the absence of plastoquinones in the isolated complex (15), DCMU binding may also occur at the Q_A site. Thus it is conceivable that the mechanism of cytochrome b559 reduction requires the pre-reduction of the added quinone at the Q_A and/or Q_B sites. According to the maximum amplitude of the light-induced signal obtained, about 30% of the cytochrome is reduced. This level of reduction was not achievable in the dark on addition of either hydroquinone, sodium borohydride or sodium ascorbate. However, when decyl plastoquinone and sodium borohydride were added together, the cytochrome could be reduced to about the same extent as the light-driven process. This observation suggests that decyl plastoquinone is directly involved in the light-induced reduction of cytochrome b559. This conclusion, together with the fact that the photoreduced cytochrome is reoxidised in the dark, even in the presence of electron donors such as $MnCl_2$, makes this light-induced signal intriguing and worthy of further investigations.

The apparent correlation between the sensitivity of DCPIP reduction to the action of herbicides and the presence of the 22 kD protein, suggest that this component functions in some way on the reducing side of PS2. Bearing in mind this observation and also the fact that this protein can be detected, albeit at a low level, as compared with that in the Ghanotakis and Yocum particle (22), in all the PS2 preparations tested, including the isolated D1/D2/cyt b559 complex, could mean that the 22 kD is functionally analogous to the H-subunit of the purple bacterial reaction centre. If this is so, then procedures need to be devised in order to obtain the isolated PS2 reaction centre still binding the 22 kD component. Such a preparation may be expected to bind the Q_A and Q_B plastoquinones and maintain a high sensitivity to PS2 herbicides.

Acknowledgements

We wish to thank the Agricultural and Food Research Council and the Science and Engineering Research Council for financial help. We acknowledge the technical help given by Kenneth Davis, John DeFelice and Jenny Nicolson, and are grateful to Dr P. Rich and Professor B. Andersson for supplying decyl plastoquinone and 22 kD antibody, respectively.

References

1. Tang, X.-S. and Satoh, K. (1985) *FEBS Lett.* 179, 60–64
2. Ikeuchi, M., Yuasa, M. and Inoue, Y. (1985) *FEBS Lett.* 185, 316–322
3. Satoh, K. and Butler, W. L. (1978) *Plant Physiol.* 61, 373–379
4. Gounaris, K. and Barber, J. (1985) *FEBS Lett.* 188, 68–72
5. Bricker, T. M., Pakrasi, H. B. and Sherman, L. A. (1985) *Arch. Biochem. Biophys.* 237, 170–176
6. Nanba, O. and Satoh, K. (1987) *Proc. Natl. Acad. Sci.* (USA) 84, 109–112
7. Marder, J. B., Chapman, D. J., Telfer, A., Nixon, P. and Barber, J. (1987) *Plant Mol. Biol.* 9, 325–333
8. Okamura, M. Y., Satoh, K., Isaacson, R. A. and Feher, G. (1987) in: *Progress in Photosynthesis Research*, Vol. 1 (ed. J. Biggins) pp. 379–381, Martinus Nijhoff Pub., Dordrecht, The Netherlands
9. Danielius, R. V., Satoh, K., van Kan, P. J. M., Plijter, J. J., Nuijs, A. M. and van Gorkom, H. J. (1987) *FEBS Lett.* 213, 241–244
10. Takahashi, Y., Hansson, O., Mathis, P. and Satoh, K. (1987) *Biochim. Biophys.* Acta 893, 49–59
11. Michel, H. and Deisenhofer, J. (1986) in: *Encycl. Plant Physiol.* III Photosynthesis, Vol. 19 (Staehelin, L. A. and Arntzen, C. J. eds.) pp. 371–381, Springer-Verlag, Berlin
12. Trebst, A. (1986) *Z. Naturforsch* 41c, 240–245
13. Barber, J. (1987) *Trends in Biochem. Sci.* 12, 321–326
14. Satoh, K. (1986) *FEBS Lett.* 204, 357–362
15. Barber, J., Gounaris, K. and Chapman, D. J. (1987) in: *ytochrome Systems: Molecular Biology and Bioenergetics* (Papa, S. *et al.* ed.) pp. 657–666, Plenum Press, New York (in press)
16. Barber, J., Chapman, D. J. and Telfer, A. (1987) *FEBS Lett.* 220, 67–73
17. Babcock, G. T. and Sauer, K. (1975) *Biochim. Biophys.* Acta 396, 48–62
18. O'Malley, P. J., Babcock, G. T. and Prince, R. C. (1984) *Biochim. Biophys.* Acta 766, 283–288
19. Berthold, D. A., Babcock, G. T. and Yocum, C. F. (1981) *FEBS Lett.* 134, 231–234
20. Arnon, D. I. (1949) *Plant Physiol.* 24, 1–15
21. Ikeuchi, M. and Inoue, Y. (1986) *Arch. Biochem. Biophys.* 247, 97–107
22. Ghanotakis, D. F. and Yocum, C. F. (1986) *FEBS Lett.* 197, 244–248
23. Ljungberg, U., Akerlund, H. E. and Andersson, B. (1986) *Eur. J. Biochem.* 158, 477–482
24. Henrysson, T., Ljungberg, U., Franzen, L. -G., Andersson, B. and Akerlund, H. -E. (1987) in: *Progress in Photosynthesis Research* (Biggins, J. ed.) Vol. 2, pp. 125–128, Martinus Nijhoff Publ., Dordrecht, The Netherlands
25. Hashimoto, F., Horigome, T., Kanbayashi, M., Yoshida, K. and Sugaro, H. (1983) *Anal. Biochem.* 129, 192–199

26. Laemmli, U. K. (1970) *Nature* 227, 680–685
27. Trebst, A. (1972) *Meth. Enzymol.* 24, 146–165
28. Hauska, G. (1977) in: *Encylcopedia of Plant Physiol.* Vol. 5 Photosynthesis (Trebst, A. and Avron, M., eds.) pp. 253–265, Springer-Verlag, Berlin
29. Satoh, K., Nakatani, H. Y., Steinback, K. E., Watson, J. and Arntzen, C. J. (1983) *Biochim. Biophys.* Acta 724, 142–150
30. Barry, B. A. and Babcock, G. T. (1987) *Proc. Natl. Acad. Sci.* (USA) in press
31. Debus, R. J., Barry, R. A., Babcock, G. T. and McIntosh, L. (1988) *Proc. Natl. Acad. Sci.* (USA) 85, 427–430
32. Trebst, A. (1987) *Z. Naturforsch* 42c, 742–750

Use of Synthetic Peptides in Studies on the Biogenesis and Phosphorylation of Photosystem II Proteins

W. E. Buvinger, H. P. Michel, A. Sutton and J. Bennett

Biology Department
Brookhaven National Laboratory
Upton NY 11973, USA

Summary

The light-harvesting chlorophyll a/b complex (LHC II) and the 8.3 kDa psbH gene product are the two most conspicuous phosphoproteins of chloroplast thylakoids. We have used synthetic peptide analogs of regions of these proteins to study their phosphorylation and biogenesis. The substrate specificity of the LHC II kinase was studied in terms of the ability of thylakoid membranes to phosphorylate several analogs of the LHC II phosphorylation site. The data indicate that the kinase can phosphorylate serine and threonine residues provided that they are located in an appropriate sequence, which includes a pair of basic amino acids located 1–3 residues to the N-terminal side of the phosphorylated residue. In another study, antibodies raised against the phosphorylation site of the psbH gene product were used to determine the photoregulation of this protein. Western blotting revealed that, like cytochrome b_{559} and the three extrinsic proteins of the oxygen-evolving complex of photosystem II, but unlike the five known chlorophyll-binding proteins of photosystem II (D1, D2, CPa-1, CPa-2 and LHC II), the psbH protein accumulates to detectable levels in both etiolated maize and maize leaves grown under far-red light (730nm), neither of which accumulate chlorophyll. This result suggests that the psbH gene product is not a chlorophyll-binding protein. Finally, a peptide analog of the first 19 residues of the transit sequence of *Arabidopsis thaliana* pre-LHC II was synthesised and used to generate antibodies specific for pre-LHC II. The peptide itself was shown to be an inhibitor of the uptake and processing of pre-LHC II by isolated intact pea chloroplasts.

Introduction

Research in this laboratory is concerned with the structure, function and biogenesis of the chloroplasts of green algae and higher plants. One of our interests is the reversible phosphorylation of chloroplast proteins, especially those phosphoproteins associated with photosystem II in thylakoid membranes. Another interest is the biogenesis of photosystem II, including such questions as the role of light in regulating assembly of the photosystem and the mechanism by which photosystem II proteins encoded by nuclear genes enter the chloroplast after synthesis as cytoplasmic precursors. In both areas of interest, we have found it valuable to use synthetic peptides, either as protein kinase substrates, or as immunogens for the preparation of monospecific antibodies, or as inhibitors of the uptake and processing of pre-LHC II by isolated intact chloroplasts. Here we summarise our experience in these applications of synthetic peptides.

Synthesis and Purification of Peptides

For the studies reported here, peptides were synthesised on a solid support either manually or in a Biosearch Model 9600 synthesiser, using t-Boc chemistry (1). Conditions of automated synthesis were determined for each peptide through the use of a programme supplied by the manufacturer. The completed peptide was deblocked and cleaved from the resin with HF. In the case of peptides containing cysteine, a portion of the product was reacted with iodoacetamide to block the sulphydryl group and was used in the subsequent purification. Following desalting through a Sephadex G15 column in 1% ammonium bicarbonate, each peptide was purified by reverse phase HPLC using a C_{18} column and a linear gradient of (a) aqueous buffer containing 100 mM sodium phosphate (pH 3.1) with 0.1% (w/v) hexane sulfonic acid and (b) 60% (v/v) acetonitrile containing 40 mM sodium phosphate (pH 3.1) with 0.1% hexanesulfonic acid. The peptide was detected by absorbance at 210 nm (Fig. 1A). The major peptide peak was collected, lyophilised and desalted by passage through a Sephadex G15 column in 0.1M acetic acid. The purity was checked by rechromatography on the C_{18} column (Fig. 1B) and on a strong cation exchange column (Synchropak S300) (Fig. 1C). Elution from the cation exchanger was done with a linear gradient of 0–1M KCl in 5 mM potassium phosphate (pH 6.5). Each peptide was finally checked for correct composition by amino acid analysis (5 nmoles) and for correct sequence by Edman degradation using an Applied Biosystems gas phase sequenator (1 nmole).

Synthetic Peptides as Protein Kinase Substrates

When isolated, intact spinach chloroplasts are incubated in the light for 15 min with [^{32}P]orthophosphate, numerous soluble and membrane-bound proteins become phosphorylated. The six most heavily labelled proteins are all components of photosystem II and have been identified as the 25 and 27 kDa apo-proteins of LHC II (2) and four photosystem II core proteins (32 kDa D1, 34 kDa D2, 44 kDa CPa-2 and the 8.3 kDa *psb*H gene product) (3,4). All six proteins are phosphorylated at or near their N-termini. The phosphorylation sites of the four photosystem II core

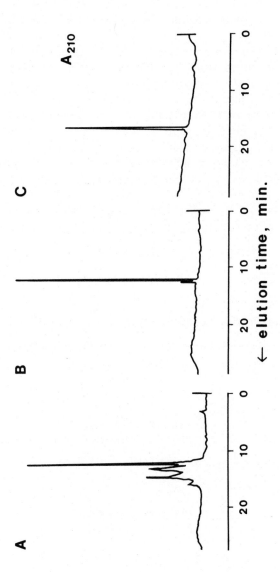

Fig. 1: Purification of a synthetic peptide analog of the LHC II phosphorylation site. The peptide (P1 in Table I) was synthesised manually (10). Purification and analysis of peptides was done with HPLC. *A:* crude peptide mixture analysed on a C_{18} reverse phase column. *B:* peak fraction after preparative reverse phase (C_{18}) HPLC re-chromatographed on a C_{18} reverse phase column. *C:* peak fraction re-chromatographed on a Synchropak 300 cation exchange column. Eluting solvents are described in the text.

proteins have been identified rigorously by sequencing of tryptic phos-phopeptides (3,4). These sequences are indicated in Table I in the standard one-letter code. The phosphorylation site of the 8.3 kDa protein was determined by Edman degradation and shown to be threonine-2 (3). The phosphorylation sites of D1, D2 and Cpa-2 were determined by mass spectrometry: all three proteins contain N-acetyl-O-phosphothreonine as N-terminal residue (4).

Although some LHC II apo-proteins are known to be phosphorylated on threonine about 5–7 residues away from their N-termini (5), there are several points of uncertainty concerning their structures in this region (Table I). First, the N-terminal residue is usually assumed to be a methionine but this point remains to be established. Second, while the N-terminus appears to be blocked against Edman degradation, the nature of the blocking group is unknown. Third, sequencing of many *cab* genes encoding LHC II has revealed that the most variable region of the protein lies between residues 4 and 15 at the N-terminus (i.e., including the phosphorylation site) and it is not yet possible to predict which genes code for phosphorylatable LHC IIs and which do not. Finally, although the phosphorylation site of one 27 kDa LHC II apo-protein of pea has been

TABLE I

Summary of Relevant Peptide Sequence Data[1]

Phosphorylation Sites of Photosystem II Core Proteins (2,3):

*psb*H gene product (8.3 kDa)	NH$_2$–ATQTVESSSR. . .
D1 (32 kDa)	acetyl–NH–TAILER. . .
D2 (34 kDa)	acetyl–NH–TIAVGK. . .
Cpa-2 (44 kDa)	acetyl–NH–TLFNGTLTLAGR. . .
analog of *psb*H site	NH$_2$–ATQTVESSSRC

Putative N-Terminal Sequence of 27 kDa LHC II of Pea (5,6):
acetyl–NH–MRKSATTKKVA. . .

analog P1	NH$_2$–MRKSATTKKAVC
analog P2	NH$_2$–MRKSASSKKAVC

First 20 Residues of Pre-LHC II Transit Peptides of A. thaliana (23):

Encoded by gene AB140	NH$_2$–MAASTMALSSPAFAGKAVKL. . .
Encoded by genes AB160/185	NH$_2$–MAASTMALSSPAFAGKAVNL. . .
synthetic analog	NH$_2$–MAASTMALSSPAFAGKAVN[Y]

---

[1]Underlined residues are known or suspected to be phosphorylated.

isolated and sequenced (6), it was not clear which of two threonyl residues at the phosphorylation site was phosphorylated. Thus, in Table I, which depicts the pea LHC II phosphorylation site, we have placed question marks against the N-terminal methionyl, the acetyl blocking group, and the phosphothreonyl group to draw attention to these uncertainties. We are currently attempting to resolve these questions by applying to LHC II apo-proteins the same techniques which we used to isolate and sequence the phosphorylation sites of the photosystem II proteins. Tandem mass spectrometry (7) will be required to identify the blocking groups and to identify which residues are phosphorylated.

In a parallel study we are using synthetic peptides as kinase substrates to assess the importance of various structural features in defining a phosphorylation site. There is now considerable evidence that LHC II is phosphorylated by a protein kinase that is distinct from the kinase which phosphorylates the four photosystem II proteins [see (8,9) for a discussion]. The LHC II kinase is activated in response to reduction of certain intersystem electron carriers [probably both plastoquinone and the cytochrome *bf* complex (8,9)] and phosphorylates LHC II within a highly basic region (6). In contrast, the photosystem II kinase can be activated by reduction of the plastoquinone pool in the absence of the cytochrome *bf* complex and phosphorylates its substrates within a rather neutral sequence right at their N-termini (2,3,8,9). We expect that at least two families of phosphorylatable peptides will emerge: those containing features recognised by the LHC II kinase and those recognised by the photosystem II kinase.

We have synthesised several analogs of the phosphorylation site of the 27 kDa LHC II of pea and have begun to analyse their effectiveness as substrates when incubated with thylakoid membranes. Peptide P1 (Table 1) has been studied most extensively. Like the 27 kDa LHC II, it is phosphorylated by the protein kinase which requires the cytochrome *bf* complex for activation (10,11). This result indicates that the substrate specificity of the kinase is determined largely or even entirely by the sequence adjacent to the phosphorylation site. Peptide P1 is phosphorylated on threonine-6, but neither serine-4 nor threonine-7 seems to be phosphorylated. When the two threonines are replaced by serines (as in peptide P2 in Table I), the kinase now phosphorylates serine. This result establishes that the failure of the enzyme to phosphorylate serine-4 in peptide P1 is due not to a requirement for threonine but to an unfavorable location of the serine. Similarly, the failure to phosphorylate threonine-7 of peptide P1 indicates that the LHC II kinase has a marked preference for phosphorylation of residue 6. Sequencing of cloned *cab* genes has established that the sequence methionyl–arginyl–(lysyl/arginyl)–(seryl/threonyl) is highly conserved in LHC II (12). However, threonyl residues are often absent from positions 6 and 7. This difference could serve to distinguish phosphorylatable apo-proteins of LHC II from non-phosphorylatable apo-proteins. Alternatively, it is possible that serine-4 or threonine-4 can be phosphorylated but only when residue 6 is not phosphorylatable, e.g., when threonine-6 is replaced by a glycine.

We are also interested in defining the roles of basic residues located near the LHC II phosphorylation site. A peptide which lacks residues 1–3 (methionyl–arginyl–lysyl) is not phosphorylated to a significant extent by thylakoids. This suggests that the LHC II kinase recognition site is defined in part by the arginyl–lysyl motif on the N-terminal side of threonine-6. It is interesting that the mammalian cyclic AMP-dependent protein kinases phosphorylate a somewhat similar site (...arginyl–arginyl-X–seryl-Y...) (13). Replacement of the second arginine by a lysine, or addition of a glycine residue between X and serine, results in a 200-fold increase in K_m and a 40–60% decline in V_{max}. Removal of X leads to a 500-fold increase in K_m and a 30% decline in V_{max}. Clearly there is great scope for similar studies on the LHC II kinase, both to establish the sequence requirements for a phosphorylation site and to relate such requirements to the sequences known to exist in LHC II apo-proteins encoded by various *cab* genes (12). In this way we shall endeavor to develop rules for deciding whether a given *cab* gene encodes a phosphorylatable or non-phosphorylatable apo-protein.

We have synthesised also a peptide analog of the phosphorylation site of the *psb*H protein (Table I). When added to thylakoid membranes, it is phosphorylated by the kinase which does not require the cytochrome *bf* complex for activation (11). From this and the above-mentioned experiments, it appears that synthetic peptides display the important characteristics of the corresponding membrane-bound proteins.

Synthetic Peptides as Immunogens

Hitherto, the *psb*H protein has been detected only as a labelled phosphoprotein. To detect the protein in unlabelled form we wanted to prepare antibodies against it. However, since the protein itself is difficult to purify from thylakoids, we tested the effectiveness of the phosphorylation site analog in eliciting production of suitable antibodies. The peptide was covalently attached to ovalbumin (14) and injected into a rabbit. Post-immunisation serum from the rabbit contained antibodies that recognised the *psb*H protein in western blots of total thylakoid proteins from maize (Fig. 2A, right panel).

The function of the *psb*H protein in photosystem II is unknown. However, on the strength of some limited sequence similarity with the N-terminal third of mature LHC II (15,16), it has been suggested to be a chlorophyll-binding protein (15). We have shown elsewhere (17) that three chlorophyll-binding proteins of photosystem II (CPa-1, CPa-2 and LHC II) fail to accumulate in maize leaves grown in darkness or under far-red light

(730 nm), presumably because of the absence of chlorophyll molecules required for the synthesis (18) and/or stabilisation (19) of the apo-proteins. To assay the level of the *psb*H protein in these leaves, we used the antibody raised against the synthetic peptide. At the same time, we assayed several other photosystem II proteins, including two additional chlorophyll-binding proteins (D1 and D2) and four known not to bind chlorophyll [cytochrome b_{559} and the 33, 23 and 17 kDa extrinsic proteins of the oxygen-evolving complex (OEC)]. No chlorophyll-binding protein (LHC II, CPa-1, CPa-2, D1 or D2) was detected by western blotting in etiolated leaves or in leaves grown under far-red light. Fig. 2A (left panel) shows the data for D2. If D2 is present in etiolated or far-red illuminated leaves, it must be below our detection limit, which corresponds to 2–5% of the white light level. In contrast, the 9 kDa subunit of cytochrome b_{559} and the three OEC proteins were readily detected by western blotting in extracts of plants grown in darkness or far-red light, and, as Fig. 2A (right panel) shows, the *psb*H protein exhibited the same behavior. Etiolated tissue contained about 10% of the white light level of this protein, while plants exposed to far-red light contained about 30%. These data suggest that the *psb*H protein is not a chlorophyll-binding protein or, if it is, that its accumulation is much less dependent on chlorophyll synthesis than that of the other chlorophyll-binding proteins of photosystem II.

The second peptide used as an immunogen in this laboratory was an analog of an LHC II transit sequence. Most chloroplast proteins are encoded in nuclear DNA and synthesised as cytoplasmic precursors, which contain 35–75 additional amino acids at their N-termini (20,21). These N-terminal extensions were originally termed transit peptides, because they were assumed to be necessary and possibly even sufficient to target a cytoplasmically synthesised protein for transport into the plastid and were removed by proteolytic cleavage during or immediately after import. It is now clear that in some cases only the leading part of the N-terminal extension is involved in transport and is removed first. The remainder of the extension appears to be involved in the subsequent targeting of the protein to a specific site within the organelle. An example of this two-stage targeting is provided by plastocyanin, which must first cross the double chloroplast envelope and then cross the thylakoid membrane to reach its site of function in the thylakoid lumen (22).

We wanted to prepare antibodies that would specifically recognise LHC II precursors and not recognise mature LHC II. In the case of a precursor for which the sequence of the transit peptide is known, the use of a synthetic peptide analog as an immunogen becomes an attractive way of raising such antibodies that are specific for the precursor protein. The particular analog that we synthesised (Table I) corresponds to the first 19 residues of the transit sequence of a pre–LHC II from *Arabidopsis thaliana* (23). A C-terminal tyrosine was also added to expand the possibilities for later chemical modification of the peptide. The peptide was attached to

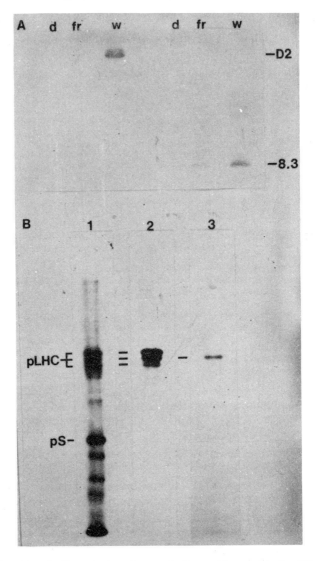

Fig. 2: Use of antibodies raised against synthetic peptides to detect *psb*H protein and pre-LHC II.

A (left panel): antibodies raised against D2 were used in western blotting with horseradish peroxidase-linked second antibody to detect the D2 protein in total membrane preparations derived from 8-day-old maize leaves (d, dark-grown; fr, grown under 730 nm far-red light for days 7 and 8; w, grown under warm white fluorescent light for days 7 and 8; see (17) for further details). *A (right panel):* duplicate blot probed with antibody raised against analog of the phosphorylation site of the 8.3 kDa *psb*H protein.

B: [^{35}S]methionine-labelled translation products [produced from *Arabidopsis thaliana* leaf RNA using the wheat germ system (34)] were separated by SDS-PAGE and autoradiographed (lane 1). The total translation mixture was also immunoprecipitated with antibodies against pea LHC II (19) (lane 2) or with antibodies against the synthetic analog of *A. thaliana* pre-LHC II transit peptide (lane 3).

ovalbumin and injected into a rabbit. As a preliminary to characterising post-immunisation serum from this rabbit, we prepared total RNA from *A. thaliana* leaves, translated it in the wheat germ system (Fig. 2B, lane 1), and identified LHC II precursors by immunoprecipitation with antibodies raised against mature pea LHC II (Fig. 2B, lane 2). Three distinct forms of pre-LHC II (30–32 kDa) were identified. The major translation product at 20 kDa (the precursor of the small subunit of Rubisco) was not precipitated. Fig. 2B (lane 3) shows the results obtained when antibodies raised against the transit peptide analog were added to the total translation products. Only the middle band of the three pre-LHC IIs was strongly selected. The most slowly migrating of the other two forms of pre-LHC II was seen only faintly on the original autoradiogram.

Our tentative explanation for the selective precipitation of the middle pre-LHC II is based on the existence of sequence heterogeneity among the LHC II transit peptides of *A. thaliana* (23). As shown in Table I, the transit sequence that we have synthesised is encoded by two out of the three *cab* genes that have been cloned from *A. thaliana*. A third gene codes for a pre-LHC II that differs from the other two in a non-conservative single residue change (asparagine→lysine) within the sequence covered by the analog. [Note that the three proteins encoded by these three genes are elsewhere identical (23).] If the region of the peptide where this substitution has occurred also contains the strongest epitopes (as a result of its hydrophilicity), these epitopes may elicit a majority of the antibodies found in the serum. As a result, pre-LHC II molecules containing the identical sequence are likely to be immunoprecipitated more efficiently than pre-LHC II molecules with a slightly different sequence in that region. As the replacement of an asparagine by a lysine might lead to slower migration on the SDS gel, we suggest tentatively that the middle pre-LHC II band in lane 2 of Fig. 2B corresponds to the asparagine-containing precursor, while the slowest band corresponds to the lysine-containing precursor. The fastest band may be encoded by a putative fourth *cab* gene that has been detected on Southern blots (23) but has not yet been cloned.

Synthetic Peptide Analog of the Transit Peptide of LHC II

There is considerable evidence that, during import into the chloroplast, cytoplasmically synthesised precursors associate with specific receptors in the envelope membrane. One early result pointing in this direction was the finding that thermolysin treatment of intact chloroplasts digests proteins on the outer surface of the organelle and diminishes the import of precursors (24). In addition, putative envelope-bound receptors can be solubilised with detergents and reconstituted into synthetic membranes which then bind precursors (25). More recently a photoaffinity labelled precursor to the small subunit of Rubisco was found to bind to a particular envelope component, estimated to have a M_r of about 67 kDa (26). Finally, anti-idiotypic antibodies raised against the second half of the transit peptide of the same precursor were found to recognise a 30 kDa protein in the zones of contact between the inner and outer envelope membranes (27).

As mentioned above, we have synthesised a peptide analog of the first 19 amino acids of the transit peptide of *A. thaliana* pre-LHC II. This particular peptide was chosen because LHC II precursors tend to have short transit peptides, the first half of transit peptides are less variable than the second half, and the transit peptide of *A. thaliana* pre-LHC II resembles very closely the consensus sequence arrived at by comparing pre-LHC II transit peptides for all species so far examined (21). Antibodies raised against the peptide can be used in a number of ways, including testing the hypothesis (28) that pre-LHC II accumulates *in vivo* under certain photoinhibitory conditions. In addition, the search for envelope-bound receptors that recognise pre-LHC II could use either anti-idiotypic antibodies prepared against the peptide or a photoaffinity derivative of the peptide. As a preliminary experiment we needed to determine whether the

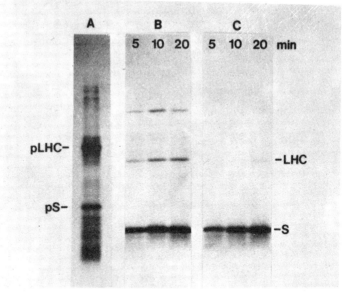

Fig. 3: Effect of LHC II transit peptide analog on import of precursor proteins into pea chloroplasts. Total leaf RNA from pea (*Pisum sativum* L. var. Alaska) was translated in the wheat germ system (34) and analysed immediately by SDS-PAGE and autoradiography (panel A), or after import into isolated chloroplasts (panels B and C). Import reactions were run for the indicated times in the absence (panel B) or the presence (panel C) of *A. thaliana* transit peptide analog (127 μM). To identify successfully imported and processed proteins, all unincorporated translation products were digested with thermolysin prior to reisolation of chloroplasts and analysis by SDS-PAGE and autoradiography.

peptide would inhibit the *in vitro* uptake and processing of pre-LHC II by isolated chloroplasts. Here we report such an experiment.

When mRNA from pea leaves is translated *in vitro* in a wheat germ S30 system, the two major products are preSSU (20 kDa) and preLHC II (32 kDa) (Fig. 3A). Addition of the translation products to intact chloroplasts leads to uptake and processing of the two precursors to give mature SSU and mature apo-LHC II (Fig. 3B). Import appears to be complete within about 10 min. When import is conducted in the presence of 127 μM peptide (Fig. 3C), the rates of pre-LHC II and pre-SSU import are reduced by at least 85% and 50%, respectively. A titration experiment indicates that, essentially, total inhibition of pre-LHC II and pre-SSU import may be achieved at 200 μM peptide. Control experiments indicate that these effects of the peptide are not due to interference with energy production or plastid integrity. Thus, transit peptide analogs appear to be a new class of inhibitor of protein import into chloroplasts and may provide a means of analysing several aspects of the process, including the sequence requirements of transit peptides, the identity of envelope-bound receptors and the substrate specificity of the proteases involved in cleaving the precursors.

Concluding Remarks

Synthetic peptides make feasible many experiments which might otherwise be difficult or perhaps even impossible. In other laboratories, the folding of D1 (29) and LHC II (30) across the thylakoid bilayer has been examined using antibodies raised against peptides corresponding to specific regions of these intrinsic membrane proteins. These studies were possible because the corresponding genes had been sequenced and the thylakoid membrane can be obtained in right-side-out and inside-out configurations. In the case of the folding of D1 (29), the results were consistent with the idea that D1 traverses the membrane five times, as found for homologous proteins in the crystallised bacterial reaction centre (31,32). In the case of LHC II, antibodies against the C-terminus of the protein coagulated only right-side-out vesicles, placing the C-terminus on the stromal side (30). However, this latter result conflicts with the finding that a tyrosine near the C-terminus of LHC II is labelled by a non-permeant reagent only in the case of inside-out vesicles (33).

We have summarised here several applications of synthetic peptides in research on photosystem II. The applications included the preparation of specific soluble substrates for the two thylakoid-bound protein kinases, the manipulation of the sequence of the LHC II kinase substrate to explore the sequence requirements of the kinase, the preparation of antibodies against proteins which are difficult to purify (the 8.3 kDa protein) or which exist only transiently (the transit peptide of LHC II), and the inhibition of protein uptake into the chloroplast by addition of an analog of the pre-LHC ll transit peptide. The success of these preliminary experiments suggests that there is considerable scope for expanding the use of synthetic peptides to other areas of photosynthesis research.

Acknowledgements

This work was performed at Brookhaven National Laboratory under the auspices of the United States Department of Energy (Division of Biological Energy Research, Office of Basic Energy Research) and with additional support from the United States Department of Agriculture (competitive grants 84–CRCR–1–1487 and 85–CRCR–1–1570). We thank C. Anderson, W. Crockett, M. Elzinga, S. Lamb, S. Lees-Miller, E. Shaw and J. Wysocki for advice and assistance in this research, and P. Nixon and J. Barber for the gift of D2 antibody.

References

1. Merrifield, R. B. (1963) *J. Amer. Chem. Soc.* 85, 2149–2154
2. Bennett, J. (1979) *Eur. J. Biochem.* 99, 133–137
3. Michel, H. P. and Bennett, J. (1987) *FEBS Lett.* 212, 103–108
4. Michel, H. P., Hunt, D. F., Shabanowitz, J. and Bennett, J. (1988) J. Biol. Chem. 263, 1123–1130
5. Cashmore, A. R. (1984) *Proc. Natl Acad. Sci.* USA 81, 2960/2964
6. Mullet, J. E. (1983) *J. Biol. Chem.* 258, 9941–9948
7. Hunt, D. F., Yates, J. R., III, Shabanowitz, J., Winston, S. and Hauer, C. R. (1986) *Proc. Natl Acad. Sci.* USA 83, 6233–6237
8. Gal, A., Shahak, Y., Schuster, G. and Ohad, I. (1987) *FEBS Lett.* 221, 205–210
9. Bennett, J., Shaw, E. K. and Michel, H. P. (1988) *Eur. J. Biochem.* 171, 95–100
10. Bennett, J., Shaw, E. K. and Bakr, S. (1987) *FEBS Lett.* 210, 22–26
11. Michel, H. P., Shaw, E. K. and Bennett, J. (1987). In: *Plant Membranes: Structure, Function, Biogenesis* (Sze, H. and Leaver, C. J., eds) pp. 85–102, Alan R. Liss, Inc., New York
12. Pichersky, E., Hoffman, N. E., Malik, V. S., Bernatzky, R., Tanksley, S. D., Szabo, L. and Cashmore, A. R. (1987) *Plant Mol. Biol.* 9, 109–120
13. Bramson, H. N., Mildvan, A. S. and Kaiser, E. T. (1984) *CRC Crit. Rev. Biochem.* 15, 93–124
14. Likhite, V. and Sehon, A. (1967). In: *Methods in Immunology and Immunochemistry*, Vol. I (Williams, C. A. and Chase, M. W., eds), pp. 150–165, Academic Press, New York
15. Allen, J. F. and Holmes, N. G. (1986) *FEBS Lett.* 202, 175–181
16. Hird, S. M., Dyer, T. A. and Gray, J. C. (1986) *FEBS Lett.* 209, 181–186
17. Sutton, A., Sieburth, L. E. and Bennett, J. (1987) *Eur. J. Biochem.* 164, 571–578
18. Klein, R. R., Mason, H. S. and Mullet, J. E. (1988) *J. Cell Biol.* 106, 289–301
19. Bennett, J. (1981) *Eur. J. Biochem.* 118, 61–70
20. Schmidt, G. W. and Mishkind, M. (1986) *Annu. Rev. Biochem.* 55, 879–912
21. Karlin-Neumann, G. A. and Tobin, E. M. (1986) *EMBO J.* 5, 9–13
22. Kirwin, P M., Elderfield, P. D. and Robinson, C. (1987) *J. Biol. Chem.* 262, 16386–16390

23. Leutweiler, L. S., Meyerowitz, E. M. and Tobin, E. M. (1986) *Nucleic Acids Res.* 14, 4051–4076
24. Cline, K., Werner-Washburne, M., Lubben, T. H. and Keegstra, K. (1985) *J. Biol. Chem.* 260, 3691–3696
25. Bitsch, A. and Kloppstech, K. (1986) *Eur. J. Cell Biol.* 40, 160–166
26. Cornwell, K. L. and Keegstra, K. (1987) *Plant Physiol.* 85, 780–785
27. Pain, D., Kanwar, Y. S. and Blobel, G. (1988) *Nature* 331, 232–237
28. Hayden, D. B., Baker, N. R., Percival, M. P. and Beckwith, P. B. (1986) *Biochim. Biophys.* Acta 851, 86–92.
29. Sayre, R. T., Andersson, B. and Bogorad, L. (1986) *Cell* 47, 601–608
30. Anderson, J. M. and Goodchild, D. J. (1987) *FEBS Lett.* 213, 29–33
31. Michel, H. and Deisenhofer, J. (1988) *Biochemistry* 27, 1–7
32. Barber, J. (1987) *Trends Biochem. Sci.* 12, 321–326
33. Bürgi, R., Suter, F. and Zuber, H. (1987) *Biochim. Biophys* Acta 890, 346–351
34. Mishkind, M., Greer, K. L. and Schmidt, G. W. (1987). In: *Methods in Enzymology* Vol. 148 (Packer, L. and Douce, R., eds.), pp. 274–294, Academic Press, New York

Functional Role for 33KDa Proteins in the Oxygen Evolution Complex (OEC)

P.M.NAIR AND N.K. RAMASWAMY

Food Technology & Enzyme Engineering Division
Bhabha Atomic Research Centre
Bombay 400 085, INDIA

Summary

Isolated thylakoid membrane, on treatment with 0.5% Triton X-100, exhibited aerobic dark oxidation of Mn^{2+} ions. However, thylakoids in which water oxidation activity was inhibited by treatments such as Tris, NH_2OH, or heating at 55°C, did not exhibit this activity. These findings, together with the detection of enzyme activity in Triton-X-100-treated PS II particles, suggest that this enzyme activity is related to the OEC. The protein catalysing this activity was purified to homogeneity and its molecular identity was established with the extrinsic, 33 KDa protein in the PS II system. The 33 KDa extrinsic protein of the OEC isolated by the method of Murata also catalysed this reaction. This enzyme activity was dependent on the presence of Cl^- ions and a sulphydryl compound like β-mercaptoethanol (β-ME). The Km for Cl^- was 20 mM and it was required for the stability of the enzyme also. The Cl^- requirement was specific and of the other anions only Br^- was somewhat (72%) effective in replacing Cl^- for the catalytic activity of Mn oxidase. So in its anion requirement also this reaction resembled water oxidation activity. Similarly, reagents like NH_2OH and NH^+_4, the known inhibitors of O_2 evolution, were strong inhibitors of this oxidase activity also. The thiol compound required for the activity of the purified enzyme is also oxidized during the reaction and Mn^{3+} formed was also detected. For continuous activity of the enzyme, thiol compounds were thus essential. There was 1:2 stoichiometry between O_2 uptake and Mn^{3+} formation and -SH disappearance. Addition of catalase in the reaction mixture did not suppress the O_2 uptake, suggesting water as the reaction product.

Abbreviations: AcA–34: Acrylamide-agarose; BTP: 1, 3-bis [Tris(hydroxymethyl)-methyl amino]-propane; BSA: Bovine serum albumin; CTAB: Cetyl trimethyl ammonium bromide; p-CMB: para-Chloromercuric benzoate; Chl: Chlorophyll; DCQ: 2, 6-dichloro-p-benzoquinone; EPR: Electron paramagnetic resonance; EDTA: Ethylene diaminetetra acetic acid; HEPES: N(2-hydroxyethyl) piperazine-N'-2-ethane sulfonic acid; β-ME: β-mercaptoethanol; MOPS: Morpholino-propane sulphonic acid; MES: 2[N-Morpholino] ethane sulphonic acid; NEM: N-ethylmaleimide; OEC: Oxygen evolution complex; PS II: Photosystem II; PAGE: Polyacrylamide gel electrophoresis; P-680: Primary electron donor in PS II; SDS: Sodium dodecyl sulphate; Tricine: N-[2-hydroxy-1, 1, bis (hydroxymethyl) ethyl] glycine; Tris: Tris (hydroxy-methyl) amino methane.

Introduction

The oxygen evolution complex (OEC) located in the thylakoid membrane has four manganese atoms at the active site, presumably acting as charge accumulator, so that stepwise 4-electron oxidation of the molecule can take place in this complex. This is represented by the 'S model' or 'Kok clock', which accounts for the periodicity of four seen in flash-induced oxygen evolution. The electrons are drawn from the complex via 'Z' having a mid-point potential over +1000 mV. Recent evidence show that tyrosine residue of one of the intrinsic protein 34 KDa is involved in this process (Vermaas, this volume). 'Z' is in turn oxidised by the PS II reaction centre providing the driving force for the oxidation reaction of S-states transitions. Thus the oxygen evolution is tightly coupled to photochemical reactions. The isolation of the D_1 and D_2 complex, which does not contain plastoquinone 9, but is highly active in photo-reversible accumulation of reduced pheophytin a, characteristic of a photosystem II reaction centre, shows that charge separation takes place in the D_1 D_2 complex (1).

The highly organised OEC spans the thylakoid membrane. There are many models proposed for the topography of this complex in the membrane. An examination of all these models reveals a common feature, that is, 4 Mn atoms are ligated between the hydrophobic cavity of 'intrinsic' and hydrophilic 'extrinsic proteins'. This type of closed configuration of OEC made it difficult to study the biochemical mechanism of oxygen evolution. The advent of new techniques such as Triton-X-100 treatment of thylakoids (2, 3, 4), and inside-out thylakoid membrane inversion by mechanical disruption (5) for the preparation of PS II particles in which active sites of oxygen evolution are exposed to the aqueous medium, have revolutionised the study of the mechanism of oxygen evolution. In such preparations, the protein components from the active site can be easily dissociated and can be reconstituted with ease under mild conditions.

Studies on Extrinsic Proteins and Their Function on OEC

The preparation of these PS II particles, free of other components of thylakoid like PS I reaction centre, cytochrome b/f or CF_1–CF_0 made it easier to study the protein composition of OEC. On treating these particles with concentrated Tris buffer (2), or subjecting them to high pH (2, 6), three proteins having molecular mass of 18 KDa, 24 KDa, and 33 KDa are released, with concomitant loss of O_2 (2, 6, 7). From this, it is inferred that these extrinsic proteins are involved in the O_2 evolution reaction. In the PS II particles, there are components of PS II reaction centre, three extrinsic proteins, and 4 Mn atoms for every 200 molecules of chlorophyll.

Among the three extrinsic proteins, the 33 KDa protein was the first to be purified and well characterised (8, 9). Subsequently, the 24 KDa and 18

KDa proteins were also purified (10, 11). 33 KDa is acidic whereas 18 KDa is basic. The polarity index of these proteins, determined on the basis of the polar amino acids, adjudicated that they are hydrophilic in nature. Dissociation and reconstitution studies of these three proteins in the PS II particle enable elucidation of the function of these proteins in oxygen evolution (12, 13, 14). These findings suggest that 33 KDa protein is necessary for both oxygen evolution and Mn binding. A high concentration of Cl^-, >100 mM, can partially substitute for the 33 KDa protein in oxygen evolution and Mn preservation. The 24 KDa protein restores O_2 evolution activity, which can be suppressed after its removal, and 5 mM Ca^{++} can substitute for this function. Thus, 24 KDa protein preserves the Ca^{++} required for O_2 evolution even when Ca^{++} concentration in the medium is low, acting as 'Ca^{++} concentrator' (15). The 18 KDa protein reduces requirement for Cl^- concentration, an essential factor for O_2 evolution. The depletion in these two proteins results in reduction of O_2 evolution which can be restored by 30 mM Cl^- and 5 mM Ca^{++}. There is evidence to show that Cl^- ions are required for transition from S_2 to S_3 (16, 17). The site of Ca^{++} ion action is not confirmed. However, there is some suggestive evidence to show that Ca^{++} is necessary for the functional coupling between photochemical reaction and oxidation of water. Ca^{++} depletion uncouples the reaction between the Z and Mn clusters in the spinach PS II particles (18, 19). The 33 KDa, 24 KDa, and 18 KDa proteins can rebind stoichiometrically to the OEC depleted of these proteins to restore O_2 evolution (12, 20, 21). The 33 KDa and 24 KDa proteins seem to bind directly to the complex. The intrinsic protein which anchors 33 KDa and 24 KDa has not yet been identified. It is suggested that, although 33 KDa is hydrophilic, a hydrophobic domain can be generated by secondary and tertiary structures, which can be associated with the intrinsic hydrophobic protein (D_1 & D_2) (22). The 18 KDa protein binds with high affinity to only the O_2 evolution complex having stoichiometric amounts of 24 KDa protein, suggesting that the binding site for 18 KDa protein is located on 24 KDa protein.

Function of 33 KDa or Manganese Stabilising Protein

The function of 33 KDa protein is further investigated by Miyao et al. (23) by removing all intrinsic proteins leaving 4 atoms of Mn bound. To this depleted particle 33 KDa can rebind. Studies using this particle suggest that a dark step is retarded by the elimination of 33 KDa. The transitions from S_1 to S_2 and S_2 to S_3 are not affected by removal and rebinding of the 33 KDa protein, but transition from S_3 to S_0, i.e. release of an O_2 molecule is slowed down and fully recovered on rebinding. Another striking feature of OEC lacking 33 KDa is the pronounced, slow deactivation of the S_2 state which in turn arises from the back reaction in PS II (24). The change in the deactivation rate upon removal and rebinding of 33 KDa may reflect the modification of oxidation–reduction potential of S-state by 33 KDa protein because of its interaction with the Mn cluster. One of the probes to the structure of the S_2 state and to the S-state turnover is an unusual multiline EPR signal originating from S_2 state (25, 26). The signal arises from the cluster of 2 or possibly 4 Mn atoms in a mixed valence state. A recent study by Styring et al (27) shows that 33 KDa could be replaced by 200 mM Cl^- for both multiline signal formation and oxygen evolution to some extent. But the absence of 33 KDa modifies the Mn cluster, leading to an altered multiline signal from S_2 state. Cl^- can also replace 24 KDa and 18 KDa with respect to oxygen evolution and the multiline signal when Ca^{++} is present. They conclude that Cl^- is very important for the formation of the multiline signal from S_2 state and suboptimal concentration of Cl^- results in the inability to form an intact multiline signal in a particle depleted of extrinsic protein. This may suggest that removal of proteins simply enhances the demand for Cl^- without any damage to the Mn cluster of OEC. However, there is evidence to show that oxygen evolution restored by supplementation of 200 mM Cl^- definitely shows some abnormalities such as low activity (28). It is evident from these findings that there is a specific function for 33 KDa in PS II electron transport which cannot be replaced by high concentration of Cl^-. Recently, using thermoluminescence studies, Inoue and coworkers (30) observed substantial stabilisation of the $S_2 Q_A^-$ charge pair upon removal of 3 extrinsic proteins in the presence of DCMU. Based on the experimental evidence of De Vault et al. (31), it is assumed that, in the dark relaxed PS II particle, the peak position of the thermoluminescence glow curve approximately manifests the stabilisation of energy of recombining charges. The particle depleted of extrinsic proteins exhibited a modified $S_2 Q_A^-$ pair evidenced from the change in peak temperature from 4°C to 30°C. Supplementation of 200 mM Cl^- which can restore the O_2 evolution does not reveal the Q band peak temperature to the original state. The reconstitution of the depleted particle with 33 KDa at stoichiometric level restored the normal peak temperature, suggesting a structural change of the donor and/or acceptor side of PS II dependent on the association with the 33 KDa protein.

Meanwhile, recent studies by Beck et al, (32) have shown that the intensity of hyperfine structure of the multiline EPR signal of S_2 produced at low temperature illumination of thylakoid PS II particle is dependent on the length of dark adaptation, suggesting that there is a change in the environment of the Mn cluster of the OEC. Their major findings are (i) the slow O_2 consumption observed in PS II particle or thylakoid after cessation of illumination correlates well with changes in hyperfine structure of the S_2 EPR signal. (ii) These changes are a result of the conversion of the OEC from an active O_2 consuming state to a rest state incapable of O_2 consumption. This means that, in its active state, an OEC can reduce O_2

to water, indicating that an active OEC is capable of reducing O_2 in the dark through a reverse S-state turnover which is thermodynamically a favourable reaction. Although direct evidence is lacking, this suggests the presence of a potential oxidase system in PS II particles. Besides this report by Beck *et al.* (32), there is no direct evidence for such an oxidase system in thylakoid, beyond that it contains a latent form of polyphenol oxidase (33).

We have undertaken experiments to detect the presence of such oxidase activity in OEC using a denatured thylakoid preparation which may be able to oxidise artificial reductant with molecular oxygen. During such studies, oxidation of Mn^{2+} ions was detected. Further investigations on this line led to the finding that isolated 33 KDa protein from ethanol-precipitated thylakoid could catalyse the oxidation of Mn^{2+} ions in the presence of an electron donor such as sulphydryl compounds in the dark. A preliminary report of this study is presented elsewhere (34). The significance of this finding in relation to the function of 33 KDa protein in an OEC will be discussed in this report.

Mn^{2+} Oxidation in Thylakoid and PS II Particles

We have initiated studies with isolated and washed thylakoid preparations from spinach. After disruption with Triton-X-100, these thylakoids exhibited 800–1000 n moles of oxygen consumption mg Chl^{-1}. min^{-1} when Mn–tricine (1:1) complex was used as substrate (Table I). The thylakoids did not show Mn^{2+} dependent O_2 uptake unless they were solubilised with

TABLE I

Effect of Different Detergents and Other Reagents on the Activation of Mn^{2+} Oxidase in Isolated Chloroplasts

Detergent added	Mn^{2+} oxidase activity O_2 uptake in nmoles/mg Chl/min
None	0
CTAB	187
SDS	317
Tween 80	0
Triton X-100	1000
Miranol S_2M-SF	0
[Perchlorate 400 mM]	0

The concentration of detergent used in each reaction mixture was 0.5%. Mn^{2+} dependent O_2 uptake was measured using a Clark-type O_2 electrode. The reaction mixture contained 5 mM Mn-Tricine complex (freshly prepared) pH 7.8, 20 mM BTP-SO_4 buffer pH 7.8, 50 mM NaCl and chloroplasts preparations equivalent to 250 µg Chl in a final volume of 1.5 ml.

β-mercaptoethanol was not added in any of the reaction mixtures used.

detergent. The effect of various detergents like CTAB, cationic, SDS, anionic, Tween 80, Triton-X-100, nonionic, Miranol S_2 M-SF, Zwitter ionic was determined on the release of this enzyme from the membrane. Among these detergents used, Triton-X-100 was found to be the best at 0.5% level. CTAB and SDS also elicited some effect (Table I). Perchlorate at 400 mM concentration did not liberate this enzyme activity. Maximum activation was dependent on the age of the leaves used for isolation of thylakoid. Leaves stored at 0–4°C for 9 days in a desiccator exhibited only 30% activity (300–400 nmoles oxygen uptake mg Chl^{-1}. min^{-1}). When thylakoid preparation was used there was no need for the addition of an external reductant like β-ME but supplementation of Cl^- ions was essential.

Treatment of thylakoids with reagents known to inhibit O_2 evolution, like treatment with NH_2OH, Tris and heating at 55°C for 7 min showed an inhibitory effect on the expression of enzyme activity (Table II). Triton-X-100 on prolonged treatment (thylakoid suspension was kept for

TABLE II

Effect of Different Treatment on Chloroplasts Mn^{2+} Oxidase

Pretreatment of chloroplasts	Mn^{2+} oxidase activity O_2 uptake in nmoles/mg Chl/min	Percentage of inhibition
None	850	–
Triton —100–15 min preincubation	380	55.3
NH_2OH	43	95.0
Tris	36	95.8
Heat	17	98.0

NH_2OH-treated chloroplasts: Hydroxylamine treatment of chloroplasts was performed as given in (35). Chloroplasts suspension (200 μg Chl/min) was incubated at 21°C for 20 min in the presence of 5 mM NH_2OH plus 1 mM EDTA at pH 7.5. The chloroplasts were then centrifuged and washed well with 10 mM HEPES pH 7.5 containing 40 mM NaCl and suspended in the same buffer with chlorophyll concentration adjusted to 2 mg Chl/ml.

Tris-washed chloroplasts: The method described in (36) was followed for the preparation of tris-washed chloroplasts. Isolated chloroplasts were incubated for 20 min at 15°C in 0.8 M tris-HCl pH 8.0. The tris-treated chloroplasts were sedimented and washed well to remove tris with 10 mM HEPES pH 7.5 containing 40 mM NaCl and finally suspended in the same buffer (2 mg Chl/ml).

Heat-treated chloroplasts: The cholorplasts were heat treated at 55°C for 7 min and later cooled in ice. After centrifugation, the sedimented chloroplasts were suspended in 10 mM HEPES pH 7.5 containing 40 mM NaCl and chlorophyll concentration adjusted to 2 mg Chl/ml.

Triton X-100-pretreatment: In this case, Triton X-100 at 0.5% concentration was added to chloroplasts suspension and incubated for about 15 min at 0–4°C and an aliquot was taken for Mn^{2+} oxidase assay. The Mn^{2+} oxidase was assayed as detailed in Table 1.

15 min at 0–4°C with Triton-X-100 at 0.5% concentration) reduced the activity considerably (55.3%).

The enzyme activity was located in PS II particles. Two types of particle preparation were tested, namely the Murata type (2) and the BBY particle (3) (Table III). In these particles, the elicitation of Mn^{2+} oxidase activity was only about 20% of that of thylakoids. The lowering of oxidase activity in these particles may be due to prolonged Triton-X-100 treatment of thylakoid for preparation of these particles. Since Triton-X-100 was used in the preparation of these particles, its addition was not necessary to detect this activity. However, supplementation of β-ME was essential for full activity. In its absence, PS II particles exhibited only 25% of maximum activity.

The expression of Mn^{2+} oxidase activity only on solubilisation of the thylakoid membrane by detergents indicates that externally added Mn^{2+} ions cannot react with the membrane-bound enzyme and it is possible that enzyme may exist as a large complex embedded in the membrane. The non-ionic detergent Triton-X-100 was most effective in releasing the enzyme from its membraneous environment. Even in the case of Triton-X-100, the activity is considerably reduced if the assay is not conducted immediately after addition of detergent (Table II). For the same reason, PS II particles prepared with Triton-X-100 treatment showed a reduced level of activity (Table III). It is worth mentioning here that, in an

TABLE III

A Comparative Study of Mn^{2+} Oxidase Activity in PS II Particles

Source of Mn^{2+} oxidase	O_2 evolution μmoles/mg Chl/hr	Mn^{2+} oxidase activity in nmoles O_2 uptake/mg Chl/min
Murata's PS II	561 (500–600)	235 (200–250)
BBY PS II	533 (500–550)	181 (170–190)

The isolated spinach chloroplasts used for preparing the PS II particles had Mn^{2+} oxidase activity of 850 nmoles of O_2 uptake/mg Chl/min or in the range of 800–1000 nmoles O_2 uptake/mg Chl/min. Parentheses values indicate the range of values possible on this assay obtained from different independent experiments carried out on separate days. The O_2 evolution rate of the chloroplasts used for this experiment was 300 μmoles O_2 evolved/mg Chl/hr or 1200 μ eq. ferricyanide reduced per mg Chl/hr when ferricyanide was used as the electron acceptor.

PS II particles were prepared as described in (2, 3).

O_2 evolution assay mixture of PS II particles contained 20 mM NaCl, 30 mM MES buffer pH 6.5, PS II particles equivalent to 20 μ Chl/ml in a final volume of 1.5 ml with 0.3 mM DCQ as electron acceptor.

For the Mn^{2+} oxidase assay, the reaction mixture (Table I) was supplemented with 1 mM β-ME along with PS II particles equivalent to 125 μg Chl in a final volume of 1.5 ml.

early study, Miles has reported consumption of oxygen in PS II catalysed by the addition of exogenous Mn^{2+}, but the reaction was not further characterised (37). Since the activity is associated with PS II particle it seems logical to assume that this enzyme protein may be a part of the water oxidation system.

Purification and Characterisation of the Protein

The first step in the purification process was to isolate chloroplast protein free of pigment and lipids. The best results were obtained with ethanol rather than acetone, because ethanol prevented complete delipidation of the thylakoid protein and from such preparations we got more active enzyme preparation. The ethanol precipitate could be stored at $-80°C$ for more than a month without loss in enzyme activity. The extraction of the enzyme from this precipitate was facilitated by grinding with high-salt buffer at pH 8.5 after freezing. The dialysed crude preparation was then subjected to gel filtration on Sephadex G-150. Fractions having Mn^{2+} oxidase activity were pooled and concentrated. Final purification of this protein was achieved by chromatography on an AcA-34 column. The pooled fractions were concentrated and used for further characterisation. The final preparation gave 87% recovery of activity on protein basis (Table IV).

TABLE IV

Purification Profile of Mn^{2+} Oxidase

Fraction	Protein yield µg/mg Chl	Specific activity nmoles/min/mg O_2 uptake protein	Activity recovery %
Salt extract from alcohol powder (crude enzyme)	420	41	100
Sephadex G-100 fractions	114	133	88
AcA-34 fraction	6	2498	87

The protein concentrations of crude enzyme, Sephadex G-150 and AcA-34 column fraction were 1.4 mg protein/ml; 378 µg protein/ml; and 12 µg protein/ml respectively. For the assay, 0.5 ml enzyme from each fractions was used for measuring activity. The protein concentrations in the fractions was determined by a modified Bradford method as detailed in (38).

Mn^{2+}-dependent O_2 uptake was measured using a Clark-type O_2 electrode. The reaction mixture contained 5 mM Tricine-Mn complex (freshly prepared) pH 7.8, 20 mM BTP-SO_4 buffer pH 7.8, 50 mM NaCl, 1 mM β-mercaptoethanol and enzyme protein as detailed above in a final volume of 1.5 ml. The initial slope was measured and activity expressed as O_2 uptake nmoles/min/mg protein. Proper blanks were included in all experiments to check the auto-oxidation of Mn^{2+}-tricine complex.

The purified enzyme had an UV absorption maximum at 278 nm with a shoulder at 228 nm. The molecular weight of the final purified enzyme was determined by gel filtration on Sephadex G-200, where it is separated as a single peak corresponding to a molecular mass of 33 KDa, when compared with a standard molecular weight marker. SDS-PAGE showed a single band having molecular weight 33 KDa (Fig. 1). Metal analysis of this homogeneous preparation showed that this protein is devoid of metal ions. An estimate of SH groups in this preparation, using Ellman's reagent, after controlled dialysis against 20 mM NaCl buffer to remove β-ME, revealed that the protein contained 2 SH groups per molecule. The purified enzyme was devoid of catecholase and catalase activities. Addition of catalase to the Mn^{2+} oxidase system did not affect O_2 uptake, indicating formation of H_2O as the reaction product.

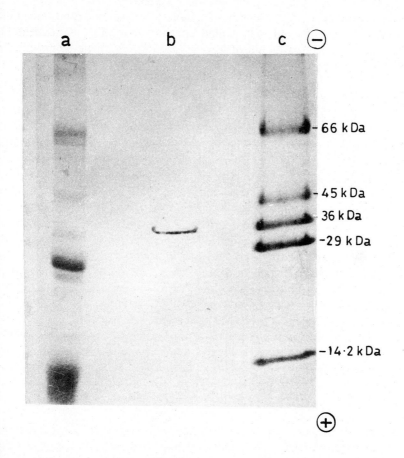

Fig. 1: SDS-PAGE pattern of purified Mn^{2+} oxidase. (a) PS II particle containing BSA; (b) Purified Mn^{2+} oxidase from AcA-34 column; (c) Molecular weight standards.

Properties of the Enzyme Activity

In dealing with the oxidation of a metal ion one has to be very careful about choosing the proper reaction mixture. Otherwise the results will be vitiated by the auto-oxidation of the metal. So the important factors are, (i) choice of the buffer and pH, (ii) stability of the metal ion in the mdium, i.e. there should not be any auto-oxidation of the metal or its precipitation. After many trial experiments, we have arrived at the following medium for best results. The buffer of choice was BTP-SO$_4$, pH 7.8, substrate was added as Mn-Tricine complex, and other requirements were enzyme protein, Cl$^-$ ions, and reducing agent β-ME. Our experience showed that use of Mn^{2+} alone, although it gave a good initial rate, resulted, after a certain period, in inactivation of the enzyme caused by the product of oxidation, and at pH 7.8 there was slight auto-oxidation Tricine-Mn complex, on the other hand, did not show any auto-oxidation at pH 7.8. It was also established, by examining the non-enzymatic oxidation rate in the absence of enzyme and with boiled enzyme, that Mn^{2+}-dependent O$_2$ uptake is truly the oxidation of Mn^{2+}. Addition of tricine alone did not show any oxygen uptake.

Requirements for Mn^{2+} Oxidation by 33 KDa Protein

The optimum pH was determined in BTP-SO$_4$ buffer at various pH values from 6.5 to 8.5. There was a sharp increase in activity after pH 7.5 and the optimum was at pH 8.0. In these experiments the rate of auto-oxidation of Mn^{2+}-tricine complex was also determined at various pH values and the auto-oxidation rate was subtracted from the enzyme reaction rate. Although activity at pH 8.0 was maximum, there was some auto-oxidation of Mn$^2+$. So pH 7.8 was selected for further experiments, where auto-oxidation was nil. The effect of Mn^{2+} concentration on the enzyme reaction was also determined and it was found that the enzyme was saturated at 5 mM Mn-Tricine concentration. Increase in enzyme protein in the reaction mixture showed linearity in the activity upto 8 μg.

Cl$^-$ ions were essential for enzyme activity. Presence of 20 mM NaCl in all the purification steps was obligatory for the retention of enzyme activity, so the reaction mixture always contained a minimum concentration of 10 mM Cl$^-$. Supplementation of Cl$^-$ in the reaction mixture showed sharp increase in activity which reached a maximum at 50 mM Cl$^-$ (Fig. 2). The apparent Km for Cl$^-$ ion was 20 mM for the routinely used assay at pH 7.8. The Cl$^-$ requirement for oxygen evolution is well recognised now (39, 40, 41).

Some of the observed properties of this enzyme strongly suggest that it is a component of OEC. This inference was further strengthened by the observation that treatments known to produce predominantly irreversible inactivation of O$_2$ evolution in the chloroplast accompanied by the release

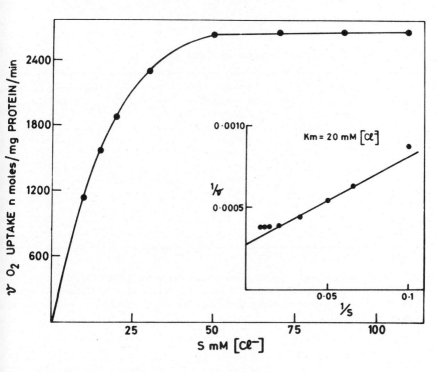

Fig. 2: Effect of Cl⁻ ions on Mn²⁺ oxidase activity. NaCl was used as source of Cl⁻ ions. The insert is the double reciprocal plot for the effect of Cl⁻ ions. The Mn²⁺ oxidase assay conditions are as detailed in Table IV.

of Mn ions like Tris, NH_2OH, or heat also inhibited the expression of this enzyme activity in thylakoids solubilised with 0.5% Triton-X-100. The similarity of this enzyme reaction to water oxidation is once more evidenced when the effect of monovalent anions like Br^-, I^-, F^-, NO_3^-, $HCOO^-$ and HCO_3^- was examined (Table V). The relative effect of these ions on enzyme activity was closely paralleled with their effect on OEC reported by Kelly and Izawa (42) on Cl⁻-depleted chloroplasts. The function of Cl⁻ ions on the enzyme activity of 33 KDa protein may be to give it proper orientation for catalytic activity and stabilisation of the oxidation product of Mn²⁺. There are innumerable reports suggesting concerted action of Cl⁻ with 33 KDa protein in OEC (13, 22, 28, 39, 43, 44). This enzyme activity was found to be susceptible to low concentration of Cu^{2+}, Cd^{2+}, Hg^{2+}, even in the presence of β-ME, and SH binding reagents inhibited the activity strongly (Table VI).

NH_2OH and NH_4^+ were found to be strong inhibitors of the enzyme activity. K_1 values for NH_2OH and NH_4^+ ions were 0.36 mM and 2.4 mM respectively (Fig. 3). The inhibition by NH_2OH and NH_4^+ is strikingly similar to that observed with OEC. It is reported that in PS II particle

TABLE V

Relative Effect of Other Anions in Place of Cl⁻

Anions	Activity O_2 uptake nmoles/mg protein/min	Relative effect %	Relative effect* on O_2 evolution
Cl⁻	2766	100	100
Br⁻	2000	72	72
I⁻	767	28	31
F⁻	500	18	7
NO_3^-	1067	39	65
HCOO⁻	667	24	27
CH_3COO^-	0	0	0
HCO_3^-	733	27	30

* Data from Kelly and Izawa (42).

The anions were used in the reaction mixtures in a concentration of 60 mM. 1 mM β-mercaptoethanol was present in the reaction mixture for the assay. 33 KDa proteins from AcA-34 eluate with 12 μg protein/ml concentration was used as the enzyme. Other details are as given in the text.

TABLE VI

Effect of Metal Ions and SH-inhibitors

Addition	Concentrations used	Activity O_2 uptake in nmoles/mg protein/min
None		2766
Cd^{2+}	100 μM	567
Cd^{2+}	250 μM	67
Cu^{2+}	250 μM	267
Hg^{2+}	250 μM	167
NEM	5 mM	233
p-CMB	5 mM	200

Metal ions and other compounds were supplemented before the addition of Mn-Tricine complex in the reaction mixture at a final concentration as indicated above. Other assay conditions are the same as given in Table IV.

NH_2OH affects the water oxidation by solubilisation of ligated Mn and mobilisation of extrinsic polypeptides (45, 46) and in chloroplasts by removal of Mn and destabilisation of S-state enzymes (47). In the present instance they may be exerting their effect by preventing the formation of the Mn-enzyme complex required for the binding of second substrate O_2 for catalytic activity.

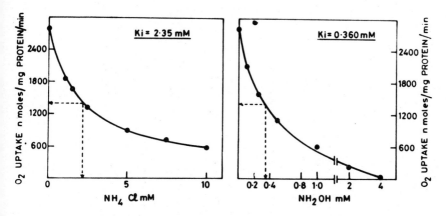

Fig. 3: Effect of NH_4^+ ions and NH_2OH on the Mn^{2+} oxidase activity. NH_4Cl was used as the source of NH_4^+ ions. Enough Cl^- were added to have the final concentration of Cl^- as 50 mM in the reaction mixture. For the experiment on NH_2OH, Cl^- ions were added at the end after adding NH_2OH.

Identity of this 33 KDa Protein with Murata Protein of PS II Particles

Our next attempt was to establish the identity of the 33 KDa protein we have isolated with the 33 KDa protein of PS II isolated by Murata (21, 48). In its physical properties, such as molecular mass, presence of 2 SH groups, UV absorption and absence of metal ion, this protein resembled the manganese-stabilising protein or Murata protein isolated by Kuwabara and Murata (8, 48), suggesting that they are the same proteins isolated by

TABLE VII

Comparative Study of Mn^{2+} Oxidase Activity on 33 KDa Protein Isolated from Different Source

Source of Mn^{2+} oxidase	Mn^{2+} oxidase activity nmoles O_2 uptake/mg protein/min
Murata protein (33 KDa)	3473 (3000–3500)
33 KDa protein from AcA-34 fraction	2766 (2400–2800)

The isolated spinach chloroplasts used for preparing the 33 KDa protein, had Mn^{2+} oxidase activity of 850 nmoles of O_2 uptake/mg Chl/min or in the range of 800–1000 nmoles O_2 uptake/mg Chl/min. Parentheses values indicate the range of values possible on this assay obtained from different independent experiments carried out on separate days.
The 33 KDa protein from Murata's PS II was isolated as described in (21, 48). Experimental details are given in the text. Enzyme was assayed as detailed in Table IV.

TABLE VIII

Evidence for Mn^{3+} Formation and SH Disappearance with Mn^{2+} Oxidase Activity in AcA-34 Column Fractions

Time duration of assay	O$_2$ uptake in nmoles	Mn^{3+} formed in nmoles	SH disappeared in nmoles	Ratios of O$_2$ uptake versus	
				Mn^{3+} formed	SH group disappeared
1 min	11.8 ± 1.79	23.4 ± 5.13	46.4 ± 8.62	1 : 1.98 ± 0.2	1 : 4 ± 0.4
2 min	22.2 ± 3.56	45.2 ± 9.39	56.2 ± 11.26	1 : 2 ± 0.2	1 : 2.4 ± 0.2
5 min	36 ± 1.83	76.5 ± 3.42	78.6 ± 8.85	1 : 2.15 ± 0.13	1 : 2.2 ± 0.2

The values are expressed as mean ± SD (standard deviation) from 10 independent experiments carried out on different days. Extra addition of β-mercaptoethanol to the assay system was avoided, only carry-over β-mercaptoethanol from enzyme preparation of about 0.5 μmoles was present in the assay medium. The reaction mixture for the assay was same as given in Table IV. 6 μg protein was used for the assay. O$_2$ uptake was measured in a Gilson oxygraph at the time intervals indicated. Mn^{3+} formation was monitored as described in (49, 50). The amount of SH group disappeared in concurrence with the oxidase activity was determined by Ellman's reagent as in (51). A reaction carried out in a 'Thunberg tube' in the absence of O$_2$ was used as the control blank for SH group estimation. For calculation of sulphydryl content, the molar absorptivity value of 13,600 M^{-1} cm^{-1} for the nitromercaptobenzoate anion at 415 mμ was employed.

different methods. Confirmation of its identity with the Murata protein was obtained by isolating Murata protein and testing whether it can catalyse the oxidation of Mn^{2+}. Under experimental conditions, Murata protein showed activity which compared well with the 33 KDa protein from the AcA-34 fraction (Table VII). In the absence of added β-ME in the assay of Murata protein there was no trace of Mn^{2+} oxidase activity. But our preparation which contained β-ME during purification showed some activity without supplementation.

Since we are measuring only oxygen uptake and claiming that it is an oxidase without demonstrating the oxidation of Mn ions, the oxidation of Mn^{2+} to Mn^{3+} and disappearance of SH groups from the reaction mixture was determined. In this assay, extra addition of β-ME was avoided. Only carry-over β-ME from an enzyme preparation of about 0.5 μ moles was present in the assay medium. Mn^{3+} formation was determined by Avron's method (49, 50). When this assay was conducted in Thunberg tubes in the absence of O_2 there was no Mn^{3+} formation. There was 1:2 stoichiometry for O_2 uptake and Mn^{2+} oxidation (Table VIII). But in the case of SH disappearance the initial rate at 1 min was very high (1:4), which got reduced with time and at 5 min it was 1:2. These experiments were repeated a number of times and also with the Murata preparation. Here also, 1 min SH values are high but there was no concomitant O_2 uptake or Mn^{3+} formation. At present we have no explanation for this observation.

Conclusions

In the elucidation of the mechanism of this reaction, knowledge of the structure of the protein molecule is essential. The 33 KDa protein has been purified by Murata and coworkers and its complete amino acid sequence was determined recently (52). The main feature of this primary structure of this enzyme which is relevant in the present context is its content of 2 cysteine residues at 26th and 51st positions. The participation of these SH groups in catalysis is inferred from the requirement of SH reagents. This idea is strengthened by the inhibition of the reaction by metal ions Hg^{2+}, Cd^{2+}, Cu^{2+}, and by SH inhibitor p-CMB and NEM. So far, there has been no report on the involvement of SH groups in the water oxidation system. But recent findings by Wada et al. (52) show otherwise. They focused their attention on the cysteine residues of 33 KDa protein. The two cysteine in the 33 KDa protein are normally in the oxidised state (-S-S-). When they are reduced to SH with β-ME, the protein loses its ability to bind to 33 KDA depleted particles and to stimulate 0_2 evolution. The β-ME-reduced 33 KDa is readily reoxidised in air to become the -S-S- form again which showed full reconstitution and 0_2 evolution. If this is true, then inclusion of β-ME or any other sulphydryl compound should inhibit O_2 evolution. In order to confirm this, an experiment was conducted. When β-ME was

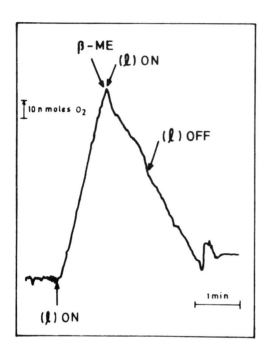

Fig. 4: Effect of β-ME on the oxygen evolving PS II particles in presence of light. PS II particles were prepared as described in (2). O_2 evolution was assayed at pH 6.5 using PS II particles equivalent to 10 μg Chl/ml along with 0.3 mM DCQ as the electron acceptor. 1 mM β-ME was added to the assay medium where indicated.

added to a PS II particle system evolving O_2, the activity was abruptly inhibited, followed by O_2 uptake (Fig. 4). This is evidence suggestive of the involvement of a thiol group for redox function in O_2 evolution. It is possible that, under reduced condition, there is a reverse turnover of the OEC in which O_2 is reduced to water. When S_0 and S_4 states are in equilibrium, binding of oxygen to the S_0 state could lead to formation of the S_4 state and, consequently, the reduction of oxygen. The details of thiol function on water oxidation is worth further investigation. In the case of a purified system, the presence of Mn enhances the redoxidation of cysteine residues of 33 KDa protein. From this evidence, a working model for Mn oxidation by 33 KDa is proposed in Fig. 5. With the purified enzyme,

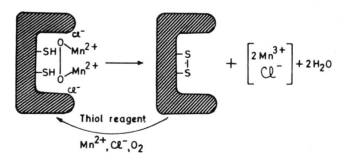

Fig. 5: A working model for Mn^{2+} oxidation.

presence of β-ME is required for continuous activity by generating SH. These findings suggest an additional functional role for 33 KDa protein giving a redox property and possibly a binding site for either O_2 or water. How this will help in O_2 evolution is matter needing further investigation.

Acknowledgements

This work was initiated in Professor S. Izawa's laboratory at the Department of Biological Sciences, Wayne State University, Detroit, Michigan 48202, USA, during PMN's tenure (July–Dec. 1983) as a Visiting Scientist. Part of this work has been presented at the 9th International Congress on Photobiology, Philadelphia, 1–5 July 1984 and also at the Japan/US Binational Symposium held in Okazaki, Japan, 17–21 March, 1987.

The authors wish to thank Professor S. Izawa for his encouragement and valuable suggestions during the initial course of these studies.

References

1. Nanba, O. and Satoh, K. (1987) *Proc. Natl Acad. Sci. USA* 84, 109–112
2. Kuwabara, T. and Murata, N. (1982) *Plant Cell Physiol.* 23, 533–539
3. Berthold, D. A., Babcock, G. T. and Yocum, C. F. (1981) *FEBS Lett.* 134, 231–234
4. Ghanotakis, D. F. and Yocum, C. F. (1986) *FEBS Lett.* 197, 244–248
5. Andersson, B. and Akerlund, H. E. (1978) *Biochim. Biophys.* Acta 503, 462–472
6. Akerlund, H. E. and Jansson, C. (1981) *FEBS Lett.* 124, 229–232
7. Yamamoto, Y., Doi, M., Tamura, N. and Nishimura, M. (1981) *FEBS Lett.* 133, 265–268
8. Kuwabara, T. and Murata, N. (1979) *Biochim. Biophys.* Acta 581, 228–236
9. Kuwabara, T. and Murata, N. (1982) *Biochim. Biophys.* Acta 680, 210–215
10. Jansson, C. (1984). In: *Advances in Photosynthesis Research* (Sybesma, C. ed.), Vol. 1, pp. 375–378, Martinus Nijhoff/Dr W. Junk Publishers
11. Kuwabara, T. and Murata, N. (1984). In: *Advances in Photosynthesis Research* (Sybesma, C., ed.), Vol. 1, pp. 371–374, Martinus Nijhoff/Dr. W. Junk Publishers
12. Miyao, M. and Murata, N. (1983) *Biochim. Biophys.* Acta 725, 87–93
13. Miyao, M. and Murata, N. (1984) *FEBS Lett.* 170, 350–354
14. Miyao, M. and Murata, N. (1983) *FEBS Lett.* 164, 375–378
15. Ghanotakis, D. F., Topper, J. N., Babcock, G. T. and Yocum, C. F. (1984) *FEBS Lett.* 170, 169–173
16. Itoh, S., Yerkes, C. T., Koike, H., Robinson, H. H. and Crofts, A. R. (1984) *Biochim. Biophys.* Acta 766, 612–622
17. Theg, S. M., Jursinic, P. and Homman, P. H. (1984) *Biochim. Biophys.* Acta 766, 636–646
18. Dekker, J. P., Ghanotakis, D. F., Plijter, J. J., Van Gorkom, H. J. and Babcock G. T. (1984) *Biochim. Biophys.* Acta 767, 515–523
19. Satoh, K. and Katoh, S. (1985) *FEBS Lett.* 190, 199–203
20. Akerlund, H. E., Jansson, C. and Andersson. B. (1982) *Biochim. Biophys.* Acta 681, 1–10
21. Ono, T. and Inoue, Y. (1984) *FEBS Lett.* 166, 381–384.

22. Camm, E. L., Green, B. R., Allred, D. R. and Staehelin, L. A. (1987) *Photosynthesis Res.* 13, 69–80
23. Miyao, M., Murata, N., Lavorel, J., Maison-Peteri, B., Boussac, A. and Etienne, A. L. (1987) *Biochim. Biophys.* Acta 890, 151–159
24. Joliot, P., Joliot, A., Bouges, B. and Barbieri, G. (1971) *Photochem. Photobiol.* 14, 287–305
25. Brudvig, G. W., Casey, J. L. and Sauer, K. (1983) *Biochim. Biophys.* Acta 723, 361–371
26. Zimmermann, J. L. and Rutherford, A. W. (1984) *Biochim. Biophys.* Acta 767, 160–167
27. Styring, S., Miyao, M. and Rutherford, A. W. (1987) *Biochim. Biophys.* Acta 890, 32–38
28. Kuwabara T., Miyao, M., Murata, T. and Murata, N. (1985) *Biochim. Biophys.* Acta 806, 283–289
29. One, T. A. and Inoue, Y. (1986) *Biochim. Biophys.* Acta 850, 380–389
30. Vass, I., Ono, T. A. and Inoue, Y. (1987) *FEBS Lett.* 211, 215–220
31. De Vault, D., Govindjee and Arnold, W. (1983) *Proc. Natl Acad. Sci. USA* 80, 983–987
32. Beck, W. F., Paula, J. C. de and Brudvig, G. W. (1985) *Biochemistry* 24, 3035–3043
33. Golbeck, J. H. and Cammarata, K. V. (1981) *Plant Physiol.* 67, 977–984
34. Nair, P. M., Ramaswamy, N. K. and Izawa, S. (1987) *Proc. Japan/US Binational Symposium* held in Okazaki, Japan, bet. March 17–21, 1987, pp. 24–25
35. Izawa, S. and Ort, D. R. (1974) *Biochim. Biophys.* Acta 357, 127–143
36. Yamashita, T. and Butler, W. L. (1968) *Plant Physiol.* 43, 1978–1986
37. Miles, C. D. (1976) *FEBS Lett.* 61, 251–254
38. Peterson, G. L. (1983) in: *Methods in Enzymology,* 'Enzyme Structure' Part I, Vol. 91 (Hirs, C. H. W. and Timasheff, S. N. eds.), pp. 95–119, Academic Press, New York and London
39. Coleman, W. J. and Govindjee (1987) *Photosynth. Res.* 13, 199–223
40. Critchley, C. (1985) *Biochim. Biophys.* Acta 811, 33–46
41. Izawa, S., Muallem, A. and Ramaswamy, N. K. (1983). In: *The Oxygen Evolving System of Photosynthesis* (Inoue, Y., Crofts, A. R., Govindjee, Murata, N. and Satoh, K. eds.), pp. 293–302, Tokyo, Academic Press
42. Kelley, P. M. and Izawa, S. (1978) *Biochim. Biophys.* Acta 502, 198–210
43. Mavankal, G., McCain, D. C. and Bricker, T. M. (1986) *FEBS Lett.* 202, 235–239
44. Miller, A. F., De Paula, J. C. and Brudvig, G. W. (1987) *Photosynth. Res.* 12, 205–218
45. Tamura, N. and Cheniae, G. (1985) *Biochim. Biophys.* Acta 809, 245–259
46. Becker, D. W., Callahan, F. F. and Cheniae, G. (1985) *FEBS Lett.* 192, 209–214
47. Radmer, R. and Cheniae, G. M. (1977). In: *Topics in Photosynthesis* (Barber, J. ed.), Vol. 2, pp. 303–348, Elsevier Science Publishers, Amsterdam
48. Kuwabara, T. and Murata, N. (1983) *Plant Cell Physio.* 24, 741–747
49. Ben-Hayyin, G. and Avron, M. (1970) *Biochim. Biophys.* Acta 205, 86–94
50. Avron, M. and Shavit, N. (1963) *Anal. Biochem.* 6, 549–554
51. Habeeb, A.F.S.A. (1972). In *Methods in Enzymology,* 'Enzyme Structure', Part B, Vol. XXV (Hirs, C. H. W. and Tamasheff, S. N. eds.) pp. 457–464, Academic Press, New York, London
52. Oh-Oka, H., Tanaka, S., Wada, K., Kuwabara, T. and Murata, N. (1986) *FEBS Lett.* 197, 63–66

Heat-shock Proteins Associated with Chloroplasts

A. GNANAM, S. KRISHNASAMY AND R. MANNAR MANNAN

School of Biological Sciences
Madurai Kamaraj University
Madurai 625 021, India

Summary

It is now well established that all living organisms, from bacteria to higher plants and animals, when exposed to non-lethal heat-shock conditions, respond by a drastic alteration of the protein synthesis pattern and the appearance of a new set of proteins termed heat-shock proteins (HSPs). The shut-down of normal protein synthesis results from translational regulation and the synthesis of HSPs begins as a result of the induction of transcription of a new set of mRNAs.

In the case of plant systems, beginning with the report on the occurrence of heat-shock response in the cultured cells of tobacco and soybean (1) there is a growing interest in the study of HSPs in plants. During the last eight years, heat-shock response has been studied in a variety of plant species (2) including unicellular green alga (3) and cyanobacterium (4). In vascular plants, HSPs fall into two size-classes: a high molecular mass group (60–110 kDa) and a complex group of polypeptides with molecular weight less than 30 kDa.

Functional Significance of HSP in Higher Plants

The functional significance of none of the HSPs in the higher plants has so far been determined. However, there is a growing body of evidence to suggest that HSPs play an important role in the induction of thermotolerance in plants. In soybean seedlings an absolute correlation has been found between the synthesis and accumulation of HSPs and the ability to survive short heat treatments at otherwise lethal temperatures (5). The ability of cotton plants to produce and accumulate HSPs in response to elevated temperatures under field conditions has been observed (6). Genetic differences in high temperature susceptibility of sorghum lines has been correlated with variations in the temporal development of the capacity of synthesised HSPs and acquired thermotolerance (7). In barley it has been shown that a short exposure to 37°C (30 min), which determines the induction of HSPs, thermoprotected the plants when they were shifted to a higher temperature of 42–45°C (8).

Localisation of HSPs in the Intracellular Organelles

To explain the molecular mechanism involved in the induction of thermotolerance through the synthesis of HSPs, it has been proposed that localisation of significant amounts of HSPs in a cellular compartment is likely to result in the non-specific lowering of the rate of irreversible loss of vital functions by an otherwise stress-labile protein in that compartment (9). Thus the acquisition of thermotolerance appears to depend not only upon the synthesis of HSPs, but also on their selective intracellular localisation. For example, treatment with arsenite, a metabolic inhibitor, mimics the effect of heat shock in inducing the synthesis of a set of HSPs even at the normal temperature. However, the arsenite-induced HSPs do not get localised in the intra-cellular organelles at room temperature, but they do become organelle-associated during subsequent heat shock (5).

HSPs Associated with Chloroplasts

In has been shown that, in plants, the high temperature inhibition of photosynthesis is not caused by a general breakdown of cellular integrity since impairment of respiration and membrane integrity cannot be detected until temperature is raised significantly above those inhibitory to photosynthesis. Also, the inhibition of photosynthesis at high temperature cannot be ascribed to either a decrease in stomatal conductance or a temperature-induced alteration of the diffusive resistance of CO_2 fixation within the chloroplasts. Thus photosynthesis at high temperature is considered to be a result of damage within the chloroplasts (10). In the light of this information, it is logical to conclude that, if the HSPs are playing a role in the induction of thermotolerance, some of them must be associated with the chloroplasts.

Kloppstech *et al.* (11) have shown that in pea and *Chlamydomonas*, a 22-kDa HSP gets localised on the thylakoid membranes. This is coded by the nuclear genome and it has been shown to be synthesised as a high-molecular-weight precursor with a molecular weight of 26 kDa. Similarly, Vierling *et al.* (12) have shown that in soybean, pea and corn, a number of HSPs between 21 and 27 kDa get localised in the chloroplasts and all of them are nuclear coded. On the other hand, Restivo *et al.*(13) have shown that, in heat-shocked *Nicotiana plumbageni-folia* protoplasts, a total of nine HSPs are found associated with the chloroplasts. Disruption of chloroplasts into membrane-bound and soluble fractions showed a tight association of 67, 22, 21, 19 and 18 kDa HSPs with the thylakoid membranes whereas 96, 74 and 26 kDa HSPs were detectable in the soluble fraction.

The results of our experiments with *Vigna sinensis* suggest that the number of HSPs associated with the chloroplasts in this species depends on the mode of heat-shock treatment given to the leaves (14, 15). When the leaves are subjected to a rapid heat-shock treatment (transferring the leaves from 25°C to 40°C for and labelling them immediately), a set of six HSPs in the molecular sizes of 96, 80, 75, 22, 20 and 15 kDa were detectable in the total leaf homogenate (Fig. 1). The chloroplast fractions isolated from these leaves showed the presence of a set of low-molecular-weight HSPs in the size range of 15–22 kDa (Fig. 2 lane B). This observation is comparable to the results obtained by Vierling *et al.* (12). However, when the leaves were subjected to gradual heat shock treatment (increasing the temperature from 25°C by 3°C over every hour and labelling the leaves after 40°C is reached), in addition to the higher-molecular-weight HSPs (96, 80 and 75 kDa) detectable in the leaves subjected to rapid heat shock, a 60 kDa HSP was detected as a distinct band in the fluorographic profile. Further, two new HSPs in the molecular size of 85 and 70 kDa were detectable as faint bands in the fluorographic profile of the total leaf homogenate (Fig. 1 lane C). In the fluorographic profile of the chloroplast preparations obtained from these leaves subjected to gradual heat-shock treatment, these three new HSPs (85, 70 and 60 kDa) appeared as distinct bands (Fig. 2 lane C). Thus these three HSPs are found associated with the chloroplasts only in leaves subjected to gradual heat-shock and are absent in leaves subjected to rapid heat-shock.

HSPs Coded by the Chloroplast Genome

Chloroplast is a semiautonomous intracellular organelle having its own genome and an independent protein synthetic machinery. So, it is possible that some of the HSPs detectable in the chloroplast fraction of the heat-shocked leaves could be coded by the chloroplast genome and synthesised within the organelle.

Kloppstech *et. al.* (16) have demonstrated that the synthesis of 70 kDa HSP associated with chloroplast in *Acetabularia mediterranea* is sensitive to chloramphenicol. Further, the synthesis of this HSP could be seen even in anucleated cells. These observations have led them to suggest that the site of synthesis of this HSP in this organism is within the chloroplasts.

Our experiments with *Vigna sinensis* and *Sorghum vulgare* indicate that, in these two species, there is a synthesis of sets of HSPs by the isolated chloroplasts at elevated temperatures (17). When the incubation temperature of the isolated chloroplasts capable of protein synthesis is increased to 40°C, there is an induction of synthesis of four HSPs in the molecular weight range of 85, 70, 60 and 24 kDa (Fig. 3). Thus, at least in these two species, the chloroplast genome seems to have a definite heat-shock response.

All of these four HSPs synthesised by the isolated chloroplasts of *Vigna* at the elevated temperature are found localised on the thylakoid

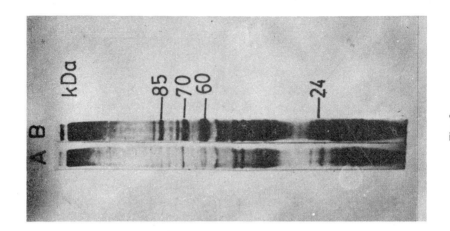

Fig. 3

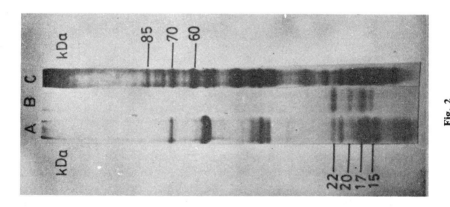

Fig. 2

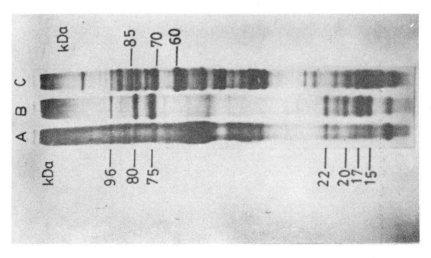

Fig. 1

membranes and their synthesis is regulated at the transcriptional level (15).

It is interesting to note at this point that three of the four HSPs synthesised by the *Vigna* chloroplasts at the elevated temperature (85, 70 and 65 kDa) are also detectable in the chloroplast fraction isolated from the leaves subjected to gradual heat shock and not in the leaves subjected to rapid shock. From these observations it can be concluded that the heat-shock response of the chloroplast genome *in vivo* is expressed only when the leaves are subjected to gradual heat shock and not when the leaves are subjected to rapid heat shock.

The explanation for the absence of *in vivo* expression of chloroplast-coded HSPs due to rapid heat–shock treatment is not immediately discernible. However, it is worth mentioning at this point the effect of rapid heat-shock treatment on the *in vivo* synthesis of the chloroplast-coded large subunit of RuBPcase in soybean cell suspension cultures. A 40°C heat-shock treatment for 2 hr, decreases the *in vivo* synthesis of this peptide by 80° and this decrease is not a result of decrease in the level of mRNA for this polypeptide (18). As an explanation for this observation, it has been suggested that, by reducing the metabolite or energy supply to the chloroplast, or by inactivating the chloroplast translation machinery, the heat shock could depress total chloroplast protein synthesis. It is most probable that, unlike the case during rapid heat-shock, during gradual heat-shock, the metabolite or energy supply between the chloroplast and cytoplasm might get adjusted in such a way that chloroplast protein synthesis is not significantly affected.

Earlier Kloppstech *et. al.* (11), based on their experiments with pea and *Chlamydomonas*, concluded that the chloroplast genome in these two species does not have any heat-shock response with the induction of synthesis of HSPs and, therefore, it is most likely that, during the course of evolution, the corresponding prokaryotic genes for the synthesis of HSPs originally present in the progenitors of chloroplast have either been lost or transferred from the prokaryotic (organelle) genome to the nucleus. Very recently, Nieto-Sotelo and Ho (19), from the observation that mitochondria and chloroplasts isolated from maize do not synthesise HSPs, have come to the conclusion that maize organelles do not have any heat-shock response of their own. They have also suggested that more studies are needed to explain the molecular basis for the absence of HSP synthesis in

Fig. 1. Fluorographic profile of the total leaf homogenate of the control and heat shocked leaves of *vigna sinensis*. A, Control leaves, B, Leaves subjected to rapid shock. C, Leaves subjected to gradual heat shock.

Fig. 2. Fluorographic profile of the chloroplast fraction obtained from control and heat shocked leaves of *Vigna sinensis*. A, Chloroplast fraction from control leaves, B, Chloroplast fraction from leaves subjected to rapid heat shock. C, Chloroplast fraction from leaves subjected to gradual heat shock.

Fig. 3. Fluorographic profile of the polypeptides synthesised by the isolated chloroplasts of *Vigna sinensis* incubated at 25°C (A) and 40°C (B).

the chloroplasts and mitochondria. On the other hand, the experiments of Kloppstech *et al.* (16) with *Acetabularia* indicate the existence of a definite heat-shock response of the chloroplast genome in this organism. Kloppstech *et al.* (16) have suggested that *Acetabularia* represents an ancient evolutionary situation in which the genes for HSPs have not yet been translocated from the organelle to the nucleus. The results of our own experiments with *Sorghum* and *Vigna* clearly indicate that the chloroplast genome in these two species have got a definite heat-shock response and this may be a case, in addition to *Acetabularia*, where the genes for HSPs are still retained in the chloroplast genome. Similarly, the experiments of Restivo *et al.* (13) on the heat-shock response of the protoplasts isolated from *Nicotiana plumbaginifolia* indicates the existence of a heat-shock response of the chloroplast genome in this species. Thus, a comparative study of the heat-shock response of the chloroplast genome in a variety of plant species could be considered as a potential approach to looking into the evolution of the chloroplast genome.

Functional Significance of HSPs Associated with Chloroplasts

At this point, nothing is known about the functional significance of the HSPs associated with chloroplasts. But it is interesting to note that all the four HSPs synthesised by the isolated chloroplasts of *Vigna* get localised on the thylakoid membranes and not in the stroma (15). The 22 kDa HSP coded by the nuclear genome, detectable in the chloroplast fraction, is also found localised on the thylakoid membranes (11). Further, when the *in vitro* translation products of poly A^+ mRNA are incubated with the intact chloroplasts from control and heat-shocked leaves, both the chloroplast preparations take up 26 kDa HSP precursor and process it to 22 kDa mature HSP. In the case of chloroplasts from control leaves, the 22 kDa HSP is detectable only in the stroma, while in the chloroplasts from heat-shocked leaves this HSP is found localised preferentially on the thylakoids (11). This observation indicates that changes induced by the heat shock is not merely a transient, temperature-dependent alteration of the fluidity of the membrane and that the localisation of HSPs on the thylakoid membrane may have some functional significance. Further, the observation that high temperature decreases the quantum yield for CO_2 fixation (10) suggests that one of the first components of the photosynthetic apparatus to be damaged by the heat shock is the thylakoid membrane. An inhibition of enzyme reactions should not affect the quantum yield when measured under strictly light-limiting conditions. Schreiber and Berry (20) have shown that the observed inhibition of photosynthesis by heat is closely related to the perturbation of the thylakoid membranes, affecting both associated enzymes and the pigment systems. In the light of these observations, the localisation of all the four chloroplast-coded HSPs in the thylakoid membranes could be taken as an indication of the functional significance of the chloroplast-associated HSPs in offering thermal tolerance to the thylakoid-membrane-bound activities.

Acknowledgements

This work was supported by a research grant to A. Gnanam from the Department of Science and Technology, Government of India [No. 21 (7) 84-STP-II].

References

1. Barnett, T., Altschuler, M., McDaniel, C. N. and Mascarenhas, J. P. (1980) *Develop. Genet.* 1, 331–340
2. Kimpel, J. A. and Key, J. L. (1985) *Trends Biochem. Sci.* 10, 353–357
3. Valliammai, T., Gnanam, A. and Dharmalingam, K. (1987) *Plant Cell Physiol.* 28, 975–985
4. Mannan, R. M., Krishnan, M. and Gnanam, A. (1986) *Plant Cell Physiol.* 27, 377–381
5. Lin, C. Y., Roberts, J. K. and Key, J. L. (1984) *Plant Physiol.* 74, 152–160
6. Burke, J. J., Hatfield, J. L., Klein, R. R. and Mullet, J. E. (1985) *Plant Physiol.* 78, 394–398
7. Ougham, H. J. and Stoddart, J. L. (1985) *Plant Science* 44, 163–167.
8. Marimiroli, N., Restivo, F. M., Stanca, M. O., Terzi, V., Giovanelli, B., Tassi, F. and Lovenzoni, C. (1986) *Genet. Agri.* 40, 9–25.
9. Minton, K .W., Karmin, P., Habb, G. M. and Minton, A. P. (1982) *Proc. Natl Acad. Sci.* 79, 7107–7111.
10. Bjorkman, O., (1975) *Carnegie Inst. Wash. Year Book* 74, 748–751
11. Kloppstech, K., Meyer, G., Schuster, G. and Ohad, I. (1985) *EMBO J.* 4, 1901–1909
12. Vierling, E., Mishkind, M. L. Schmidt, G. W. and Key, J. L. (1986) *Proc. Natl. Acad. Sci.* 83, 361–365
13. Restivo, F. M., Tassi, F., Maestri, E., Lorenzari, C., Puglizi, P. P. and Marimiroli, N. (1986) *Curr. Genet.* 11, 145–149
14. Kirshnasamy, S., Mannan, R. M. Krishnan, M. and Gnanam, A. (1988a) *Proc. Natl. Acad. Sci. India* (in press)
15. Krishnasamy, S., Mannan, R. M., Krishnan, M. and Gnanam. A. (1988b) *J. Biol. Chem.* (in press)
16. Kloppstech, K., Ohad, I. and Schweiger, H. G. (1986) *Eur. J. Cell. Biol.* 42, 239–245
17. Krishnan, M., Krishnasamy, S., Mannan, R. M. and Gnanam, A. (1987) In: *Progress in Photosynthesis Research*, Vol. IV (Biggins, J.ed) pp. 593–597, Martinus Nijhoff Publishers, Dordrecht, The Netherlands
18. Vierling, E. and Key, J. L. (1985) *Plant Physiol.* 78, 155–162
19. Nieto-Sotelo, J. and Ho, T. H. D. (1987) *J. Biol. Chem.* 262. 12288–12292
20. Schreiber, U. and Berry, J. A. (1977) *Planta* 136, 233–238

The Role of Chloride in Oxygen Evolution

GOVINDJEE*

Department of Physiology and Biophysics
University of Illinois at Urbana-Champaign
524 Burrill Hall
407 South Goodwin Avenue
Urbana, IL 61801 USA

* Correspondence:
Department of Plant Biology
University of Illinois at Urbana-Champaign
289 Morrill Hall
505 South Goodwin Avenue
Urbana, IL 61801 USA

Summary

A brief review on the role of chloride in oxygen evolution is presented here. This paper deals with the discovery of the phenomenon, the site of action, specificity and effectiveness, possible binding sites and a possible mechanism of action. The phenomenon was discovered by Otto Warburg and has been studied extensively since the pioneering work of Sei Izawa and coworkers. The site of action is on the electron donor side of the Photosystem II (PSII) between the so-called water oxidation complex (WOC) and the electron donation sites of external donors such as NH_2OH, diphenylcarbazide and catechol. The effectiveness of the anions in supporting water oxidation, protecting O_2 evolution against mild heating and in preventing the dissociation of the extrinsic polypeptides of WOC, follows the series: $Cl^->Br^->NO_3^->I^->ClO_4^-$. Bicarbonate ($HCO_3^-$), that appears to be required on the electron acceptor side of PSII, is ineffective in replacing Cl^-. Sulphate (SO_4^{2-}) and PO_4^{3-} are totally ineffective, and F^- and OH^- might even be inhibitory. The size, ionic field and hydration energies of the anions seem to play significant roles in their effectiveness. It appears that several (4 to 40) Cl^- ions may be weakly and reversibly bound per WOC. In all likelihood, there are two binding domains: (1) intrinsic: on the PSII reaction centre proteins D_1 and D_2 (perhaps, on the lumenal histidines): and (2) extrinsic: on the extrinsic 33 kDa polypeptide (perhaps, on lysines and/or arginines). Removal of the extrinsic polypeptides seems to increase the [Cl^-] requirement for water oxidation activity. It is suggested that the intrinsic sites are the catalytic sites and the extrinsic sites are the regulatory sites. In the absence of Cl^-, the WOC is stuck in an abnormal state, and there is no O_2 evolution activity. Our hypothesis for Cl^- action is: it facilitates the abstraction of H^+s from H_2O during its oxidation to O_2 ($2H_2O = 4H^+ + e^- + O_2$). We have suggested that Cl^- binds to a positively charged amino acid (N^+) which is close to a negatively charged amino acid (B^-) in the neighbourhood of the substrate H_2O molecules. Binding to N^+ changes the pKa of B^- such that its affinity to H_2O protons increases. This facilitates H^+ abstraction from H_2O during its oxidation to O_2. In this hypothesis, release of H^+s is accompanied by the release of Cl^- ions.

I. Discovery

Forty-four years ago, Warburg and Lüttgens (1; also see 2) discovered that various anions (Cl^-, Br^-, I^- and NO_3^-) stimulate oxygen evolution during Hill reaction in water-washed broken chloroplasts. Chloride was the most effective anion, but rhodanide, SO_4^{2-} and PO_4^{3-} were ineffective in stimulating the Hill reaction. Gorham and Clendening (3) provided the first clear demonstration that Cl^- is specifically involved in O_2 evolution during the Hill reaction, and that the Cl^- effect is not just the reversal of injurious effects of exposure to light. Gorham and Clendenning further noted that (a) Cl^- shifts the pH optimum of the Hill reaction from 6.5 to 7.5; and (b) this Cl^- effect has similarity to that in dialyzed α-amylase. It is now generally accepted that Cl^- is a necessary cofactor for O_2 evolution, based mainly on the pioneering work of S. Izawa and coworkers [see e.g., Hind et al. (4); Izawa et al. (5); Kelley and Izawa (6); Muallem et al. (7) and Izawa et al (8)]. After 44 years of research, that occurred mostly in spurts, the molecular mechanism by which Cl^- activates O_2 evolution remains unknown. Readers are encouraged to consult earlier reviews by Homann et al. (9), Govindjee et al. (10, 11), Critchley (12), Homann (13), Coleman and Govindjee (14,15) and Coleman (16).

II. Site of Action

Figure 1 shows a current Z-scheme for electron flow from H_2O to $NADP^+$. This figure is a modification of that presented earlier by Govindjee and Eaton Rye (17); all the symbols are defined in the legend of the figure. The site of Cl^- action is shown here to be on the electron donor side of photosystem II (PSII) between water oxidation and the electron donation by Z of PSII (see left top of Fig 1); this site is suggested to involve the charge accumulator "M", of the water oxidation complex (WOC).

Figure 2 shows a current model of the thylakoid membrane with four major protein complexes: photosystem II, plastoquinol-plastocyanin oxidoreductase or cytochrome b_6/f complex, photosystem I, and the coupling factor (CF_1-CF_0) or ATPase; this figure is also a modification of that presented earlier (17). The site of Cl^- action is on the lumenal portion of photosystem II. The diagram shows Cl^- ions bound to the extrinsic 33 kilodalton polypeptides; Cl^- ions bound to the intrinsic polypeptides are not shown due to lack of space.

Figure 3 shows a scheme of the reaction centre of PSII, except that cytochrome b559 is not shown because its function is unknown. The reaction center of PSII was first isolated by Nanba and Satoh (25) and shown to contain two polypeptides of 32–34 kDa, lebelled as D_1 and D_2, and the two subunits of cytochrome b559 (4 and 9 kDa). This preparation contains 4 chlorophyll and 2 pheophytin molecules, but lacks plastoquinones Q_A and Q_B. The diagram shows the membrane portion of D_1 nd D_2 polypeptides arranged in the same fashion as the L and M polypeptides of bacterial reaction centres [see Michel and Deisenhofer (26)]; the lumenal portions of D_1 and D_2 as suggested by Coleman and Govindjee (15); the herbicide niche (see amino acids, as solid circles) as shown by Trebst (27); and the bicarbonate sites as suggested by Blubaugh and Govindjee (28) The *intrinsic* sites of chloride binding are suggested to be on histidines; specifically, H–337 on D_2 and H-332 and H-337 on D_1 are possible sites (15); there are no lysines on the lumenal portion of D_1 and and only one on D_2. Since 4Mn are involved in O_2 evolution, chloride binding sites may depend on where these manganese atoms are located. We expect the chloride site to be close to the Mn site. Coleman and Govindjee (15) proposed 4 possible Mn binding sites on the 4 lumenal loops; these are suggested to be bound to acidic amino acids, as shown in Fig. 3. Distances between Mn atoms larger than 3Å may be contradictory to the existing data [see e.g., reviews by Babcock (29), Brudvig (30) and Renger (31)]. If 'Z' is in fact tyrosine-160 (see Y on D_1-III in Fig. 2), it would be 'convenient' for the electron flow if the charge accumulator Mn were nearby. Thus, the possible sites of Mn could be on the lumenal portions connecting D_1-I and D_1-II (G. Brudvig, personal communication) and/or on the two loops with COOH ends (see D_1-V). Since the X-ray structure and the distances within the reaction centre of PSII are totally unknown, none of the possibilities can be ignored.

Bové *et al.* (32) showed that photophosphorylation associated with NADP-dependent non-cyclic electron flow, but not that associated with cyclic flow, requires chloride. Further localisation of the site of Cl^- action was made by chlorophyll *a* flourescence studies. Chlorophyll *a* fluorescence is an excellent monitor of PSII reactions (see reviews in 33). Upon illumination of a dark-adapted sample, fluorescence rises instantly to O (Fo) level, and then further increases to a 'P' (F maximum or Fp) level. This O→P rise is a monitor of the reduction of Q_A to Q_A^-. This fluorescence rise can be abolished either by stopping electron flow from the electron donor side of PSII, or by accelerating the electron flow out of Q_A^-, for example, by adding an efficient electron acceptor such as methylviologen. However, if the fluorescence rise can be restored by the addition of electron donors to Z or P680, such as hydroxylamine, diphenylcarbazide or catechol, then we know that the absence of a fluorescence rise was because of inhibition of electron flow from the electron donor side. Health and Hind (34, 35) and Critchley *et al.* (36 showed that chloride-deficient thylakoid membranes did not have any O→P fluorescence rise, but the addition of electron donors such as NH_2OH, catechol or diphenylcarbazide, or chloride (or bromide) restored the fluorescence rise (35,36; see Fig. 4). Thus, these data show that the chloride effect is on the electron donor side of PSII. Kelley and Izawa (6) showed restoration of electron flow in chloride-deficient thylakoid

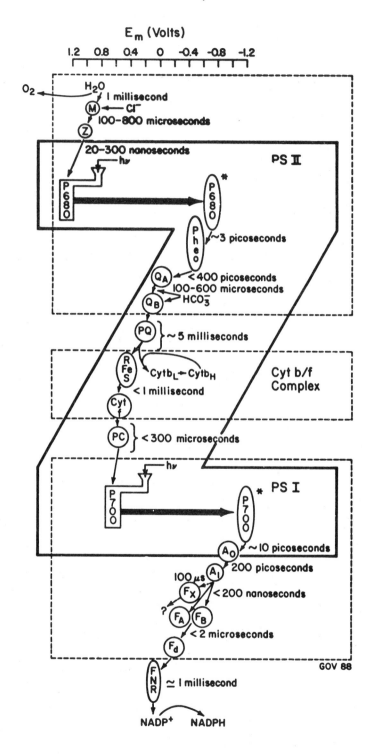

membranes by the addition of such electron donors or sufficient chloride, thus, establishing the site of chloride action.

III. Specificity and Effectiveness

The effectiveness of the anions in stimulating electron flow follows the series: $Cl^- > Br^- > NO_3^- > I^- > ClO_4^-$; F^- and OH^- are somewhat inhibitory, and SO_4^{2-} and PO_4^{3-} are ineffective. Sometimes F^- is found to be more effective than ClO_4^-, and NO_3^- is less effective than suggested

---

Fig. 1: *The 'Z' scheme of electron transport in photosynthesis*

Dashed retangles: Three major multiprotein complexes (PSII Photosystem II; Cyt b/f, cytochrome b_6/f complex; and PSI, Photosystem I), located in the thylakoid membrane, containing the photosynthetic components required for electron flow from H_2O to $NADP^+$.
Primary reactions: The electron carriers are placed horizontally according to their midpoint redox potentials (Em, 7). Electron flow is initiated when a photon or exciton reaches the reaction centre chlorophyll *a* P680 (in PSII) and P700 (in PSI) (see hν going into the funnel). P680* and P700* (see ovals) indicate the first singlet excited states of P680 and P700. The *first* reaction of P680* is the conversion of excitonic energy into chemical energy *i.e.* charge separation, with the formation of the cation $P680^+$ and the anion pheophytin$^-$ (Pheo$^-$) within $\sim$ 3ps. However, the first reaction of P700*, the charge separation into $P700^+Ao^-$, may occur in $\sim$ 10ps. Here, A_o is a special chlorophyll *a* molecule. The $P680^+$ recovers its 'lost' electron from Z, now thought to be tyrosine-160 of the D$_1$ polypeptide of PSII.
Secondary reactions of PSII. The positive charge on Z is then transferred to the charge accumulator 'M', or the water oxidizing complex (WOC). It is suggested that 'M' is nothing else but a Mn-cluster located on the lumenal portions of the D$_1$ and D$_2$ polypeptides of PSII (see Fig. 3). The *intrinsic* binding site of Cl^- lies somewhere in this region. Water oxidation seems to require another polypeptide, an extrinsic 33 kDa polypeptide, and, this is where *extrinsic* binding sites of Cl may exist. Four positive charges must accumulate before an O_2 molecule is evolved. The Pheo$^-$ delivers the extra electron to a primary (plastoquinone) electron acceptor Q_A located on the D$_2$ polypeptide of PSII; Q_A^- delivers its electron to a secondary (plastoquinone) electron acceptor Q_B, located on the D$_1$ polypeptide of PSII (see Fig. 3). Bicarbonate ions seem to be involved in the Q_A–Fe–Q_B region, where Fe is an iron atom (Fe^{2+}) between Q_A and Q_B. After reduction to plastoquinol, that is after two turnovers of the reaction centre P680, Q_B (H$_2$) exchanges with a mobile plastoquinone (PQ) molecule.
Reactions of the Cyt b/f complex. Plastoquinol (PQH_2) delivers one electron to the Rieske Fe-S centre (R-Fe-S), and the other to a cytochrome b (b_L). The electron on R-Fe-S reduces cytochrome f(Cyt f) and the one on Cytb$_L$ is transferred to Cytochrome b$_H$ (Cyt b$_H$), returning in a cyclic process (called the 'Q'-cycle).
Secondary reactions of PSI. Reduced Cyt f delivers its extra electron to a copper protein plastocyanin (PC) which delivers the electron to $P700^+$ (produced in the PSI primary reaction). On the other hand, A_o passes its electron to A_1 (perhaps, a phylloquinone). The rest of the electron carriers are: F_x (an iron-sulphur centre X), F_B (an iron-sulphur centre B), F_A (an iron-sulphur centre A), Fd (ferredoxin) and FNR (Ferredoxin-NADP$^+$ reductase).
Reaction times. The diagram shows either measured or estimated times of the various reactions in the Z-scheme, except for the production of P680* and P700* that occur on a femtosecond time scale. The bottleneck reaction is of the order of 5 ms and it involves the total time involved in the exchange of $Q_B(H_2)$ with PQ, diffusion of PQH_2 to Cyt b/f complex, and the reoxidation time of PQH_2. The electron flow from A_0 to A_1 may occur in 50 ps, rather than 200 ps. (The diagram does not show the steps involved in H$^-$ uptake and release.)

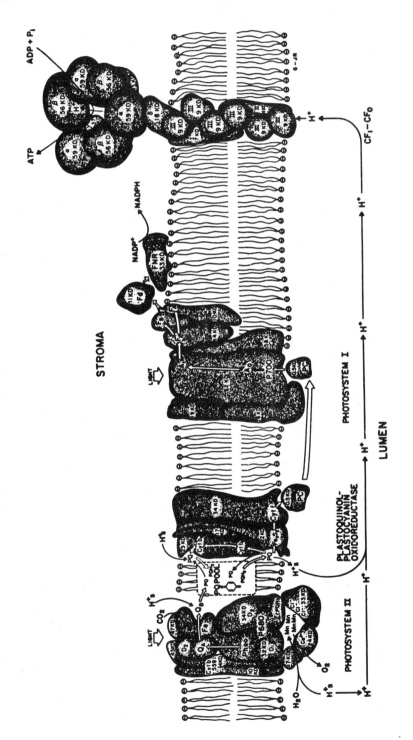

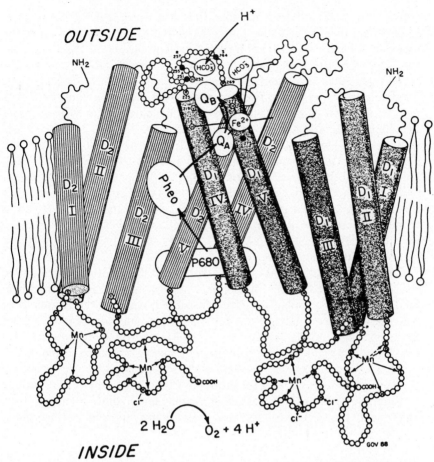

Fig. 3: *A highly-speculative working model of the reaction centre of photosystem II.*
(Contd.)

Fig. 2: *A stylised model of the thylakoid membrane with four major protein complexes.*

The light-harvesting pigment-protein complexes have, however, been omitted. The depiction of photosystem II is adapted from Mathis (18) and the organisation of the cytochrome $b_{6/f}$ complex is based on that by Mansfield and Anderson (19) and by Ort (20). The organisation of PSI is adapted from Ort (20) and Malkin (21). The organisation of the H+ -ATPase (CF_1–CF_0) is highly schematic; each of the subunits in the membrane, perhaps, span the membrane. The hydrophobic CF_0 appears to contain 4–6 copies of the DCCD (N,N' dicyclohexylcarbodiimide) binding protein or subunit III but CF_0 has not yet been purified [see McCarty and Nalin (22)]. A model for isolated hydrophilic CF_1 (see the portion sticking out of the membrane) has been proposed by Tiedge *et al.* (23). The subunit stoichiometry shown here is $3\alpha\ 3\beta:1\gamma:1\delta:1\epsilon$ (22). The symbols, used in the figure have the same meaning as in the legend of Fig. 1; Fe_2S_2 stands for Rieske iron sulphur centre. Molecular masses of the various polypeptides are denoted on the figure in kilodaltons (KD). Details of photosystem II, without the extrinsic polypeptides (17, 24 and 33 KD) and the 43 and 47 KD antenna proteins, and their associated symbols, are shown in Figure 3 and its legend. The figure is modified from that presented earlier by the author in (17); the modification is mainly in photosystem II and was first used by Eaton-Rye (24).

above. A similar, although not identical, effectiveness of Cl^- substitutes against heat damage [Coleman *et al.* (37)] and against dissociation of the extrinsic polypeptides [Homann (38)] has been observed. Hind *et al.* (4) suggested that the ionic volume may be the critical factor; that is, the smaller the anion the greater the activity. Perhaps a 'gate' or 'pocket' exists for the entry of the anion. Figure 5 shows that the Cl^- volume of 25Å is optimal for activity in spinach thylakoids. Critchley *et al.* (36) suggested that ionic volume is not the only factor because HCO_3^-, F^-, OH^-, and Acetate$^-$ that have smaller volume than Cl^+ gave low or no stimulation of electron flow (for function of HCO_3^- on electron acceptor side, see Ref. 17). A plot of the effectiveness of the anion as a function of ionic field (= $Q_A/e_s r_A^2$, where Q_A is the anion charge, r_A is the Pauling radius and e_s is the differential dielectric constant of water) showed that the effectiveness of anions decreases with decreasing ionic field. However, anions with large ionic field (HCO_3^-, SO_4^{2-}, Ac^-, PO_4^{3-}, OH^- and F^-) were also relatively ineffective in stimulating electron flow.

Coleman *et al.* (37) had suggested that the similarity of the specificity of the anions for stimulating electron flow and protection against mild heat damage is due to similar selective reversible binding of the anions to or near the O_2-evolving complex. Additionally, Homann (38; also see 39) has found that the anions also influence the binding of the 17 and 23 kDa extrinsic polypeptides to the membrane, implying a structural role of Cl^- proposed 36 years ago by Gorham and Clendenning (3). Any proposed mechanism of the Cl^- or substitute function must, therefore, include consideration of the anion volume, ionic field and hydration energy of the anion, and the ability to organise the structure (protein surface etc.)

IV. Binding Sites

Three major points are [see (15)]: (1) there are several (ranging from 4 to

Fig. 3 (*Contd.*)

D_1-I though V, equivalent to the L-subunit of the bacterial reaction centre, is the 32 kilodalton rapidly turning over herbicide-binding protein. D_2-I through V, equivalent to the M-subunit of the bacterial reaction centre, is the 34 kilodalton polypeptide. Approximate locations of the functional chlorophyll *a* P680, the functional pheophytin (Pheo), bound plastoquinone Q_A (on D_2), Fe^{2+}, and bound plastoquinone Q_B (on D_1) are shown. Q_A is *not* buried as shown, but is located there for the convenience of drawing. By analogy with photosynthetic bacteria, it is thought to be located at the same distance in the membrane as Q_B. The N-terminals of D_1 and D_2 face the outside, whereas, the C-terminals face the inside of the membrane. Suggested sites of HCO_3^- on Arginine-257, and on Fe^{2+} are also shown. Amino acids that are known to be involved in the herbicide niche are shown by solid dots; mutations at these sites are known to create herbicide resistance. Possible sites for Mn and Cl^- binding are suggested to be on the lumenal portions of the D_1 and D_2 polypeptides. In this model, H^+s are brought in via HCO_3^- from the outside and are released on the inner side when Cl^- is released. This composite model is based on concepts presented in references (15, 26, 27 and 28).

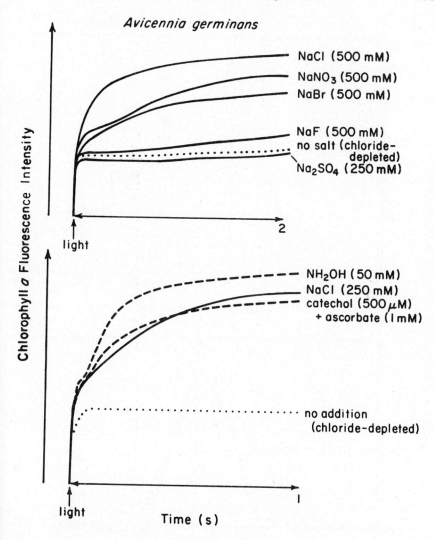

Fig. 4: *Chlorophyll a fluorescence transients in halophyte thylakoids.*

Top curves: in the absence of added salt (no salt) and in the presence of NaCl, NaNO₃, NaBr, NaF, and Na₂SO₄. *Bottom curves:* in the absence of added chloride ('no addition'), in the presence of NaCl ('NaCl') and in the presence, without added chloride, of electron donors to PSII: NH₂OH and reduced catechol. [After Critchley *et al.* (36).]

40) binding sites; (2) there are two types of binding sites: (a) catalytic sites on D_1 and D_2 polypeptides; and (b) regulatory on 33 kDa extrinsic polypeptide; and (3) removal of extrinsic polypeptides (e.g., 17 and 23 kDa) increases the $[Cl^-]$ requirement by ten times—further removal of 33 kDa increases, by another ten fold, the $[Cl^-]$ requirement.

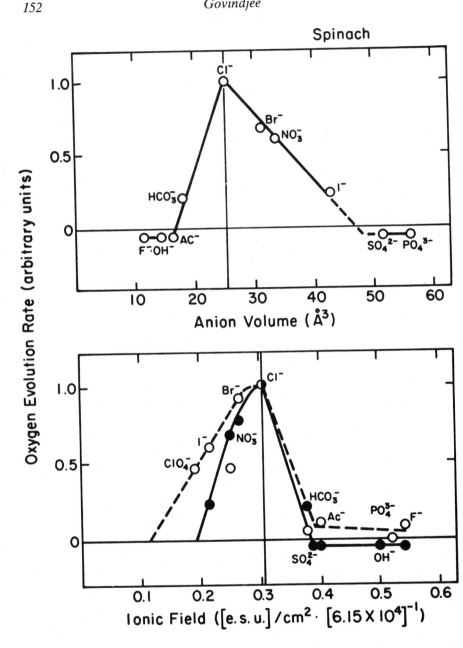

Fig. 5: *Hill reaction rates.*

Rate of O_2 evolution, with ferricyanide as electron acceptor, versus the anion volume and ionic field for spinach thylakoids. Data from Kelley and Izawa (6). [After Critchley *et al.* (36).]

Number of Binding Sites. Using radioactive ^{36}Cl, Theg and Homann (40) obtained a value of 10 Cl$^-$/600 Chl in spinach thylakoids. Izawa (41) obtained a value that ranged from 4 to 40 Cl$^-$/600 Chl. Using ^{35}Cl NMR results, Baianu *et al.* (42) and Govindjee *et al.* (10) estimated a value of 20–40 Cl$^-$/600 Chl in thylakoids from halophytes, suggesting that multiple binding sites exist for Cl$^-$.

Multiple Binding Sites. The existence of multiple sites of Cl$^-$ was most dramatically demonstrated by ^{35}Cl-NMR measurements on photosystem II and thylakoid membranes from spinach [see Coleman *et al.* (43–46); Fig. 6]. Plots of ^{35}Cl NMR linewidth, corrected for that in chloride solution, show several maxima that occur at roughly the same [Cl$^-$] as intermediary plateaus in Hill activity (at low light intensities) versus [Cl$^-$] plot, indicating a possible correlation between the two measurements (44).

These maxima were obtained from data collected by an NMR instrument at the University of Illinois at Urbana (NSF-250; 250 MHZ homebuilt NMR spectrometer; spectra obtained at 24,508 KHz using a 33μs 90° pulse and 360 ms recycle time; signals detected in quadrature with 32K data points and a spectral width of ± 25,000 Hz; signals were then block averaged (500 scans per block) and transferred to a Nicolet 1180E computer [for further details, see Ref. (44)].

No such maxima were obtained in the NMR linewidths of (a) Cl$^-$/buffer solutions containing 0.1, 1.0 and 10mM Cl$^-$; these linewidths were found to have exactly the same values (differences of Hz or less); thus, there was no apparent concentration dependence of the ^{35}Cl-NMR linewidth in the absence of proteins; and (b) highly purified bovine serum albumin (a Cl$^-$ binding protein) in the same [Cl$^-$] range as used for PSII membranes [see Ref. (44)].

Furthermore, (1) NMR linewidths for PSII at the 'maxima' depended directly on the activity of the preparations (44), suggesting that whatever structures contribute to the linewidth maxima also contribute to the overall activity; (2) each maximum appears to have a different degree of sensitivity to the inactivation of the Hill reaction (43–46), suggesting that each maximum reflects a discrete structural entity; (3) mild heating (46); Tris (to remove Mn and 18,24 and 33kDa polypeptides), salt-washing (to remove 18 and 24 kDa; and 33 kDa polypeptides) and hydroxylamine (to remove Mn) (45) treatments show different effects on the different NMR linewidth maxima; (4) the NMR linewidth maxima are altered by removing 18 and 24 kDa polypeptides by NaCl washing (as mentioned above) but 3 of the 4 maxima were restored by adding 2mM CaSO$_4$; this also demonstrates another correlation with Hill activity, which is also partially restored by adding Ca^{2+}; (5) ^{35}Cl-NMR line-bradening in the presence of PSII membranes in the [Cl$^-$] range between 0.1 mM and 0.3 mM was eliminated by a very low concentration of added Br$^-$ (0.1 mM); the observed reduction in the height of the first linewidth maximum is also consistent with the competition between the anions for a limited set of binding sites.

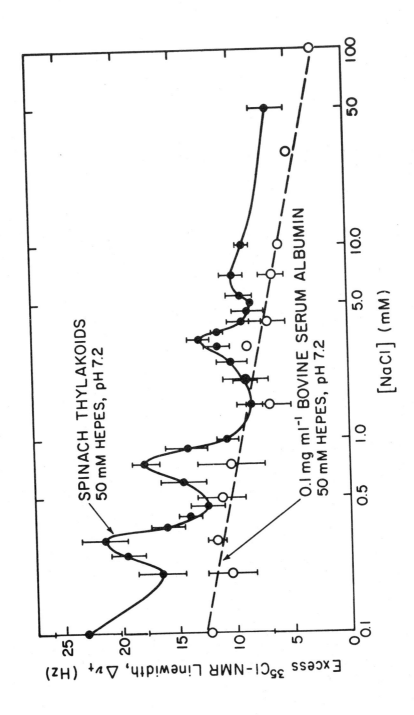

Two Major Binding Domains. Since several of the ^{35}Cl-NMR maxima, discussed above, occur at higher [Cl$^-$] than that needed for activating O$_2$ evolution, it is suggested that these may be related to the *regulatory*, rather than the catalytic, function of Cl$^-$. Mild heating (3 min. 30°C) eliminates the high [Cl$^-$] 'kinks' of maxima and minima in both the [Cl$^-$] dependence of the ^{35}Cl$^-$NMR linewidth and the Hill reaction (46). Furthermore, these are the ones that are dramatically altered by the removal of the regulatory proteins (18 and 24 kDa) and partially restored by the addition of Ca^{2+} (45).

The [Cl$^-$] dependence of excess ^{35}Cl-NMR linewidth data in halophytes [Baianu *et al.* (42)] and in spinach (44–46) show a general pattern: there is a large excess linewidth at low [Cl$^-$] that decreases with increasing [Cl$^-$]. Such a curve suggests that the tightly binding sites are filled at low [Cl$^-$]. We suggest that this tightly binding site at the lowest [Cl$^-$] is related to the catalytic function. However, Cl$^-$ affinity at this site decreases when the 33kDa extrinsic polypeptide is also removed: a peak appears in the [Cl$^-$] dependence of ^{35}Cl-NMR linewidth at ~ 0.5 mM. This is qualitatively consistent with the idea that the [Cl$^-$] requirement for the O$_2$ evolution activity of such preparations is extremely high. Further research is necessary to correlate ^{35}Cl binding with the O$_2$ evolution since binding is required for O$_2$ evolution, but binding is not a sufficient condition for O$_2$ evolution. It can be easily seen that removal of Mn by NH$_2$OH, that eliminates O$_2$ evolution, only slightly alters the [Cl$^-$] dependence of the ^{35}Cl-NMR linewidth (45).

V. Mechanism of Cl$^-$ Action

We have already mentioned (see section II above) that Cl$^-$ functions on the electron donor side of PSII between or at the charge accumulator M and 'Z' where several artificial electron donors donate electrons. Now, the question is: What is its mechanism of action? First of all 'M' exists in several redox states, labelled as S-states [see (29-31) and Wydrzynski (47)].

Fig. 6: *^{35}Cl-NMR binding curve for thylakoid membranes and for bovine serum albumin (BSA).*
The buffer was 50 mM HEPES at pH 7.2; the chlorophyll concentration wa 0.5 mg. ml^{-1}. BSA was dissolved at a concentration of 0.1 mg. ml^{-1} in 50 mM HEPES at pH 7.2. $\Delta \nu_t$ (Hz) = excess line width; it is equal to $\Delta \nu_{obs}$ (observed linewidth at half-maximum intensity) minus $\Delta \nu f$ (linewidth at half-maximum intensity for Cl$^-$ solution). $\Delta \nu_{obs} = \Delta \nu_{bound}$ f$_{bound}$ + $\Delta \nu f$ (1-f$_{bound}$), where $\Delta \nu_{bound}$ is the weighted average of the contributions from Cl$^-$ in the bound state, $\Delta \nu f$ is the same in the free state, and f$_{bound}$ is the fraction bound. Since $\Delta \nu_b$ (approximately 10kHz) is very much larger than $\Delta \nu_f$ (approximately 12–30 Hz, depending on the viscosity) and f$_{bound}$ <<1 (for a dilute protein solution), $\Delta \nu_t = \Delta \nu_{bound}$ f$_{bound}$. The details of the NMR instrument and its settings are given in the text. [After W. J. Coleman, unpublished; also see ref. (44).]

In darkness, 'M' is in state 'S_1', and the following sequence takes place in the first cycle in a dark-adapted sample:

$$S_1 \xrightarrow{h\nu_1} S_2 \xrightarrow{h\nu_2} S_3 \xrightarrow{h\nu_3} S_4 \xrightarrow[2H_2O \quad O_2]{} S_0$$

And, the following sequence occurs in the subsequent cycle(s):

$$S_0 \xrightarrow{h\nu_4} S_1 \xrightarrow{h\nu_5} S_2 \xrightarrow{h\nu_6} S_3 \xrightarrow{h\nu_7} S_4 \xrightarrow[2H_2O \quad O_2]{} S_0$$

Here, $h\nu_n$ stands for a light flash, and S_n for a redox state of 'M', the higher n's representing higher oxidation states. Thermoluminescence [see Sane and Rutherford (48)] is a powerful tool to monitor the S-states because it represents the back reaction of PSII involving the S-states and the electron acceptor side. Thermoluminescence is suggested to appers as follows:

$$S_2 \cdot Z \cdot P680 \cdot Pheo \cdot Q_A \cdot Q_B^- \text{ (created by light reaction)} \xrightarrow{\text{warm}}$$

$$S_1 \cdot Z^+ \cdot P680 \cdot Pheo \cdot Q_A^- \cdot Q_B \xrightarrow{\text{warm}} S_1 Z \cdot P680^+ \cdot Pheo^- \, Q_A \, Q_B \xrightarrow{\text{warm}}$$

$$S_1 \cdot Z \cdot P680^* \cdot Pheo \cdot Q_A \cdot Q_B \xrightarrow{\text{spontaneous}} S_1 \cdot Z \cdot P680 \cdot Pheo \cdot Q_A \cdot Q_B + light$$

This light appears as a glow peak 'B' in thermoluminescence curves and is said to originate from an S_2Q_B back reaction. On the basis of their thermoluminescence results, Homann *et al.* (49) and Vass *et al.* (50) suggested that Cl^- depletion generates an abnormal 'S_2' state. It was, however, Muallem *et al.* (7) who had first suggested that Cl^- depletion causes the formation of defective oxidant storage. Several investigators, using a variety of experimental approaches, have confirmed this conclusion [see *e.g.* (50–53)]. Flash illumination of dark-adapted Cl^- depleted preparations generates an abnormal S_2-state (labelled as 'Σ_2 state') which has an abnormally long lifetime and has a lower than normal oxidation potential [Homann *et al.* (49)]. This Σ_2 state, unlike the normal S_2 state, does not produce the normal multiline ESR signal for manganese [see Damodar *et al.* (54) and Imaoka *et al.*(55)]. However, the g = 4.1 signal of the S_2 state is still present in Cl^- depleted samples [Ono *et al.* (56)]. Addition of Cl^- restores the normal S_2 signal. ^{35}Cl-NMR linewidth broadening, suggestive of Cl-binding, appears to be dependent upon the S-state transitions. Preston and Pace (57) showed its relationship to the $S_2 + S_3$ states.

    In order to formulate a theory for the mechanism of Cl^- action, it is useful to keep in mind the following points [Homann (13); Coleman and

Govindjee (58)]: (1) Inactivation by Cl^- depletion is reversible; it appears that only electrostatic interactions are involved; (2) Cl^- depletion is accelerated by incubating the thylakoids at high pH; (3) Cl^- binding is pH dependent and reversible; (4) activation of the Hill reaction by added Cl^- shows hyperbolic kinetics, indicating saturation; (5) activation of the Hill reaction by anions is relatively, but not exclusively, specific for Cl^-, the selectivity does not resemble the lyotropic Hofmeister series; and there is a competitive interaction of anions; (6) the pH optimum of the Hill reaction is shifted to more alkaline pH by Cl^- binding; this suggests that Cl^- addition is equivalent to an acidification; Cl^- may be involved in splitting of H_2O and protonation steps; and (7) an interaction of Ca^{2+} with the Cl^- effect exists.

To this check list, we may add [also see Homann (13)] the following unique properties of Cl^- in the water oxidising steps: (1) there are two types of Cl^- binding/interactions (see Section IV); (2) Cl^- binding is controlled by a group with rather low pKa; (3) there is modulation of Cl^- binding by the extrinsic polypeptides [see Section IV; also see Homann (59)]; (4) prevention of binding of other anions and chemicals by Cl^-; (5) effect of sulphate ions on Cl^- action suggesting that its interaction involves deprotonation, and (6) the variation of Cl^- binding during S-state transitions [see (57)].

There are two major effects of Cl^-: a regulatory (a structural) role; and a catalytic role. We shall discuss here only the latter. Govindjee *et al.* (10) suggested that the function of Cl^- was to act as a counterion to positive charges on the oxygen-evolving complex. Since these charges arrive after each light flash, it was suggested that Cl^- binding occurred after each S-state transition. The release of Cl^- was associated with the release of a positive charge, i.e. a proton (H^+). Since there is no proton release during $S_1 \rightarrow S_2$ transition, the unbinding of Cl^- was suggested not to occur after this transition. Thus, the absence of Cl^- could specifically affect the S_2 state created. In this hypothesis, the function of Cl^- was to bind to the system when a positive charge arrived on the water oxidation complex, stabilising the system, and to lead the H^+s away from the system by simultaneous unbinding from the system. The reversible binding of Cl^- (high exchange rate) and the low binding energy ($\sim$ 3Kcal) [see Baianu *et al.* (42)] supported this picture.

Coleman and Govindjee (58) proposed a detailed picture as to how Cl^- activates the base catalysis, i.e. how it activates the H^+ removal from water (see Fig. 7). Here, a proton accepting group B^- functions to extract H^+s from H_2O, whereas Mn functions to extract an electron from H_2O, and Cl^- functions to bind transiently (and thus, reversibly) to a positively charged group N^+ that raises the pKa of B^-. The N^+ may represent an amino acid histidine (see Fig. 3), and B^- may represent a proton accepting group (Fig. 3). Thus, as Cl^- binds to N^+, the pKa of the nearby B^- changes, and this increases it affinity to catalytic H_2O protons at the

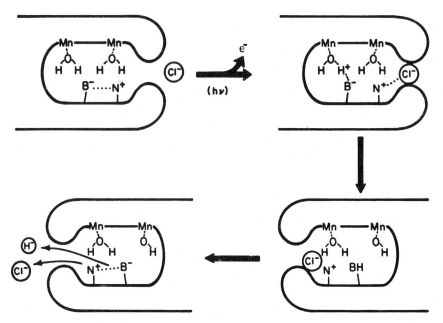

Fig. 7: *A working model for Cl⁻ activation of O₂ evolution.*
Here binding of the anion to a basic amino acid (N⁺) shifts the pKa of a reactive group (B⁻; an acidic amino acid) on the lumenal side of D_1 and D_2 polypeptides (also see Fig. 2). This makes B⁻ more reactive in 'pulling' water protons from water molecules bound at the active site. As soon as BH is made, H⁺ (bound at B⁻) and Cl⁻ (on N⁺) leave the binding sites restoring the system to its original state. In this hypothesis, Cl⁻ is reversibly bound and unbound. [Modified after Coleman and Govindjee (58).]

Mn-site making it easier to 'pull' H⁺s out of water, as 'Mn' pulls electrons from H_2O. Both the H⁺s and Cl⁻s leave the catalytic site at the same time. H⁺s may then be conducted through the 33 kDa Cl⁻ -binding extrinsic polypeptide to the outside, i.e., into the lumen (for details, see [15]). For yet another, but somewhat similar picture, see Homann (13).

The pH profile of the anion requirement, and the correlation of the anion volume and the hydration energies of the activating anions with their effectiveness, suggest [see Govindjee and Homann (60)] that protonation of unique functional groups are critical for the mechanism of water oxidation. This supports the belief of Homann (13) [also see Coleman and Govindjee (15, 58)] that one function of the anions is to organise the protein surfaces and suitable H⁺ acceptor/donor groups at the active site of the water oxidase, thereby facilating the proton abstraction from substrate water during its oxidation to molecular O_2. Further research is needed before we can understand the molecular mechanism of Cl⁻ action; only a modest beginning has been made thus far.

Acknowledgements

I am grateful to the McKnight Foundation for an interdisciplinary research grant to the University of Illinois. I thank Dr. William Coleman for providing the data for Figure 6, and Dr. Julian Eaton-Rye for editing the text.

References

1. Warburg, O. and Lüttgens, W. (1944) *Naturwissenschaften*, 32, 301
2. Warburg, O. and Lüttgens, W. (1946) *Biofyzika* 11, 303–321
3. Gorham, P. R. and Clendenning, K. A. (1952) *Arch. Biochem. Biophys.* 37, 199–223
4. Hind, G., Nakatani, H. Y. and Izawa, S. (1969) *Biochim. Biophys.* Acta 172, 277–289
5. Izawa, S., Heath, R. L. and Hind, G. (1969) *Biochim. Biophys.* Acta 180, 388–f398
6. Kelley, P. M. and Izawa, S. (1978) *Biochim. Biophys.* Acta 502, 198–210
7. Muallem, A., Farineau, J., Laine-Boszormenyi, M. and Izawa, S. (1981) In: *Photosynthesis II. Electron Transport and Photophosphorylation* (Akoyunoglou, G., ed.), pp. 435–443, Balaban International Science Services, Philadelphia, PA
8. Izawa, S., Muallem, A. and Ramaswamy, N. K. (1983) In: *The Oxygen Evolving System of Photosynthesis* (Inoue, Y., Crofts, A. R., Govindjee, Murata, N., Renger, G. and Satoh, K., eds.), pp. 293–302, Academic Press Japan, Tokyo
9. Homann, P. H., Johnson, J. D. and Pfister, V. R. (1983) In: *The Oxygen Evolving System of Photosynthesis* (Inoue, Y., Crofts, A. R., Govindjee, Murata, N., Renger, G., and Satoh, K., eds.), pp. 283–292, Academic Press Japan, Tokyo
10. Govindjee, Baianu, I. C., Critchley, C., and Gutowsky, H. S. (1983) In: *The Oxygen Evolving System of Photosynthesis* (Inoue, Y., Crofts, A. R., Govindjee, Murata, N., Renger, G., and Satoh, K. eds.), pp. 303–315, Academic Press Japan, Tokyo
11. Govindjee, Kambara, T. and Coleman, W. (1985) *Photochem. Photobiol* 42, 187–210
12. Critchley, C. (1985) *Biochim. Biophys.* Acta 811, 33–123
13. Homann, P. H. (1987) *J. Bioenerg. Biomemb.* 19, 105–123
14. Coleman, W. J. and Govindjee (1987) In: *Current Trends in Life Sciences, XIII Biomembranes: Structure, Biogenesis, and Transport* (Rajamanickam, C., ed.), pp. 215–220, Today and Tomorrow's Printers and Publishers India, New Delhi
15. Coleman, W. J. and Govindjee (1987) *Photosynthesis Research 13*, 199–223
16. Coleman, W. J. (1988) *Photosynthesis Research*, in press
17. Govindjee and Eaton-Rye, J. J. (1986) *Photosynthesis Research* 10, 365–379
18. Mathis, P. (187) In: *Porgress in Photosynthesis Research* (Biggins, J., ed.), Vol. I, pp. 151–160, Martinus Nijhoff, Dordrecht
19. Mansfield, R. W. and Anderson, J. M. (1985) *Biochim. Biophys.* Acta 809, 435–444
20. Ort, D. (1986) In: *Encyclopedia of Plant Physiology: Photosynthetic Membranes* (Staehlin, A. and Arntzen, C. J., eds.), New Series, Vol. 19, pp. 143–196, Springer-Verlag, Berlin
21. Malkin, R. (1986) *Photosynthesis Research* 10, 197–200
22. McCarty, R. E. and Nalin, C. M. (1986) In: *Encyclopedia of Plant Physiology: Photosynthetic Membranes* (Staehlin, A. and Arntzen, C. J., eds.), New Series, Vol. 19, pp. 576–583, Springer-Verlag, Berlin
23. Tiedge, H., Schafer, G. and Mayer, F. (1983) *Eur. J. Biochem.* 132, 37–45
24. Eaton-Rye, J. J. (1987) Ph.D. Thesis in Biology at the University of Illinois at Urbana-Champaign, see pp. 4 and 5
25. Nanba, O. and Satoh, K. (1987) *Proc. Natl Acad. Sci. USA.* 84, 109–112

26. Michel, J. and Deisenhofer, J. (1988) *Biochemistry* 27, 1–7
27. Trebst, A. (1987) Z. *Naturforsch*. 42c, 742–750
28. Blubaugh, D. and Govindjee (1988) *Photosynth. Res.*, in press
29. Babcock, G. T. (1987) In: *Photosynthesis* (J. Amesz, ed.), Chap. 6, Elsevier Science Publishers B. V. (Biomed. Div.), pp. 125–158
30. Brudvig, G. (1987) *J. Bioenerg. and Biomemb.* 19, 91–104
31. Renger, G. (1987) *Photosynthetica* 21, 203–224
32. Bové, J. M., Bové, C., Whatley, F. R. and Arnon, D. I. (1963) Z. *Naturforschg*. 18b, 683–688
33. Govindjee, Amesz, J. and Fork, D. C. (editors) (1986) *Light Emission by Plants and Bacteria*. Academic Press, Orlando, FL
34. Heath, R. L. and Hind, G. (1969) *Biochim. Biophys.* Acta 172, 290–299
35. Heath, R. L. and Hind, G. (1969) *Biochim. Biophys.* Acta 180, 414–416
36. Critchley, C., Baianu, I. C., Govindjee and Gutowsky, H. S. (1982) *Biochim. Biophys.* Acta 682, 436–445
37. Coleman, W. J., Baianu, I. C., Gutowsky, H. S. and Govindjee (1984) In: *Advances in Photosynthesis Research* (C. Sybesma, ed.), Vol. I, pp. 283–286. Martinus Nijhoff: Dordrecht, the Netherlands
38. Homann, P. H. (1988) *Plant Physiology*, in press
39. Homann, P. H. (1986) *Photosynth. Res.* 10, 497–503
40. Theg, S. M. and Homann, P. H. (1982) *Biochim. Biophys.* Acta 679, 221–224
41. Izawa, S. (1984) Reported by Homann, P. H. (1985) In: *Photobiology 1984* (Longworth, J. W., Jagger, J. and Shropshire, W. Jr., eds.), Praeger Scient, New York, pp. 249–253
42. Baianu, I. C., Critchley, C., Govindjee and Gutowsky, H. S. (1984) *Proc. Natl Acad. Sci. USA* 81, 3713–3717
43. Coleman, W. J., Govindjee and Gutowsky, H. S. (1987) In: *Progress in Photosynthesis Research* (Biggins, J., ed.), Vol. I, pp. 629–632. Dordrecht: Martinnus Nijhoff
44. Coleman, W. J., Govindjee and Gutowsky, H. S. (1987) *Biochim. Biophys.* Acta 894, 443–452
45. Coleman, W. J., Govindjee and Gutowsky, H. S. (1987) *Biochim. Biophys.* Acta 894, 453–459
46. Coleman, W. J., Govindjee and Gutowsky, H. S. (1988) *Photosynth. Res.*, 16, 261–276 16 pages, in press
47. Wydrzynski, T. (1982) In: *Photosynthesis. I. Energy Conversion by Plants and Bacteria* (Govindjee, ed.). Academic Press, N. Y., pp. 469–506
48. Sane, P. V. and Rutherford, A. W. (1986) In: *Light Emission by Plants and Bacteria* (Govindjee, J. Amesz and D. C. Fork, eds.). Academic Press. Orlando. pp. 329–360
49. Homann, P. H., Gleiter, H., Ono, T. -A. and Inoue, Y. (1986) *Biochim. Biophys* Acta 850, 10–20
50. Vass, I., Ono, T.-A. Homann, P. Gleiter, H. and Inoue, Y. (1987) In: *Progress in Photosynthesis Research* (J. Biggins, ed.) pp. 649–652. Martinus Nijhoff, Dordrecht
51. Itoh, S., Yerkes, C. T., Koike, H., Robinson, H. H. and Crofts, A. R. (1984) *Biochim. Biophys.* Acta 766, 612–622
52. Theg, S., Jursinic, P. A., and Homann, P. H. (1984) *Biochim. Biophys.* Acta 766, 636–646
53. Ono, T.-A., Conjeaud, H., Gleiter, H., Inoue, Y. and Mathis, P. (1986) *FEBS Lett.* 230, 215–219
54. Damoder, R., Klimov, V. V. and Dismukes, G. C. (1986) *Biochim. Biophys.* Acta 848, 378–391
55. Imaoka, A., Akabori, K., Yanagi, M., Izumi, K., Toyoshima, Y., Kawamori, A., Nakayami, H. and Sato, J. (1986) *Biochim. Biophys.* Acta 848, 201–211
56. Ono, T.-A., Zimmermann, J. L., Inoue, Y. and Rutherford, A. W. (1986) *Biochim. Biophys.* Acta 851, 193–201
57. Preston C. and Pace, R. J. (1985) *Biochim. Biophys.* Acta 810, 388–391

58. Coleman, W. and Govindjee (1985) *Proceedings of the 16th FEBS Congress*, edited by S. Ovchinikov, Part B, pp. 21–28, VNU Science Press, Utrecht
59. Homann, P. H. (1988) *Photosynth. Res.* 15, 205–220
60. Govindjee and Homann, P. H. (1988) Abstract, *14th International Congress of Biochemistry*, July 10–15, 1988, Prague, Czechoslovakia; full paper, in press

Isolation of a Highly Active PS II Preparation from Wheat *Triticum Aestivum* and the Chemical Modification of Tyrosine Residues with 4-chloro-7-nitrobenzofurazan (NBD-C1)

RANJANA PALIWAL, JAI PARKASH AND G. S. SINGHAL

School of Life Sciences
Jawaharlal Nehru University
New Delhi 110 067, INDIA

Summary

A highly active photosystem II (PS II) preparation was obtained from 8-day old laboratory grown wheat seedlings. Oxygen evolving activity of the PS II preparations was 954 μmoles O_2/mg Chl/hr. The SDS-PAGE showed the presence of 34, 32, 29, 18, 14, 10, 9 and 6 kD proteins. The LHC having M_r = 29 kD was depleted from PS II during isolation as shown by gel analysis. The fluorescence emission spectrum of the LHC obtained after β-octylglucoside treatment had peaks at 683, 693 and 735 nm at 77 K, however, the PS II preparation showed only one emission peak at 685 nm at 77 K. PS II particles from spinach showed slight differences in oxygen evolving activity and relative intensities of bands of different polypeptides in the gel as compared to PS II from wheat.

The 4-chloro-7-nitrobenzofurazan (NBD-Cl) modified PS II preparations showed 45 to 55 per cent inhibition in oxygen evolving activity as compared to control. NBD-Cl modified tyrosyl groups of PS II particles also caused significant decrease in F_m and F_v and in light-induced absorbance changes at 680 nm. F_o level was also decreased by this modifier. Dithiothreitol could reverse the inhibition of fluorescence kinetics in NBD-Cl treated PS II particles. The decrease in oxygen evolving activity, light-induced absorbance change at 680 nm and F_v would suggest that treatment of PS II particles with NBD-Cl resulted in the modification of the tyrosine residues of polypeptide(s) involved in electron transport from water to P 680.

Introduction

Recently various procedures have been reported for the isolation of the reaction centre core complex of photosystem II (PS II) from higher plants (1–4). The PS II reaction centre consists of D-1, D-2 polypeptides and the apoprotein of cyt b-559, and shows functional similarities to the photochemical reaction centre of purple photosynthetic bacteria. The crystal structures of the reaction centre complexes from *Rhodopseudomonas viridis* (5) and *Rhodobacter sphaeroides* (6) have been determined at high resolution by X-ray crystallography. The L and M subunits from the bacterial reaction centre have closely related aminoacid sequences (7) which share a high percentage of sequence homologies with the two thylakoid membrane proteins D-1 and D-2.

The water oxidising side of PS II has two redox components (Z and D) which exhibit identical EPR spectra. The component Z which is the immediate electron donor to P 680 gives rise to EPR signal II_{vf} and II_f (8). The other component D does not seem to participate in the steady state electron transport from H_2O to P 680, but has been shown to be involved in redox reactions with S states. It is responsible for the EPR signal II_s, which is stable in the dark (9–11). Earlier it had been speculated that Z and D could be semiquinones (12) but increasing evidence favours the possibility that they are tyrosine (Tyr) residues. Iodide labelling experiments had suggested that Tyr residues in D-1 and D-2 are involved in electron donation to the PS II reaction centre (13, 14). Iodide specifically labelled the D-1 protein (apparent molecular mass of 30 kD) upon illumination. The D-2 protein (apparent molecular mass of 32 kD) in dark was exclusively labelled by iodide (15). The inhibition of signal II_s and the amount of $[^{125}I^-]$-iodide incorporated into the D-2 protein are proportional to each other (16). In the Rps. *viridis* reaction centre (P 960) a tyr residue is located between P 960 and the cytochrome complex indicating a possible role in electron transport (17). Strong evidence also comes from site-directed mutagenesis experiments in which the Tyr-160 of D-2 polypetide was changed by mutation to phenylalanine. This mutation led to a 3 to 4 times reduction in photosynthetic growth and loss of EPR signal II_s (18).

In this work we present another approach to investigate the involvement of Tyr residues in electron transport. Highly active oxygen evolving photosystem II particles were isolated from wheat (and spinach) and the Tyr residues were specifically modified by using the peptide modifing reagent 4-chloro-7-nitrobenzofurazan (NBD–C1). This reagent at pH lower than 8 (in our case at pH 6), specifically reacts with the phenolic OH^- group of residues of a protein. The results of light induced absorbance change (at 680 nm), oxygen evolution and fluorescence induction would suggest that Tyr groups are involved in electron transport from H_2O to P 680.

Materials and Methods

Thylakoid membranes were isolated from 8-day old laboratory grown seedlings of wheat (*Triticum Aestivum*, var. HD 2329 obtained from Indian Agricultural Research Institute, New Delhi). PS II particles were prepared from the thylakoid membranes by following the procedure of Ghanotakis et al. (19). PS II particles from spinach (obtained from local market) were also prepared by the same procedure.

The purity of the isolated PS II preparations was analysed by using sodium dodecyl sulfate-polyacrylamide gel electrophoresis (SDS-PAGE) and 77 K fluorescence emission spectra. The SDS-PAGE of the PS II preparations was carried out by following the buffer system of Laemmli (20) with an acrylamide gradient of 10 to 20 per cent. The gels were stained with Coomassie brilliant blue G-250 (Sigma Chemical Co., USA) and photographed on fine granule NP 22 Panchromatic film (ORWO, Panchromatic). The standard molecular weight polypeptides obtained from Pharmacia Fine Chemicals (Uppsala, Sweden) were used for comparing the relative molecular mass (M_r) of PS II polypeptides.

The 77 K fluorescence emission spectra of the PS II preparations were determined with the help of our laboratory fabricated fluorometer. The excitation wavelength was 440 nm and the 90° fluorescence emission was scanned from 650 to 750 nm. The photomultiplier tube was protected from the actinic light with the help of a Corning 2–73 filter. To protect the PS II membrane against chilling induced injury at 77 K, glycerol (final concentration 50 per cent) was added to the PS II preparation. The fluorescence induction curve of the PS II preparation at room temperature was determined on the same instrument except that a very fast strip chart recorder (Riken Denshi, Japan) was used in this case. The excitation and emission wavelengths were fixed at 440 and 685 nm respectively.

The chemical modification with NBD-Cl, of the PS II particles both from wheat and spinach was carried out by following the procedure of Deters *et al.* (21). PS II particles were incubated with varying concentrations of NBD-Cl for 60 min in dark at room temperature. The reaction mixture consisted of 0.4 M sucrose, 10 mM $CaCl_2$, 15 mM NaCl and 50 mM MES-NaOH, pH 6.0 in a final volume of 3 ml and PS II particles containing 25 μg Chl/ml. After the incubation the preparations were centrifuged at 40,000 × g for 30 min and the pellet was suspended in the same medium.

Fresh stock solution of NBD-Cl was made in methanol and the final concentration of methanol in the reaction mixture for chemical modification did not exceed 1 per cent. Wherever required, 4 mM DTT was added immediately after the dark incubation period.

The oxygen evolving activity of the PS II preparations was determined polarographically with a Clark type electrode (Model YS 1, Yellow Spring, Ohio, USA). The 3 ml of the assay mixture contained 0.4 M sucrose, 10 mM $CaCl_2$, 15 mM NaCl, 2 mM $K_3Fe(CN)_6$, 500 μM para-phenyl benzoquinone (PPBQ), 50 mM MES-NaOH pH 6.00 and PS II particles equivalent to 7 μg Chl/ml. The samples were illuminated with white saturating light (light intensity $> 2 \times 10^6$ lux) during the assay. The temperature of the suspension was maintained at 25°C with the help of a constant temperature water bath circulator.

The light induced absorbance changes of PS II preparations at 680 nm were determined on a Double beam/Dual wavelength spectrophotometer provided with a cross illumination set up at 90° to the measuring beam (UV-3000, Shimadzu, Japan). The actinic light provided by a 50 W tungsten-halogen lamp (intensity $\approx 3 \times 10^4$ lux at the position of reaction vessel) was passed through a Corning 4–96 filter before being focused on to the reaction mixture. The assay mixture consisted of 50 mM MES-NaOH pH 6.00, 0.4 M sucrose, 10 mM $CaCl_2$, 50 mM NaCl, 5 mM $K_3Fe(CN)_6$ in a final volume of 0.2 ml and PS II particles equivalent to 90 μg Chl. The optical path length was 1 mm. The photomultiplier tube was protected from blue actinic light by placing a Corning 2–62 filter.

Results and Discussion

The oxygen evolving activity of the PS II preparation from wheat was 954 μmoles O_2/mg Chl/hr. The rattio of chlorophyll *a* to chlorophyll *b* was 2.0 for the PS II preparation from wheat as compared to 2.7 in wheat thylakoid. However, the same isolation procedure when used for spinach yielded a PS II preparation with oxygen evolving activity of 503 μmoles O_2/mg Chl/hr. The difference in the oxygen evolving activity may be attributed to relatively more susceptibility of wheat thylakoids to the detergents (Triton X-100 and β-octyl glucoside) as compared to that of spinach thylakoid membranes.

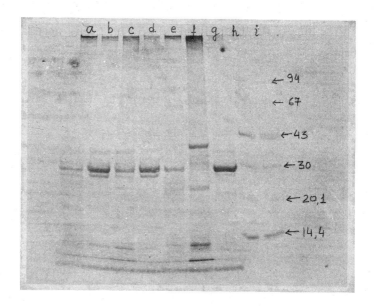

Figure 1

Figure 1 shows the polypeptide pattern on the SDS-PAGE of different fractions obtained during isolation of PS II particles from wheat and spinach chloroplasts. The PS II enriched fraction from spinach obtained after treatment with Triton X-100 and β-octylglucoside had main polypeptides with M_r 34–32, 29, 27, 24–23, 18 and 10 kD (Fig. 1, lane c). The corresponding PS II enriched fraction from wheat had main polypeptides having M_r 34–32, 29, 24–23, 18, 14, 10, 9 and 6 kD (Fig. 1, lane f). A comparative analysis of the intensities of different bands in the SDS-PAGE shows an insignificant quantity of 29 kD light-harvesting pigment-protein complex in wheat PS II (Fig. 1 lane f) whereas the same polypeptide was significantly present in spinach PS II (Fig. 1, lane c). The intensities of the bands at 34–32, 24–23 and 10 kD were more dense in wheat PS II than in PS II from spinach. In addition to this, polypeptides with M_r 14, 9 and 6 kD were present in wheat PS II (lane f) while in spinach PS II particles 14 and 6 kD were absent. These differences in the polypeptide pattern on the SDS-PAGE suggest that the isolated PS II particles from wheat seems to be different in composition and structure from the PS II particles of spinach which may arise as a result of greater susceptibility of wheat to detergent action as compared to spinach. These structural differences in PS II particles in turn may lead to differences in the exygen evolving activity of the PS II preparations (see above).

The LHC pellets obtained after β-octylglucoside treatment of spinach and wheat Triton X-100 particles i.e. BBY particles, showed a major band with M_r of 29 kD on the SDS-PAGE (Fig. 1, land d for spinach, and lane g

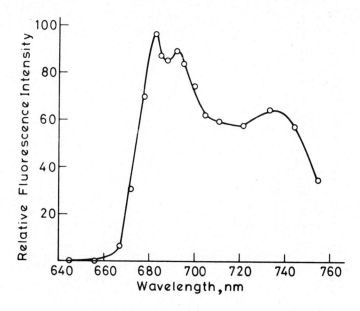

Figure 2

for wheat). However, minor amounts of other polypeptides were also present in the gel. Our result on the SDS-PAGE of this LHC fraction is in agreement to the earlier reports (22) though these workers have used only spinach as the starting material.

The low temperature (77 K) fluorescence emission spectrum of 29 kD wheat LHC fraction (lane g in Fig. 1) is shown in Figure 2. It shows three peaks at 683, 693 and 735 nm. The ratio of F 693 nm to F 683 nm was 0.81 and the ratio of F 735 nm to F 683 nm was 0.67. However, the corresponding fraction from spinach had a peak at 685 nm and a small shoulder at 693 nm in the low temperature fluoresence emission spectrum (Fig. 3). The ratio of F 693 nm to F 685 nm was 0.96. The chlorophyll concentration used in our low temperature fluorescence emission experiments was 8 μg Chl/8 ml. Although the 77 K emission spectrum of spinach LHC fraction is in agreement to the earlier reported values, our results on the corresponding fraction from wheat having an emission band at 693 nm do not coincide with the hypothesis that 693 nm peak originates from the reaction centre of PS II. As a matter of fact, our data on the low temperature emission spectra of oxygen evolving PS II core complex enriched fraction both from wheat and from spinach showed only one emission peak at 685 nm and no peak at 695 nm (see Figs. 4 and 5). This is in good agreement to the values reported by Barber *et al.* (see Fig. 3 of Ref. 4) and contradicts a previous assumption by Klimov and co-workers

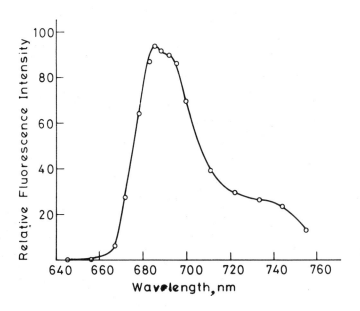

Figure 3

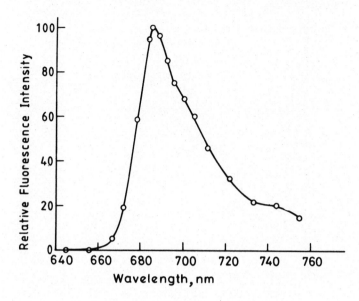

Figure 4

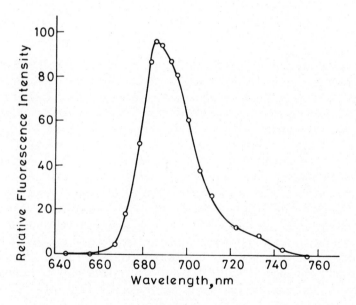

Figure 5

(23) that low temperature fluorescence at 695 nm arises from pheophytin within the PS II reaction centre.

The chemical modification of Tyr residues of PS II membranes from spinach and wheat with NBD-Cl at pH 6.0 resulted in inhibition of the

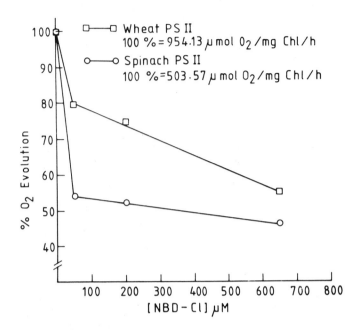

Figure 6

oxygen evolving activity (see Fig. 6). The spinach PS II particles seemed to be more prone to inhibition caused by NBD-Cl treatment than the wheat PS II particles. The inhibition of oxygen evolving activity in case of spinach PS II was around 46 per cent even at the lowest concentration of NBD-Cl used (50 μM) whereas wheat PS II membranes required 800 μM NBD-Cl to give a 45 per cent inhibition of the oxygen evolving activity. Our data suggest (i) the involvement of Tyr residue(s) of polypeptides in the electron transport process in the PS II particles and (ii) that the structural variations in PS II particles probably leads to differential susceptibility of these particles to NBD-Cl treatment. As pointed out above, the polypeptide pattern on SDS-PAGE gel and the oxygen evolving activity of spinach and wheat PS II were different, thus indicating that they are probably structurally different. Our suggestion (i) is in agreement with (a) the results obtained by Vermaas and co-workers (18) regarding site-directed mutagenesis of Tyr 160 of D2 protein to Phe, (b) [^{125}I$^-$]-iodide labelling studies of Takahashi *et al.* (15,16) wherein iodination of the Tyr residues of D2 protein led to inhinition of EPR signal II$_s$ and (c) results obtained by Michel *et al.* (17) on the reaction centre of Rps viridis.

Our results as shown in Table I indicate that chemical modification of Tyr residue(s) of PS II particles with NBD-Cl caused 88 and 83 per cent inhibition of light minus dark induced absorbance change at 680 nm (i.e. $\triangle A_{680}$) in wheat and spinach PS II respectively. The inhibition of $\triangle A_{680}$ by

TABLE 1

Light-minus-dark $\triangle A_{680}$ with NBD-Cl modification.

	$\triangle A.10^2$	$\triangle A$ (%)
Wheat		
Control	3.00	100
NBD-Cl treated	0.36	12
Spinach		
Control	1.80	100
NBD-Cl treated	0.32	17

NBD-Cl suggests that the rate of electron transport from H_2O to P 680, the reaction centre of PS II, is significantly inhibited when Tyr residue(s) are chemically modified. Thus supports our data on inhibition ofoxygen evolving activity by NBD-Cl (Fig. 6).

The treatment of PS II particles with NBD-Cl also affected the fluorescence induction parameters F_o, F_m and F_v (Figs. 7 and 8). A continuous decrease in F_m and F_v was seen with increasing concentrations of NBD-Cl both in spinach and in wheat PS II preparation. The effect of

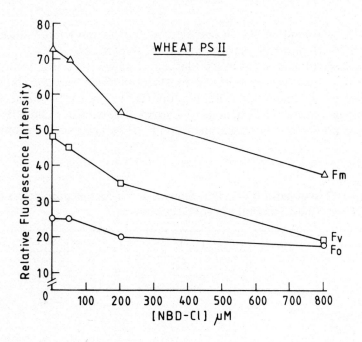

Figure 7

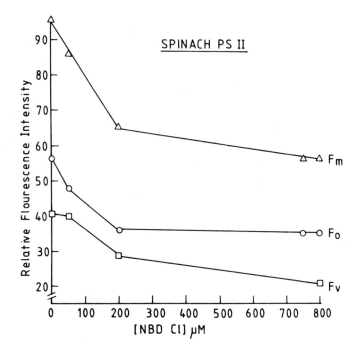

Figure 8

NBD-Cl treatment of PS II particles on F_o was more pronounced in spinach PS II particles than in wheat PS II particles.

Khananshvili and co-workers (24) have shown that inactivation of CF_1-ATPase activity by NBD-Cl in *Rhodospirilium rubrum* could be reversed on treatment with dithiothreitol (DTT) since DTT could break the linkage between the phenolic-OH group of Tyr residue and NBD-Cl. This was observed in our experiments on PS II particles also. On addition

TABLE 2

Reversal effect of 4mM DTT on the flourescence induction parameters of control and NBD-Cl treated wheat PS II particles

Sample	−DTT			+DTT		
	F_o	F_m	F_v	F_o	F_m	F_v
Control PS II particles	22	46	24	24	44	20
800 μM NBD-Cl treated PS II particles	11	18	7	24	44	20

of 4 mM DTT to the PS II preparation treated with 800 μM NBD-Cl, a reversal of inhibition of F_o, F_m and F_v due to NBD-Cl was seen (Table 2).

Our observations on the decrease in F_v with increase in the concentration of NBD-Cl in the PS II preparation indicate that the Tyr moiety which gets modified by this chemical modifier is the one which is involved in the electron transport from H_2O to P 680 on the donor side of PS II. Similar results with another chemical modifier, tetranitromethane (TNM) have been reported previously (25). However, these authors (25) have obtained multiple sites of TNM modification and moreover TNM can modify cysteine as well as Tyr. The participation of Tyr moiety on the donor side of PS II has not been excluded by these experiments.

Our observations on the involvement of Tyr residues in the electron transport activity on the donor side of PS II are further substantiated by the fact that tetrapheylboron (TPB), an exogenous electron donor to P 680 (26) when added to either control or NBD-Cl treated PS II particles caused almost same amount of decrease in F_m (Table 3). Similarly, $t_{1/2}$ (half-time for the rise of variable part of the fluorescence) decreased on addition of 10 μM TPB to both control and NBD-Cl treated PS II particles (Tables 3).

TABLE 3

Effect of TPB on E_m and $t_{1/2}$

Sample	Per cent Inhibition of F_m	$t_{1/2}$ for rise time of F_v m sec
Control PS II	–	75.0
Control PS II + TPB	21.3	65.6
NBD-Cl treated PS II	–	70.3
NBD-Cl treated PS II + TPB	29.1	65.6

The final concentration of TPB was 10 μM.

In conclusion we would like to suggest that the decrease in the oxygen evolving activity, light induced absorbance change at 680 nm and F_v on treatment of PS II particles with NBD-Cl indicates that Tyr residue(s) of polypeptides on the donor side of PS II are involved in electron transport from H_2O to P 680. Further experiments particularly with [^{14}C]-labelled NBD-Cl and ESR analysis are required to delineate this Tyr moiety.

Acknowledgements

The authors would like to acknowledge the financial support provided by USDA grant No. FG-IN-678, project No. IN-ARS-401.

References

1. Nanba, O. and Satoh, K. (1987) Proc. Natl. Acad. Sci. USA 84, pp. 109–112
2. Ghanotakis, D. F., Demetriou, D. M. and Yocum, C. F. (1987) *Biochim. Biophys. Acta.* 891, 15–21
3. Yamada, Y., Tang X-S., Itoh, S. and Satoh, K. (1987) *Biochim. Biophys.* Acta 891, 129–137
4. Barber, J., Chapman, D. J. and Telfer, A. (1987) *FEBS Lett.* 220, 67–73
5. Diesenhofer, J., Epp, O., Miki, K., Huber R. and Michel, H. (1985) *Nature* 318, 618–624
6. Allen, J. P., Feher, G., Yeates, T., Komiya, H. and Rees, D. C. (1987) *Proc. Natl. Acad. Sci.* USA 84, 6162–6166
7. Youvan, D. C., Bylina, E. J., Alberti, M., Bergusch, H. and Hearst, J. E. (1984) *Cell*, 37, 949–957
8. Blankenship, E. E., Babcock, G. T., Warden, J. T. and Sauer, K. (1975) *FEBS Lett.* 51, 287–293
9. Babcock, G. T. and Sauer, K. (1973) *Biochim. Biophys.* Acta 325, 483–503
10. Velthuys, B. K. and Visser, J. W. M. (1975) *FEBS Lett.* 55, 109–112
11. Styring, S. and Rutherford, A. W. (1987) *Biochemistry* 26, 2401–2405
12. Ghanotakis, D. F., O'Malley, P. J., Babcock, G. T. and Yocum, C. F. (1983) In: *Oxygen Evolving System of Photosynthesis* (Inoue *et al.*, eds) pp. 87–97. Academic Press, Tokyo.
13. Takahashi, Y., Takahashi, M. and Satoh, K. (1986) *FEBS Lett.* 208, 347–351
14. Ikeuchi, M. and Inoue, Y. (1987) *FEBS Lett.* 210, 71–76
15. Takahashi, M. and Asada, K. (1985) *Plant Cell Physiol.* 26, 1093–1106
16. Takahashi, Y. and Styring, S (1987) *FEBS Lett.* 223, 371–375
17. Michel, H., Epp, O. and Diesenhofer, J. (1986) *EMBO J.* 5, 2445–2451
18. Vermaas, W. (1988) These proceedings
19. Ghanotakis, D. F., Demetriou, D. M. and Yocum, C. F. (1987) In: *Progress* Vol. I (Biggins, J. ed.) pp. 681–684, Martinus Nijhoff, Dordrecht
20. Laemmli, U. K. (1970) *Nature*, 227, 680–685
21. Deters, D. W., Nelson, N., Nelson, H. and Racker, E. (1975) *J. Biol. Chem.* 250, 1041–1047
22. Ghanotakis, D. F. and Yocum, C. F. (1986) *FEBS Lett.* 197, 244–248
23. Klimov, V. V., Kelvanik, A. V., Shuvalov, V. A. and Krasnovsky, A. A. (1977) *FEBS Lett.* 82, 183–186
24. Khananshvili, D. and Gromet-Elhanan, Z. (1983) *J. Biol. Chem.* 258, 3714–3719
25. Walczak, C., Kumar, S. and Warden, J. T (1987) *Photosyn. Res.* 12, 145–154
26. Hanssum, B., Renger, G. and Weiss, W. (1985) Biochim. Biophys. Acta 808, 243–251.

Phosphorylation of Photosystem Two Reaction Centre Polypeptides

ALISON TELFER, JONATHAN B. MARDER AND
JAMES BARBER

AFRC Photosynthesis Research Group,
Department of Pure & Applied Biology,
Imperial College of Science & Technology,
London SW7, UK

Summary

The D1 and D2 proteins, which are important components of the photosystem two reaction centre, are reversibly phosphorylated by a membrane-bound kinase. Proteolytic mapping techniques indicate that the site of phosphorylation in both D1 and D2 is a conserved threonine at position 2.

Introduction

Reversible phosphorylation of thylakoid membrane proteins by a redox-activated membrane-bound kinase/phosphatase system was originally discovered by Bennett (1). Since then a role for the rapidly-reversible phosphorylation of the light harvesting chlorophyll a/b protein (LHC2) in the control of excitation energy distribution between the two photosystems (PS) has been generally accepted (see 2,3). The function of reversible phosphorylation of the other proteins is still unclear. These proteins are associated with PS2-enriched fractions and have been implicated in the regulation of PS2 electron transfer (e.g. 4,5). A clear demonstration of the identity of these phosphopeptides and of the amino acids which are phosphorylated is necessary before the reason for their reversible phosphorylation can be fully understood. The gene for one of the PS2-associated proteins, the 9 kD phosphoprotein, has been sequenced (6) and although its site of phosphorylation has been shown to be threonine 2 (7), as yet the function of this protein is not known. Michel and Bennett (7) suggested that the other phosphoproteins in PS2-enriched fractions were the 43 kD, chlorophyll a binding protein, and the D1 and D2 proteins although Millner et al (8), using a proteolytic technique, had concluded that a phosphorylated band running at approx. 32 kD was not the D1 protein.

Recently the D1 and D2 proteins, products of the *psb*A and *psb*D gene respectively, have been demonstrated to be major components of a complex displaying features of the PS2 reaction centre (9,10). This means that definitive proof that the D1 and D2 protein are reversibly phosphorylated is of extreme interest and suggests a strong correlation between their reversible phosphorylation and the known effects of thylakoid membrane phosphorylation on both the oxidising and reducing activities of PS2.

In this paper we demonstrate positively that both D1 and D2 are phosphorylated (confirming our previous work, ref. 11) and deduce the most likely sites of phosphorylation, using amino acid sequence data and proteolytic mapping techniques. The results are discussed not only in relation to the importance of reversible phosphorylation of the PS2 reaction centre but also are shown to highlight the phosphate label as a useful probe in proteolytic mapping.

Materials and Methods

[^{35}S]-methionine labelling of wheat leaves was carried out using methods described previously (12). Thylakoid isolation from pea and wheat, membrane phosphorylation with [γ^{32}P]ATP and isolation of PS2 reaction centres were carried out as described by Telfer *et al.* (11). Thylakoid and reaction centre preparations from both pea and wheat were subjected to defined enzymatic digestion and analysed by SDS-polyacrylamide gel electrophoresis, transfer to nitrocellulose and immunoblotting with D1- and D2-specific antibodies as described by Marder *et al* (10). The phosphoproteins and distribution of ^{35}S-label were visualised by autoradiography and fluorography of either dried gels or the nitrocellulose-fixed electrophoretic profiles.

Results

Phosphorylation of PS2 Polypeptides

Phosphorylation of pea thylakoids after incubation in the dark with [γ^{32}P]ATP and dithionite (to activate the kinase) was found to yield a number of phosphorylated polypeptides (Fig. 1, lane 1) very similar to the pattern of labelling reported previously (2,8). In addition to the LHC2 polypeptides (26–27 kD) there are major phosphopeptides running in this gel system at apparent M_r of 9, 30, 32 and 40 kD. The autoradiograms in Fig. 1 show the fate of these ^{32}P-labelled polypeptides during the isolation of the PS2 reaction centre by a method similar to that of Nanba and Satoh (9). The phosphopeptides seen in thylakoids are also present in PS2-enriched membranes prepared by the method of Berthold, Babcock and Yocum

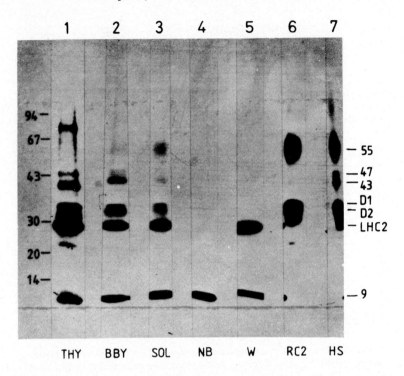

Fig. 1: Autoradiograph of SDS-PAGE (7–17%) of phosphorylated pea thylakoids and various fractions isolated during preparation of the PS2 reaction centre complex. See text for identification of lanes. *Left*: Apparent M_r of molecular weight markers. *Right*: Suggested identification of phosphorylated proteins.

(13), see lane 2 (BBY), and in the same material after solubilisation with 4% Triton-X100 (lane 3). The appearance of a phosphorylated band running at approx. 55 kD is related to the detergent treatment and is also sometimes seen in 'BBY' preparations (e.g. 8). Lanes 4–7 represent various fractions obtained after passage of the detergent-solubilised material through an anion exchange column, in the presence of 30 mM NaCl, and elution with a salt gradient (30–200 mM NaCl). Lane 4 shows that the 9 kD phosphoprotein is substantially non-binding at 30 mM NaCl as is the non-phosphorylated LHC2 (detected by Coomassie blue staining, not shown). Extensive washing with 30 mM NaCl elutes the phosphorylated LHC2 (lane 5) and the tail of the 9 kD protein. Lane 6 shows the fractions which elute at approx. 130 mM NaCl and lane 7 a higher salt fraction. The lane 6 fraction has the 30, 32 and 55 kD phosphorylated bands. The 'contaminating' tail has additional bands in the 40–45 kD region and some phosphorylated LHC2. The lane 6 fraction is highly enriched in the D1/D2/cytochrome b-559 polypeptides and we conclude that it is the phosphorylated PS2 reaction centre.

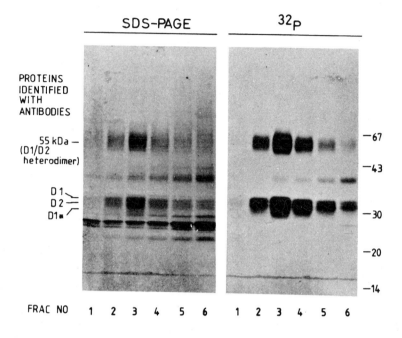

Fig. 2: SDS-PAGE and corresponding autoradiographs of fractions eluted by 110 to 180 mM NaCl during chromatography of Triton X-100 solubilised phosphorylated pea membranes.

Fig. 2 shows the Coomassie blue stained bands of the PS2 reaction centre-enriched fractions produced during purification of PS2 reaction centres from phosphorylated pea thylakoids. These fractions eluted from the column over the 110 to 180 mM NaCl range and are shown with their corresponding autoradiograms. Fraction 3 represents the peak of the PS2 reaction centre which eluted during this preparation. It consists of strongly staining bands at 55, 32 and 30 kD. The cytochrome b-559 bands (approx. 9 and 4 kD) were not resolved in this gel system which was designed to obtain a clear separation of the bands above 30 kD from those of the LHC2 bands. We found that when the normal technique used for preparation of non-phosphorylated reaction centres is applied to phosphorylated material it results in reaction centre preparations contaminated with LHC2. We can however usually remove all LHC2 if the column is washed more extensively with 30 mM salt before application of the salt gradient. The autoradiogram in Fig. 2 shows that ^{32}P-label copurifies with all the stained bands which are enriched in fraction 3 (except the cytochrome b559 bands, not shown).

In a previous paper (11) we concluded, after using a lysine-specific proteolysis technique, that both D1 and D2 in pea PS2 reaction centres are phosphorylated and speculated that the phosphorylation sites are near the amino termini of the polypeptides. In this paper we have extended these experiments to compare preparations from wheat and pea in order to show more clearly the correlation between ^{32}P-labelling and specific D1 and D2 proteolytic fragments and hence to deduce the probable phosphorylation sites.

Identification of Lys-C Fragments of D1 and D2

Fig. 3 shows the position of lysine residues in both D1 and D2 from pea and wheat as determined from their gene sequences. D1 from pea has no lysine residues whereas D2 has four (at positions 7, 10, 265 and 318). This difference in lysine content was exploited previously (14) to confirm the reaction of antibodies raised against the *psb*A and *psb*D gene products to the D1 and D2 proteins observed in various preparations from the thylakoid membranes. The amino acid sequences of D1 and D2 of wheat show two differences in lysine content: D1 has a single lysine at residue 238 (arginine in pea D1) and in D2 lysine 7 of pea is replaced by arginine. Fig. 3 also shows the proteolytic fragments which would be expected after digestion with the lysine-specific endoprotease (Lys-C) provided that all the lysine residues were equally susceptible to this enzyme. D2 of both pea and wheat should yield a large fragment (residue 11 to 265) from which both the amino and carboxy termini have been digested. However,

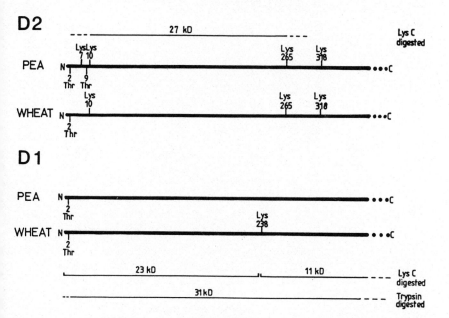

Fig. 3: A scheme of the D1 and D2 protein sequences from pea and wheat (pea D2, ref. 19 and other sequences, T. A. Dyer, personal communication) indicating lysine residue distribution and various protease digestion fragments. The threonine residues present in the first 10 amino acid region only are also indicated.

whereas D1 of pea should be resistant to digestion (shown previously, e.g. 14), wheat D1 would be expected to yield two fragments: a large amino terminal fragment and a smaller carboxy terminal fragment.

Initially it was necessary to identify the various Lys-C digestion products of pea and wheat PS2 reaction centres using D1-and D2-specific antibodies (Fig. 4). The polypeptide composition of the undigested wheat reaction centres appears to be essentially the same as that of pea. The D2 immunoblot shows the main D2 band at approx. 30 kD (the slight reaction to a lower molecular weight component running below D2 is probably due to a breakdown product of D2, P. J. Nixon, personal communication). The D1 blot shows the main D1 band at approx. 32 kD and a lower molecular weight band which Greenberg *et al.* (15) have indicated to be a conformer of D1. The 55 kD band reacts with both antibodies which is to be expected of a D1/D2 heterodimer (see 10).

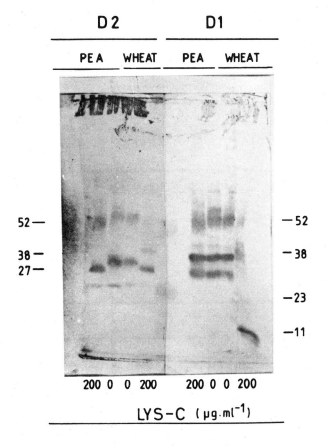

Fig. 4: Immunoblots with D1- and D2-specific antibodies of pea and wheat PS2 reaction centre preparations before and after digestion with Lys-C. The Lys-C digestion products of D2 (left) and D1 (wheat only, right) are indicated.

Lys-C digestion of pea reaction centres yields the results demonstrated previously for spinach (e.g. 10), i.e. D1 is resistant to the enzyme but digestion of D2 gives an approx. 27 kD fragment. Fig. 4 shows that the wheat digestion patterns are as predicted in Fig. 3. Wheat D2 yields a similar 27 kD fragment to that of pea whereas D1 is digested to give two fragments, one at 11 kD which reacts strongly with the antibody to D1 and another weakly reacting band at 23 kD. Because of the poor response to antibody of this latter fragment we have carried out Lys-C digestion of [^{35}S]-methionine pulse-labelled wheat thylakoids. In this experiment only the rapidly turned over D1 protein is labelled with ^{35}S: thus the label provides an alternative means of identifying D1 fragments (12). Fig. 5 shows that in [^{35}S]-methionine labelled wheat thylakoids D1 is virtually the only radioactive band. Digestion with Lys-C yields a strongly ^{35}S-labelled 23 kD fragment and one at 11 kD which is more weakly labelled. The unequal degree of labelling is consistent with the known ratio of

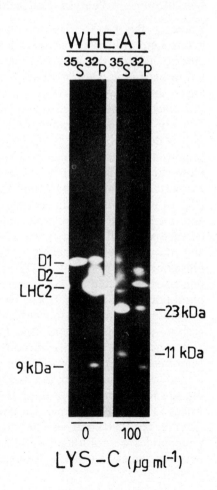

Fig. 5: ^{35}S-and ^{32}P-labelled digestion products yielded by lysine-specific proteolysis of [^{35}S]-methionine-labelled and phosphorylated wheat thylakoids respectively.

methionine residues in the two predicted Lys-C cleavage products (3 per 11 kD and 9 per 23 kD). The appearance of a phosphorylated Lys-C product corresponding to the 23 kD band will be discussed later.

Comparison of the D1 and D2 proteolysis patterns of pea and wheat PS2 reaction centres (see Fig. 4) also confirms the previous conclusion that the 55 kD band is a heterodimer of D1 and D2. In pea where D1 itself is not digested, the anti-D1-reacting 55 kD band is slightly decreased in size, indicating that the reacting species is aggregated with lysine-containing D2. The approx. 52 kD band is presumably a heterodimer of D1 and the 27 kD D2 fragment. In wheat, where the main D2 band is still only decreased in size by a few kD, the 55 kD band is digested to yield D2 reacting bands in the 38–50 kD region. These bands also react with the D1 antibodies and must therefore consist of heterodimers containing the 27 kD D2 fragment and one or other of the two D1 fragments.

Identification of Phosphorylation Sites on D1 and D2

Knowing the Lys-C digestion patterns for D1, D2 and the D1/D2 heterodimer from both pea and wheat PS2 reaction centres, we have been able to identify the probable phosphorylation sites on D1 and D2 by following the ^{32}P-label distribution during proteolytic degestion as follows. Fig. 6 shows the SDS-PAGE and corresponding autoradiogram of a pea PS2 reaction centre-enriched preparation and the effect of digestion with a range of Lys-C concentrations. The strong ^{32}P-labelling of D2 disappears as it is degraded to the 27 kD fragment but the loss of label is obscured somewhat by the resistance to Lys-C of the ^{32}P-labelled D1 conformer. However, there is a slight difference in apparent molecular weight between the weakly ^{32}P-labelled band and the strongly stained D2 27 kD fragment (Fig. 6) which is consistent with the former being due to the D1 conformer and not D2. As expected, the undigested D1/D2 heterodimer is heavily labelled with ^{32}P and Lys-C reduces both its apparent molecular weight and the degree of labelling. This is consistent with the loss of a phosphorylated fragment of D2 to leave a heterodimer of D1 and the non-phosphorylated 27 kD fragment of D2. These data show that the phosphorylation site on D2 in pea is not on the 27 kD fragment and therefore it must be on either of the relatively short amino- or carboxy-terminal fragments (see Fig. 3).

Because of the resistance of pea D1 to Lys-C no conclusion can be drawn from this data as to the position of its phosphorylation site. We therefore turned to a comparison of pea and wheat D1 in order, initially, to localise the site to one or other of the Lys-C fragments produced on digestion of wheat D1 (see Figs. 4 and 5). These experiments could be carried out directly on thylakoids as the digestion products of the other phosphopeptides were distinguishable from those of D1 by direct comparison between pea and wheat digestion patterns (Fig. 7). In wheat, Lys-C yields

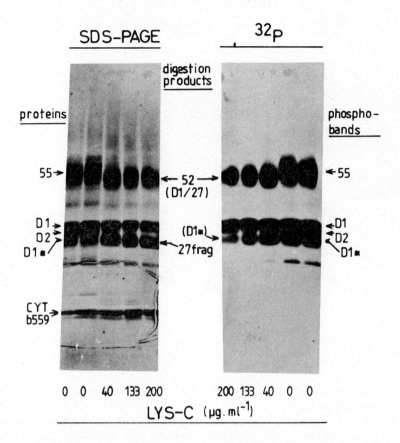

Fig. 6: Lysine-specific proteolysis of a phosphorylated pea PS2 reaction centre preparation. *Left*: Coomassie blue stain. *Right*: ^{32}P-labelling.

a strongly ^{32}P-labelled band at approx. 23 kD which appears concomitantly with the loss of label from the D1 band. This phosphorylated band is not seen on digestion of pea thylakoids and, as shown in Fig. 5, corresponds directly with the ^{35}S-labelled 23 kD fragment. From the data of Figs. 5 and 7 we conclude that the larger amino-terminal fragment of D1 is phosphorylated.

The location of the phosphorylation site in D1 was studied further by trypsinisation. It has been shown previously in Spirodela thylakoids that mild trypsin digestion removes a small fragment of D1 to yield a 31 kD band designated T31 (12). This cleavage was reproduced in thylakoids from pea and wheat (Fig. 8) using D1-specific antibodies to follow the digestion. The observed slight reduction in size of D1 must be due to removal of a few amino acid residues from either or both termini of the protein (see Fig. 3). Since T31 is not phosphorylated and considering the previous conclusions from lysine-specific proteolysis (Figs. 5 and 7), the

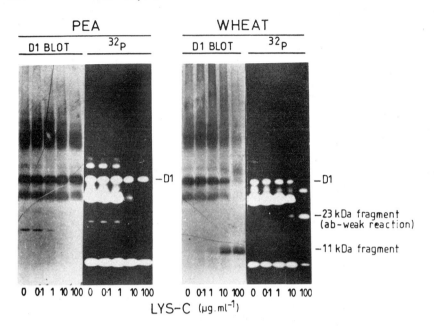

Fig. 7: Lysine-specific proteolysis of phosphorylated pea and wheat thylakoids showing D1 proteolytic fragments (D1 immunoblot) and phosphorylated peptides (^{32}P).

phosphorylation site must be located on a short, trypsin digestible amino-terminal fragment.

These results were confirmed by examining the distribution of ^{32}P-label in digested wheat reaction centres (Fig. 9). As stated previously, the expected Lys-C digestion products of wheat D1 (23 kD and 11 kD) can be seen by Coomassie blue staining and are absent in the corresponding gel lanes from pea. The autoradiogram shows clearly that, of the two D1 cleavage products in wheat, only the 23 kD fragment is labelled. No ^{32}P-labelling in the 23 kD region was seen in digested pea PS2 reaction centres (Fig. 5). Thus, considering both the thylakoid and PS2 reaction centre data on proteolysis of D1, we conclude that the phosphorylation site must be very close to the amino terminus of this protein.

As demonstrated earlier, the data on proteolysis of pea reaction centres locates the phosphorylation site of D2 to an end fragment but does not distinguish whether it is the amino or carboxy terminus. In Fig. 9 the autoradiogram of the wheat PS2 reaction centre digestion products shows the appearance of a strongly ^{32}P-labelled band at approx. 27 kD which, as can be seen in Fig. 5, was not present in digested pea reaction centres. The Coomassie blue stain in Fig. 9 and the D2 blot in Fig. 4 show, on digestion with Lys-C, the appearance of an approx. 27 kD D2 fragment in both wheat and pea. Thus lysine-specific proteolysis yields a very similar sized fragment of D2 which is phosphorylated in wheat but not in pea.

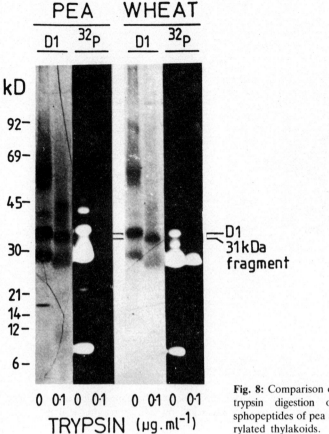

Fig. 8: Comparison of the effect of mild trypsin digestion on D1 and phosphopeptides of pea and wheat phosphorylated thylakoids.

Consideration of the lysine residue distribution in D2 of wheat compared to pea shows a single difference at position 7 (lysine in pea, arginine in wheat, see Fig. 3). The difference in phosphorylation status of the two approx. 27 kD Lys-C fragments in pea and wheat could therefore result from a difference in Lys-C susceptibility at residue 7 which would require also a resistance of lysine 10 to digestion by Lys-C in both species. This would locate the phosphorylation site on residues 1 to 7. The results could also be explained by a species difference in susceptibility of lysine 10 to Lys-C but this would still locate the phosphorylation site of D2 within the first 10 residues of the amino terminus.

Discussion

The results presented in this paper show that both the D1 and D2 proteins, which are the two major polypeptides of the PS2 reaction centre, are

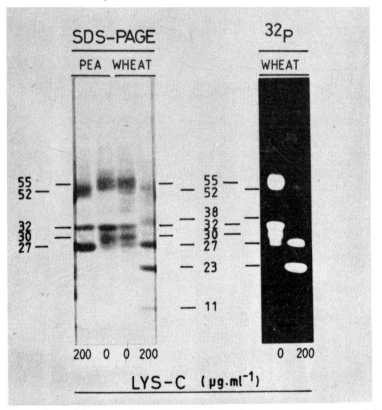

Fig. 9: Comparison of Coomassie blue stained peptide and phosphopeptide patterns yielded by lysine-specific proteolysis of pea and wheat PS2 reaction centre preparations.

reversibly phosphorylated by a redox-activated kinase. They also show that the phosphorylation sites on both proteins must be within a few residues of the amino terminus. These results confirm and extend the findings presented in an earlier paper (11).

Michel and Bennett (7) have shown previously that phosphorylation of PS2 polypeptides is essentially confined to threonine residues. In all plants, algae and cyanobacteria, for which D1 has been sequenced, a threonine residue is conserved at position 2 except in *Euglena gracilis* (16). The next conserved threonine in the sequence is at position 22. This is not consistent with the trypsinisation data which suggests that the phosphorylation site is very near the amino terminus. A threonine residue is also conserved at position 2 in the related D2 polypeptide whereas a threonine at position 9 which is present in spinach and pea D2 is not conserved in wheat. We therefore deduce that the most likely site of phosphorylation in D1 and D2 are the threonines at position 2. This is consistent with the proposition that the amino-terminal portions of these two proteins are exposed on the stromal side of the membrane in both proteins (see Trebst's folding models for D1 and D2, ref. 17).

Michel and coworkers have recently reached a similar conclusion as to the phosphorylation site of D1 and D2 based on analysis of phosphorylated peptides from PS2; amino-terminal fragments of D1 and D2 were identified by mass spectrometry and found to contain N-acetyl-O-phosphothreonine (18). The physiological role for this phosphorylation is as yet unclear, but it could regulate the functional activity of PS2 either by a direct effect on electron transfer and/or, as suggested by Bennett and colleagues (7), by changes in the organisation and assembly of the complex in the membrane. So far we have found no effects on the electron transport activities of the D1/D2/cytochrome b-559 containing PS2 reaction centre complex (unpublished observations). However, as this complex lacks both the electron donors to $P680^+$ and acceptors from $Pheo^-$, phosphorylation of D1 and D2 may yet turn out to have a role in control of electron transfer reactions in the PS2 reaction centre.

Acknowledgements

This work was supported by grants from the Science and Engineering Research Council and the Agricultural and Food Research Council.

References

1. Bennett, J. (1977) *Nature* 269, 344–346
2. Bennett, J. (1983) *Biochem. J.* 212, 1–13
3. Barber, J. (1986) *Photosyn. Res.* 10, 243–253
4. Horton, P. and Lee, P. (1984) *Biochim. Biophys.* Acta 767, 563–567
5. Packham, N. K. (1987) *Biochim. Biophys.* Acta 893, 259–266
6. Hird, S. M., Dyer, T. A. and Gray, J. C. (1986) *FEBS Lett.* 209, 181–186
7. Michel, H. P. and Bennett, J. (1987) *FEBS Lett.* 212, 103–108
8. Millner, P. A., Marder, J. B., Gounaris, K. and Barber, J. (1986) *Biochim. Biophys.* Acta 852, 30–37
9. Nanba, O. and Satoh, K. (1987) *Proc. Natl. Acad. Sci.* USA 84, 109–112
10. Marder, J. B., Chapman, D. J., Telfer, A., Nixon, P. J. and Barber, J. (1987) *Plant Mol. Biol.* 9, 325–333
11. Telfer, A., Marder, J. B. and Barber, J. (1987) *Biochim. Biophys.* Acta 893, 557–563
12. Marder, J. B., Mattoo, A. K. and Edelman, M. (1986) *Methods in Enzymol.* 118, 384–396
13. Berthold, D. A., Babcock, G. T. and Yocum, C. J. (1981) *FEBS Lett.* 134, 231–234
14. Nixon, P. J., Dyer, T. A., Barber, J. and Hunter, C. N. (1986) *FEBS Lett.* 209, 83–86
15. Greenberg, B. M., Gaba, V., Mattoo, A. K., and Edelman, M. (1987) *EMBO J.* 6, 2865–2869
16. Karabin, G. D., Farley, M. and Hallick, R. B. (1984) *Nuc. Acids Res.* 12, 5801–5812

17. Trebst, A. (1985) *Z. Naturforsch.* 41c, 240–245
18. Michel, H. P., Hunt, D. F., Shabanowitz, J. and Bennett, J. (1988) *J. Biol. Chem.* 263, 1123–1130
19. Rasmussen, O. F., Bookjans, G., Stummann, B. M. and Henningsen, K. W. (1984) *Plant Mol. Biol.* 3, 191–199

Trafficking and Distribution of the Photosynthetic Reaction Centre Proteins in the Chloroplast Membranes

AUTAR K. MATTOO, FRANKLIN, E. CALLAHAN, SUDHIR K. SOPORY[1] AND MARVIN EDELMAN[2]

Plant Molecular Biology and Hormone Laboratories, USDA/ARS,
Beltsville Agricultural Research Center,
Beltsville, MD 20705
U.S.A

[1] On leave from Jawaharlal Nehru University, New Delhi, India
[2] On leave from Weizmann Institute of Science, Rehovot, Israel

Summary

We have investigated relative distribution of proteins constituting different multi-enzyme complexes in chloroplast membranes by Western blot analyses. In addition, pulse-chase experiments revealed lateral diffusion of some of these chloroplast proteins between stromal and granal membranes of some of these chloroplast proteins. Two classes of chloroplast proteins became evident. One class constitutes those proteins that undergo lateral diffusion within the thylakoids and show dual location. This class includes the 32kDa protein, D2 protein and cytochrome $^b559$ which are protein components of the photosystem II reaction centre, and the light harvesting chlorophyll a/b apoprotein. The second class of proteins do not appear to translocate between stromal and granal membranes, and are associated exclusively with one of the membrane types. To this class of proteins belong the 43kDa and 51kDa photosystem II core proteins (granal membranes), alpha and beta subunits of the ATPase and subunit 2 of the photosystem I reaction centre (stromal membranes).

Proteins in the photosynthetic organelle (i.e., the chloroplast) are encoded by both the nuclear and plastid genomes (1). Accordingly, nuclear encoded and cytoplasmically synthesised proteins, viz., small subunit of Rubisco, the light harvesting chlorophyll a/b (LHC) protein and ferredoxin are post-translationally imported into the chloroplast (2, 3). In general, proteins imported into the chloroplast contain an N-terminal extension called transit peptide (4, 5),that is required for protein uptake (6, 7). In addition, there are specific binding sites and receptors on the chloroplast membranes that appear to recognise the transit peptide (8, 9) and assist in integration of membrane proteins (10, 11).

Relatively little is known about the processes and mechanisms that direct proteins synthesised within the chloroplast to their functional sites in the organellar compartments. Our studies are concentrated on a chloroplast protein that is a constituent of the photosystem II reaction centre. This protein, the 32 kDa protein (also known as the D1 protein, QB protein), is coded for by the psbA gene located in the chloroplast genome (12). Using an aquatic angiosperm, *Spirodela oligorrhiza*, which readily incorporate radioactive precursors, we have followed the appearance and distribution of the 32kDa protein and other chloroplast-membrane proteins.

Approach

Experimental design for studying *in vivo* transport of proteins to stromal and granal membranes was the following: Spirodela plants were pulsed for 1 to 3 min with [^{35}S] methionine. The plants were washed to remove unincorporated radioactivity, transferred to fresh growth medium containing 1 mM non-radioactive methionine and the radioactivity chased for up to 24 h. Samples were removed at different times during the chase and whole thylakoids were isolated (13). These thylakoids were then fractionated into stromal and granal membranes by a double detergent solubilisation and differential centrifugation method (13). Purity of the two membrane types was checked by electron microscopic observations (14) and stained protein pattern on sodium dodecyl sulfate polyacrylamide gels (15). In addition, Western blot analysis using antibodies raised against several different thylakoid proteins provided data on specific location, and thus distribution of these proteins in stromal versus granal membranes.

32kDa Protein, D2 and Cytochrome [b]559

The 32kDa protein is synthesised as a precursor of 33.5–34.5kDa (16). The precursor protein appears exclusively on the stromal membranes where it is post-translationally processed (16) at the carboxy terminus (17) to the mature 32kDa form. Following processing, the mature protein translocates to the appressed regions of the thylakoids, the grana, where the protein acts as a component of the photosystem II reaction centre (14). The mechanism of translocation is not yet understood. The 32kDa protein on the stromal membranes is integrated in an orientation that seems similar to the protein integrated within the granal membranes (18). Any explanation put forward for the lateral diffusion of the protein has to reconcile this fact. These data point out that the site of synthesis of the protein is spatially separated from that where the protein is functional.

The half-life of the newly synthesised 32kDa protein on stromal membranes, in cultures grown at $30\mu mol.m^{-2}.s^{-1}$ of white light, is between 9 and 18 min while the newly translocated granal 32kDa protein has a half-life of 6–12 hours (14). This difference in the life times of its presence in stromal versus granal membranes is also reflected in Western blot analysis of steady state levels of protein at these two distinct locations, viz., the distribution of 32kDa protein in the stromal and granal membranes was found to be 11% and 89%, respectively (19).

Besides 32kDa protein, the photosystem II reaction centre is comprised of two other proteins, viz., D2 protein and cytochrome [b]559 (20) which are also encoded within the chloroplast (21, 22). Together, they carry out the process of light induced charge separation. The D2 protein and cytochrome [b]559 were also found to be distributed between stromal and granal membranes in similar proportions as the 32kDa protein (19). The equivalent relative abundances of the three photosystem II reaction centre proteins between stromal and granal membranes raise the possibility that D2 and cytochrome ^{b}559 might also translocate from stromal membranes during their integration into grana. Preliminary results support such an intramembrane translocation of the D2 protein (23).

These data lead to questions of whether the photosystem II reaction centre is assembled on stromal membranes and then translocated as a unit to granal membranes to form a functional photosystem II. Experiments need to be designed to test this possibility.

Other Photosystem II Proteins

Western blot analysis of LHC protein using monoclonal antibodies (24) indicate an abundance of 22% of the 25–26kDa LHC proteins in the stromal membranes compared to 78% in the granal membranes (19). In contrast to the 32kDa and D2 proteins, the LHC protein, once imported into the chloroplast, is directly integrated within the granal membranes (25). However, LHC protein is known to disengage from the granal membranes upon phosphorylation (26). Detection of significant amounts of this protein in the stromal membranes (19) is consistent with the contention that the mobile form of LHC protein might migrate from the granal membranes to stromal membranes.

In contrast to the dual location of the photosystem II reaction centre and LHC proteins, the 43kDa and 51kDa photosystem II core proteins were found selectively localised in the granal membranes (19). These results suggest that the core proteins are immobile and might act as anchors for proper integration and functioning of the photosystem II reaction centre proteins in the granal membranes. In this context, it is known that the photosystem II located in the stromal membranes is not fully functional (27) and does not bind the herbicides atrazine or diuron (28).

CF_α and CF_β (ATPase) and Photosystem I Subunits

The alpha and beta subunits of the ATPase were found exclusively associated with the stromal membranes (19). In pulse-chase experiments as well, these polypeptides could be detected only on the stromal membranes (14). A similar situation was found for subunit 2 of photosystem 1 (19).

In contrast to the nuclear-coded subunit 2 (30), subunit 1 of photosystem I was found to be present in both membranes, 17% being present on granal membranes and 83% on stromal membranes (19). The significance of this dual location for this photosystem I subunit is hard to comprehend unless this protein also undergoes translocation. Currently no evidence

TABLE I

Localisation and Intramembrane Diffusion of Thylakoid Proteins*

Proteins	Location Membrane Type	Intramembrane Translocation
51kDa	Granal	No
43kDa	Granal	No
CFα	Stromal	No
CFβ	Stromal	No
PSI (subunit 2)	Stromal	No
32kDa (D1)	Granal and Stromal	Yes
D2	Granal and Stromal	Yes
Cyrochrome ^{b}559	Granal and Stromal	?
LHC protein	Granal and Stromal	Yes

*Results are based on references 14, 19, 26 and our unpublished data.

exists for the lateral diffusion of any subunit of photosystem I.

Table I gives an overview of results presented above. Two classes of thylakoid proteins become apparent. Class 1 proteins are associated exclusively with either granal or stromal membranes, while Class 2 proteins are largely localised to one membrane domain but are significantly present also in the other, topologically distinct, region. Consistent with this localisation, only Class 2 proteins appear to undergo lateral diffusion between stromal and granal regions of the thylakoid.

Acknowledgement

The results reported in this paper were supported in part by a U.S.-Israel BARD grant to Marvin Edelman and Autar K. Mattoo.

References

1. Ellis, J. (1981) *Ann. Rev. Plant Physiol.* 32, 111–137
2. Chua, N.-H. and Gilham, N. W. (1977) *J. Cell Biol.* 74, 441–452
3. Cashmorè, A., Szabo, L., Timko, M., Kansch, A., VandenBroeck, G., Schreie, P., Bohnert, H., Herrera-Estrella, L., VanMontagu, M., and Schell, J. (1985) *Bio/ Technology* 3, 803–808.
4. Schmidt, G. W., Devilliers-Thiery, A., Desruisseaux, H., Blobel, G. and Chua, N.-H (1979) *J. Cell Biol.* 83, 615–622
5. Cashmore, A. R. (1984) *Proc. Natl. Acad. Sci.* (USA) 81, 2960–2964
6. Schmidt, G. W. and Mishkind, M. L. (1986) *Ann. Rev. Biochem.* 55, 879–912
7. Schreier, P. N., Seftor, E. A., Schell, J., and Bohnert, M. J. (1985) *EMBO J.* 4, 25–32
8. Cornwell, K. L., and Keegstra, K. (1987) *Plant Physiol.* 85, 780–785

9. Pain, D., Kanwar, P. S. and Blobel, G. (1988) *Nature* 331, 232–237
10. Smeekens, S., Bauerle, C., Hageman, J., Keegstra, K. and Weisbeek, P. (1986) *Cell* 46, 365–375
11. Chitnis, P. R., Nechushtai, R. and Thornber, N. P. (1987) *Plant Mol. Biol.* 10, 3–11
12. Zurawski, G., Bohnert, M.-J., Whitfeld, P. R. and Bottomley, W. (1982) *Proc. Natl. Acad. Sci.* (USA) 79, 7699/7703
13. Leto, K. J. Bell, E. and McIntosh, L. (1985) *EMBO J.* 4, 1645–1653
14. Mattoo, A. K. and Edelman, M. (1987) *Proc. Natl. Acad. Sci.* (USA) 84, 1497–1501
15. Mattoo, A. K., Pick, U., Hoffman-Falk, H. and Edelman, M. (1981) *Proc. Natl. Acad. Sci.* (USA) 78, 1572–1576
16. Reisfeld, A., Mattoo, A. K. and Edelman, M. (1982) *Eur. J. Biochem.* 124, 125–129
17. Marder, J. D., Goloubinoff, P., Edelman, M. (1984) *J. Biol. Chem.* 259, 3900–3908
18. Mattoo, A. K., Callahan, F. E., Greenberg, B. M. and Edelmen, M. (1988) In *Biotechnology of Crop Protection*, (ed. Hedin, P.) ACS Symposium Series, ACS Books, Washington, D.C. (in press)
19. Callahan, F. E., Wergin, W. P., Nelson, N., Edelman, M. and Mattoo, A. K. (1988) (submitted)
20. Nanba, O., and Satoh, K. (1987) *Proc. Natl. Acad. Sci.* (USA) 84, 109–112
21. Murata, M. and Miyao, M. (1987) In *Progress in Photosynthesis Research*, Vol. 1, (ed. Biggins, J.) pp. 454–462, Martinus Nijhoff Pub., Dordrecht
22. Bottomley, W. and Bohnert, M. J. (1982) In *Encyclopedia of Plant Physiology*, New Series, Vol. 14B (eds. Parthier, B. and Boulter, D.) pp. 531–596, Springer Verlag, Berlin
23. Callahan, F. E., Edelman, M. and Matoo, A. K. (1987) In *Progress in Photosynthesis Research*, Vol. III (ed. Biggens, J.). pp. 799–802, Martinus Nijhoff Publ. Dordrecht
24. Dorr, S. C., Arntzen, C. J. and Somerville, S. C. (1987) In *Progress in Photosynthesis Research*, Vol. II (ed. Biggens, J.) pp. 141–144, Martinus Nijhoff Publ., Dordrecht
25. Kline, K. (1988) *Plant Physiol.* 86, 1120–1126
26. Staehelin, L. A. and Arntzen, C. J. (1983) *J. Cell Biol.* 97, 1327–1337
27. Melis, A. (1985) *Biochim. Biophys.* Acta 808, 334–342
28. Wetterm, M. (1966) *Plant Science Lett.* 43, 173–177
29. Vallon, D., Wellman, F. A. and Olive, J. (1986) *Photochem. Photobiophys.* 12, 203–220
30. Nechustai, R., Nelson, N., Mattoo, A. and Edelman, M. (1981) *FEBS Lett.* 125, 115–119

PART 2
Energy Transduction

Linear-dichroic Triplet-minus-Singlet Absorbance Difference Spectra of Borohydride-treated Reaction Centres of *Rhodobacter sphaeroides R26*

A. Angerhofer,[1] D. Beese,[2] A. J. Hoff,[3]
E. J. Lous[3] and H. Scheer[2]

[1]Universität Stuttgart,
Physikalisches Institut Teil 3,
Pfaffenwaldring 57,
D-7000 Stuttgart 80, FRG
Present address
Chemistry Division,
Argonne National Laboratory,
Argonne/IL 60439, U.S.A.

[2]Botanisches Institut der Universität,
Menzinger Str. 67,
D-8000 München 19, FRG

[3]Department of Biophysics
Huygens Laboratory of the
State University
P.O. Box 9504
2300 RA Leiden, The Netherlands

Summary

Linear-dichroic triplet-minus-singlet absorbance difference spectra of borohydride-treated reaction centres of *Rhodobacter sphaeroides R26* have been measured by the technique of absorbance detected magnetic resonance (ADMR) in zero field in the wavelength region between 450 and 950 nm. They were compared with spectra of untreated reaction centres. Besides the absence of the absorbance of the removed B800$_M$ 'accessory monomer' pigment a blue shift of about 7 nm of the remaining band in the 800 nm region occurs.
Key words: bacterial photosynthesis, linear dichroism, ADMR, reaction centre, pigment modification, *Rhodobacter sphaeroides*.

1. Introduction

The photosynthetic reaction centre (RC) of *Rhodobacter (Rb.) sphaeroides R 26* consists of 3 protein subunits (designated H, L and M according to their apparent molecular weight) (1), 4 bacteriochlorophyll (BChl) molecules and 2 bacteriopheophytins (BPh) as functional pigments (2). The geometrical positions of the RC pigments have been deduced from x-ray crystallographic data of RC-crystals of *Rhodopseudomonas (Rps.)*

List of abbreviations: BChl = Bacteriochlorophyll; BPh = Bacteriopheophytin; *C.* = *Chloroflexus*; LD = linear dichroism; *Rb.* = *Rhodobacter*; RC = reaction centre; *Rps.* = *Rhodopseudomonas*; T−S = triplet-minus-singlet.

viridis at 3 Å resolution (3). A similar geometry was suggested for the RC of *Rb. sphaeroides* (4,5). Two of the BChl molecules, in dimeric configuration, form the primary electron donor P870, whereas the other two so-called 'accessory monomers' $B800_L$ and $B800_M$ as well as the two BPh monomers (BPh_L and BPh_M) are located at specific distances and orientations to the dimer. Electron transfer, the primary step of photosynthesis, seems to take place only on the pigment-chain which is located in the L-part of the RC protein, whereas the function of the pigments of the M-chain is still under discussion (6–9).

By treatment with borohydride, the $B800_M$ -molecule is reduced at the 3-acetyl group and can be dissociated from the remaining complex (10,11). From kinetic and spectroscopic evidence (7,11–13) it has become clear that the electron transfer as well as the 'special pair' geometry remain practically unchanged by the borohydride treatment. Nevertheless, in a side reaction the reagent attacks the protein which results in a change of the apparent molecular weight of the M-unit from 23 to 15.5 kD (13). The homogeneity of the preparation concerning spectroscopic features and the pigment content was questioned by Shuvalov *et al.* (14). These authors discuss a product mixture in which $B800_M$ is partly pheophytinised and BPh_M is partly reduced, too. On the basis of optical and ODMR work on our somewhat differently prepared RC sample (13) we believe that the present preparation is homogeneous and consists of RCs from which the $B800_M$-pigment is removed.

Borohydride treated RCs are of great theoretical and experimental interest. They can be used to elucidate the complex interactions which the pigments undergo with each other in the RC. Under the assumption that the accessory BChl somehow plays a role in charge separation, the fact that their electron transfer in treated RCs seems hardly changed supports the idea that the primary charge separation only takes place on the 'L-chain'. Distinct changes in absorption (10–15), Raman resonance (13), circular dichroism (13, 15) and triplet-minus-singlet (T–S) spectra (13) reflect the extensive excitonic interactions among the RC pigments. Theoretical simulations of the different spectroscopic properties on the basis of the known geometry of *Rps. viridis* seem helpful to reveal the various eigenstates of the whole complex (8,16–20). Linear dichroic (LD) T–S spectra have also been used in such calculations (16–19). They are especially useful to reveal the angles between the optical transition moments of the different pigments and the fine-structure tensor of the RC triplet state (18,19,21,22).

In this contribution we present LD-(T–S) as well as isotropic T–S spectra in the wavelength range between 450 and 950 nm and compare them with spectra of untreated RCs.

2. Materials and Methods

Reaction centres from *Rb. sphaeroides R26* were prepared as described earlier (1). The RCs were modified with solid sodium borohydride according to the original method of Ditson *et al.* (10) in a buffer system containing the detergent lauryl dimethyl aminoxid (20 mM Tris, 0.08% LDAO, pH 8). In order to replace the LDAO by the detergent Triton X100 (TX), the RCs (15 ml with $A_{870} = 0.5$ cm^{-1}, 48 nmoles) were dialysed against a Iris buffer (20 mM, pH 8) with 0.1% TX and then washed on a DE 52 column (2×6 cm) with the tenfold column volume of the same buffer. The RCs were eluted with the same buffer containing 190 mM NaCl and concentrated by centrifugation (4000 rpm) in a centricon 10 (Amicon).

The first quinone acceptor was reduced by adding 10mM sodium ascorbate to the RCs and freezing the sample in the light after dilution with ethylene glycol to 66 % v/v (OD 0.21 at 801 nm).

The linear dichroic ADMR apparatus is described in (21,23); the (T−S) spectra were recorded at 1.2 K.

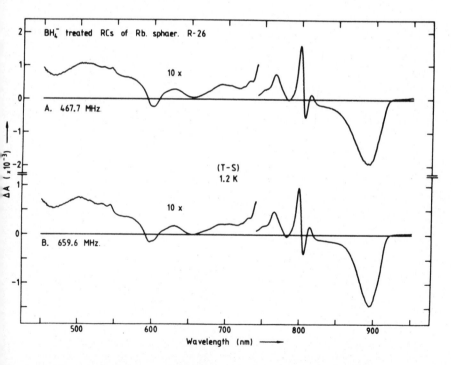

Fig. 1: Isotropic T−S spectra of borohydride-treated RCs of *Rb. sphaeroides R26* at T = 1.2 K. The short wavelength region is enhanced 10 times in intensity.
A: Detected with microwaves at 457.7 MHz.
B: Detected with microwaves at 659.6 MHz.

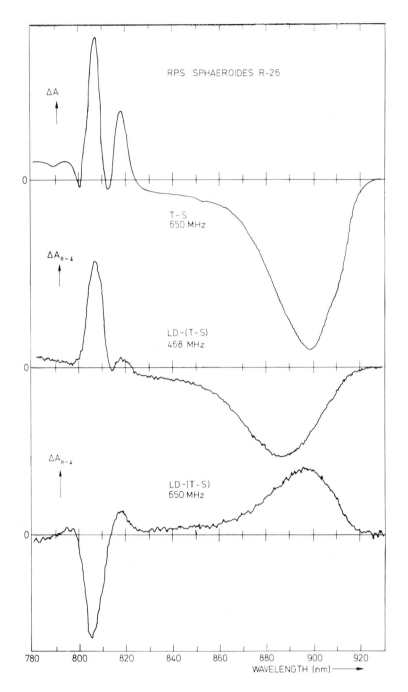

Fig. 2: T−S and LD-(T−S) spectra of native RCs of *Rb. sphaeroides R26* at T=1.2 K. Upper trace: T−S spectrum detected at 650 MHz. Middle and lower trace: LD-(T−S) spectra detected at 468 and 650 MHz, respectively. The two LD-(T−S) spectra are drawn on the same vertical scale but are not calibrated to the T−S spectrum. The T−S spectrum taken at 468 MHz was virtually indistinguishable from that taken at 650 MHz. From Ref. 22.

3. Results

T−S spectra of borohydride treated RCs taken on the |D|−|E|- and |D|+|E|-transition at microwave frequencies of 457.7 MHz and 659.6 MHz, respectively, are depicted in Fig. 1. The main peak in the infrared is located at 892 and 895 nm in the two cases, just as in untreated RCs (Fig. 2) (22). The sharp positive band which is visible in untreated RCs at 818 nm is missing. Instead we find a small positive band at 811 nm and a negative trough at 807 nm. The second sharp positive band, at 807 nm in untreated RCs, is preserved but appears blue-shifted to 799 nm. The BPh-bands at 762 and 748 (shoulder) nm are surprisingly intense. At shorter wavelengths we find a negative band at 600 nm which we ascribe to the Q_x-absorption of the primary donor. The Q_x-absorption of the BPh-molecules is visible at 531 and 542 nm. The broad, unstructured positive signal between 450 and 720 nm is mainly due to triplet-triplet absorption of the primary donor (24,25).

The corresponding LD-(T−S) spectra, depicted in Fig. 3, show the same spectral features as the T−S spectra with different signs and intensities

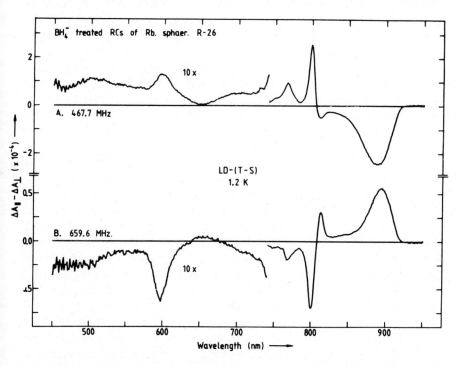

Fig. 3: Linear dichroic T−S spectra of borohydride-treated RCs of *Rb. sphaeroides R26* at T=1.2 K. The short wavelength region is enhanced 10 times in intensity.
A: Detected with microwaves at 457.7 MHz.
B: Detected with microwaves at 659.6 MHz.

with the exception of the trough at 807 nm. The near infrared bands are located at 889 and 893 nm for the $|D|-|E|$-and the $|D|+|E|$-transition, respectively. The main difference as compared with unmodified RCs (Fig. 2) (22) is the blue-shift of 7–8 nm of the bands around 800 nm. We also wish to point to differences in the shape of the band at 811 nm. In the spectrum in Fig. 3B this band appears unusually strong, whereas it is weak and negative in the spectrum of the $|D|-|E|$-transition (Fig. 3A). In unmodified RCs, where the corresponding band is located at 818 nm, it appears in both cases weakly positive.

4. Discussion

Removal of $B800_M$ by borohydride treatment of the RC causes an absorption decrease around 800 nm (10–15). The separate assignment of absorption bands to different pigments is generally not possible because of the complex structure of the RC and the various interactions between its pigments. The situation with T–S or LD-(T–S) spectra is even more complicated because of the differential structure of such spectra which involve twice the number of eigenstates as compared to normal absorption, since one has to take into account also the triplet state and its excitations. Nevertheless, in a rough estimate, we provisionally assign a major part of the sharp positive band which is visible at 818 nm in the T–S spectra of untreated RCs to a triplet–triplet transition of ^3P that acquires oscillator strength from $B800_M$, because it is largely removed and shifted to 811 nm in the spectra of modified RCs. In the LD-(T–S) spectra this is reflected in a change in intensity, which indicates a change in orientation of the transition moment. A more detailed correlation of the spectral features with the pigment eigenstates through a simulation of the spectra with exciton theory is currently in progress and will be published elsewhere.

Another prominent feature is the blue-shift of the second sharp intensive band from 807 to 799 nm, which is visible in both the T–S and the LD-(T–S) spectra. This bandshift is surprising because the spectral location of this band seemed to be preserved in all bacterial RC-preparations investigated so far, irrespective of the excitation of different sites (26,27). Even in RCs of *Chloroflexus aurantiacus*, which contain a monomeric BPh instead of $B800_M$, this band appears at 807 nm (27). This indicates that the remaining BChls that contribute to the 799 nm band may sense a change in the local environment. Qualitative simulations of the T–S spectrum of modified RCs with the theory of Knapp *et al.* (16,28) as well as the observation of a partial proteolytic cleavage of the M-protein during treatment with borohydride (13) seem to support this explanation.

Since the origin of the spectral features around 800 nm are still under discussion (16–19), we are not yet able to assign the bandshift to a particular one of the remaining 3 BChl molecules. Whether the cause of the blue-shift lies in the breakdown of excitonic interaction of B800$_M$ with B800$_L$ or in a local variation of the protein environment resulting from a small geometry variation of one of the 3 remaining BChls is not yet clear.

The correspondence of the T-S and LD- (T-S) spectra of RCs of *C. aurantiacus* and those of borohydride treated RCs of *Rb. sphaeroides R26* is remarkable. The only differences are the presence of the 811 nm band in the T–S spectrum of the latter and some changes in the BPh region, especially in the LD-(T–S) spectrum of the |D|–|E| transition. The latter changes are due to the weakly interacting BPh$_M$ that replaces B800$_M$ in *C. aurantiacus*.

Acknowledgements

This work was supported by the Deutsche Forschungsgemeinschaft, 5300 Bonn (SFB 143, project A1). We are indebted to Dr Reng at the Gesellschaft für Biotechnologische Forschung, 3301 Stöckheim, for mass culture of *Rb. sphaeroides*. We acknowledge the continuing support of Professor W. Rüdiger.

The work in the Leiden laboratory was supported by the Netherlands Foundation for Chemical Research (SON), financed by the Netherlands Organisation for the Advancement of Pure Research (ZWO).

AA wants to thank Professor H. C. Wolf and Dr J. U. von Schütz for their support.

References

1. Feher G. and Okamura M. Y. (1978) in: *The Photosynthetic Bacteria* (Clayton R. K. and Sistrom W. eds), pp. 349–386, Academic Press, New York
2. Reed D. W. and Peters G. A. (1972) *J. Biol. Chem.* 247, 7148–7152
3. Deisenhofer J., Epp O., Miki K., Huber R. and Michel H. (1984) *J. Mol. Biol.* 180, 385–398
4. Chang C. -H., Tiede D., Tang J., Smith U., Norris J. and Schiffer M. (1986) *FEBS Lett.* 205, 82–86
5. Allen J. P., Feher G., Yeates T. O., Rees D. C., Deisenhofer J., Michel H. and Huber R. (1986) *Proc. Natl, Acad. Sci. USA* 83, 8589–8593
6. Shuvalov V. A. and Klevanik A. V. (1983) FEBS Lett. 160, 51–55
7. Breton J., Martin J. -L., Petrich J., Migus A. and Antonetti A. (1986) *FEBS Lett.* 209, 37–43
8. Kuhn H. (1986) *Phys. Rev.* A 34, 3409–3425
9. Chang C. -H., Schiffer M., Tiede D., Smith U. and Norris J. (1985) *J. Mol. Biol.* 186, 201–203
10. Ditson S. L., Davis R. C. and Pearlstein R. M. (1984) *Biochim. Biophys.* Acta 766, 623–629
11. Maroti P., Kirmaier C., Wraight C., Holten D. and Pearlstein R. M. (1985) *Biochim. Biophys.* Acta 810, 132–139
12. Holten D., Kirmaier C. and Levine L. (1987) in: *Progress in Photosynthesis Research*

(Biggins J. ed) Proceedings of the VIIth International Congress on Photosynthesis Vol. I, pp. 169–176, Martinus Nijhoff Publishers, Dordrecht, Boston, Lancaster

13. Beese D., Steiner R., Scheer H., Robert B., Lutz M. and Angerhofer A. (1988) *Photochem. Photobiol.* 47, 293–304

14. Shuvalov V. A. and Duysens L. N. M. (1986) *Proc. Natl Acad. Sci.* USA 83, 1690–1694

15. Shuvalov V. A., Shkuropatov A. Ya., Kulakova S. M., Ismailov M. A. and Shkuropatova V. A. (1986) *Biochim. Biophys.* Acta 849, 337–346

16. Knapp E. W., Scherer P. O. J. and Fischer S. F. (1986) *Biochim. Biophys.* Acta 852, 295–305

17. Scherer P. O. J. and Fischer S. F. (1987) *Biochim. Biophys.* Acta 891, 157–164

18. Lous E. J. and Hoff A. J. (1987) in: *Progress in Photosynthesis Research* (Biggins J. ed) Proceedings of the VIIth International Congress on Photosynthesis Vol. 1, pp. 169–176, Martinus Nijhoff Publishers, Dordrecht, Boston, Lancaster

19. Lous E. J. and Hoff A. J. (1987) *Proc. Natl. Acad. Sci. USA* 84, 6147–6151

20. Vasmel H., Amesz J. and Hoff A. J. (1986) *Biochim. Biophys.* Acta 852, 159–168

21. den Blanken H. J., Meiburg R. F. and Hoff A. J. (1984) *Chem. Phys. Lett.* 105, 336–342

22. Hoff A. J., den Blanken H. J., Vasmel H. and Meiburg R. F. (1985) *Biochim. Biophys.* Acta 806, 389–397

23. Hoff A. J. (1985) in: *Antennas and Reaction Centers of Photosynthetic Bacteria* (Michel-Beyerle M. E. ed), pp. 150–163, Springer, Berlin

24. Parson W. W., Clayton R. K. and Cogdell R. J. (1975) *Biochim. Biophys.* Acta 387, 265–278

25. den Blanken H. J. and Hoff A. J. (1982) *Biochim. Biophys.* Acta 681, 365–374

26. den Blanken H. J., Jongenelis A. P. J. M. and Hoff A. J. (1983) *Biochim. Biophys.* Acta 725, 472–482

27. den Blanken H. J., Vasmel H., Jongenelis A. P. J. M., Hoff A. J. and Amesz J. (1983) *FEBS Lett.* 161, 185–189

28. Scherer P. O. J. and Fischer S. F. (1987) *Chem. Phys. Lett.* 137, 32–36.

Ca^{++} Gating of Proton Fluxes in Thylakoid Membranes: Regulation of Localised and Delocalised Energy Coupled Proton Gradients

RICHARD A. DILLEY, GISELA G. CHIANG AND WILLIAM A. BEARD

Department of Biological Sciences
Purdue University
West Lafayette, IN 47907, USA

Summary

This report gives an overview of recent developments in the study of how proton gradients are generated and utilised for energising ATP formation in spinach or pea chloroplasts. The evidence is reviewed that supports the hypothesis for there being energisation either by membrane domain-localised proton gradients or delocalised proton gradients. Delocalised H$^+$ gradients competent to drive ATP formation form under several different conditions—all of which appear to displace Ca^{++} ions from binding sites in the thylakoid or on the lumen side of the membrane. A localised proton gradient coupling mode is observed when Ca^{++} ions are not displaced in the first place (such as by pretreatments which maintain low K$^+$ concentrations or by not over-acidifying the thylakoids) or if Ca^{++} ions are added to thylakoids which were treated so as to displace Ca^{++} ions from thylakoid binding sites. Treatments such as 100 mM KCl incubation, over-acidifying the thylakoid by basal electron flow-dependent H$^+$ accumulation, or those in which a lipid-soluble Ca^{++} chelator is added, lead to delocalised H$^+$ gradients. Ca^{++} ions added to the first and third treatments listed above reverse the delocalising tendency, and maintain the localised gradient coupling mode.

Intact chloroplasts respond to such treatments in the same manner as isolated thylakoids. Therefore, we conclude that the evidence favours the hypothesis that thylakoid energy coupling proton gradients are regulated by Ca^{++} binding to a 'gating structure' so that either localised or delocalised H$^+$ fluxes can provide the energetic H$^+$ gradient needed to drive ATP formation.

Introduction

Background on Localised and Delocalised Energy Coupling

This paper will present experiments which led to the hypothesis, and various tests of the hypothesis, concerning a Ca^{++} controlled gate for regulating H$^+$ ion diffusion pathways between localised and delocalised energy coupling gradients in thylakoid membranes. The arguments about whether photophosphorylation could be energised by localised proton gradients or only by delocalised gradients were unresolved for some time and had proponents on each side (see ref. 1 for a thorough review of this question as of 1985). The issue is clearer now, in the light of recent results (see ref. 2 for a review), and our present view is that both types of energy coupling proton gradients occur in thylakoids. This is not to say that we understand what may be involved in the mechanism of localised proton diffusion, for this remains a mystery, although there are suggestions as to how it may occur in membranes. The most explicit idea is that of the Hydrogen Bonded Chain (3), which uses as a model the notion that proton movement along adjacent hydrogen bonded water molecules occurs in ice or ice-like water. The model was extended to include acid–base donor and acceptor groups of proteins (3). However, it remains for future work to provide experimental evidence directly testing this hypothesis for relevance to biological membranes. Indirect support for the model is, at present, limited to arguments such as those based on still hypothetical amino acid arrangements in the Fo polypeptides of the ATPase, which span the *E. coli* membranes (4). It is an attractive hypothesis, especially if one allows for membrane-associated water to participate in the Hydrogen Bonded Chain structures.

Between 1976 and 1986 the controversy raged about the validity of localised protonic coupling. The problem was that no single laboratory could demonstrate both localised and delocalised H^+ gradient coupling in photophosphorylation [except for the case of delocalised gradient-driven post-illumination phosphorylation, which occurred in thylakoids showing localised gradient coupling in a flash-excitation ATP formation protocol (6)]. Thus, while abundant experimental evidence made it easy to accept the notion of a delocalised proton gradient as the driving force, it was not easy to accept the data for localised coupling, and there was a tendency to suspect that 'something was wrong' with the experimental results put forward in support of localised coupling. A much more believable situation would accrue if, in a given laboratory, the results would be such as to dispel the nagging doubts that any membrane system used to 'demonstrate' localised coupling was not being irreparably switched from a normal state (i.e., a comfortably acceptable Mitchellian chemiosmotic system) to some altered state, or that the methods used for the experiment were not artifactually slanting the results. It must be remembered that, by 1976, the Mitchell delocalised chemiosmotic hypothesis had had stunning support (1), and the dictums of that conceptual scheme had, for many workers, taken on the aspects of sacred dogma. After all, in a variety of systems the predictions of the chemiosmotic energy coupling hypothesis were borne out by experiments (1,7), and that conceptual scheme fits our understanding of biological systems too well to be discarded as a paradigm.

It was, perhaps, inevitable that the resolution of the polemic would occur when experimental results with a given thylakoid preparation would show that either (a) no localisation of energy coupling proton gradients could occur, or (b) a system could be shown to demonstrate both localised and delocalised coupling, depending on some regulatory factor(s). Clearly, negative evidence has no weight, so it was with some diligence that we sought to resolve the experimental question as to whether our techniques could demonstrate both coupling modes in a flash excitation phosphorylation experiment. The clue to this issue was found in reports from two laboratories (8,9) which indicated that high KCl concentrations caused thylakoids to show delocalised proton gradient bahaviour.

Methods and Predictions Based on Theory

Our experimental approach for detecting localised or delocalised proton gradient energy coupling was to measure the lag time (number of single-turnover flashes) required to reach the ATP formation threshold energisation level, as influenced by the permeable buffer pyridine (6, 10–12). Provided the buffering range of the pyridine sufficiently overlaps the internal acidic pH required to energise ATP formation [in the presence of valinomycin and K^+ to collapse the $\Delta\psi$ component of the protonmotive force (6) and with the particular reactant and product concentrations we use a $\Delta pH \approx 2.3$ units is required to reach the energisation threshold], a delocalised coupling mode should show a detectable and predictable increase in the threshold energisation flash number when 5 mM pyridine is present ($pK_a = 5.44$). Localised proton gradient coupling should show no effect of pyridine on the length (number of flashes) of the ATP formation onset lag. Similarly, for delocalised coupling, a post-illumination phosphorylation ATP yield should show predictable permeable buffer effects, either increasing the ATP yield when the pK of the pyridine is near or slightly more acidic than the threshold pH_{in}, or decreasing the ATP yield if the energisation threshold pH_{in} is considerably below the pyridine pK (11).

These predictions were thoroughly tested using spinach (13) or pea (14) thylakoids stored in either a 'high salt' (100 mM KCl) containing buffer, or a 'low salt' buffer (200 mM sorbitol and no KCl). The phosphorylation assay utilised a real-time ATP detection system which measured ATP-dependent luminescence using the LKB luciferin-luciferase kit (10). The assay utilised single turnover flashes, usually at 5 Hz, at 10°C, either pH 8 or 7 and with or without 5 mM pyridine. See references (10–12) for complete details.

Experimentally Demonstrating Reversible Switching between Localised and Delocalised Coupling Modes

Fig. 1 shows typical ATP formation onset lags without and with valinomycin (val), K^+. The greater flash number at ATP onset with val and

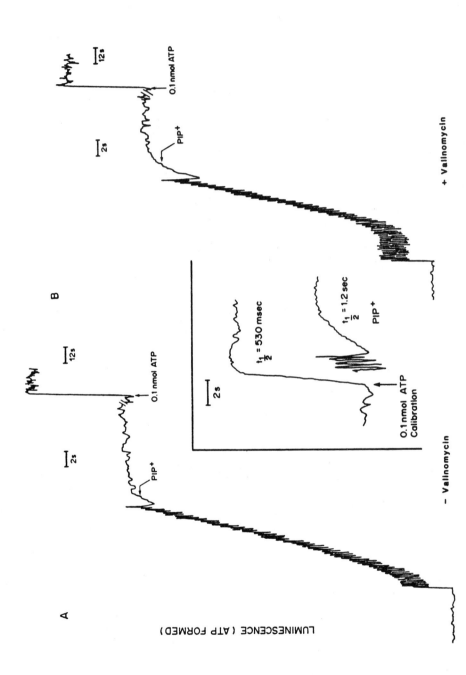

K^+ present is the expected result for a system with the ΔpH as the sole energy component of the protonmotive force, ΔP. The $\Delta \psi$-collapsing val and K^+ assure this condition in later $\pm$ pyridine experiments; if protonated pyridine builds up in the lumen, its positive charge would not contribute to the ΔP through the $\Delta \psi$ effect. Also shown is the post-illumination phosphorylation (PIP) ATP yield, identified as PIP^+ in the figure, the $+$ sign meaning that ADP and Pi were present during the flash train. In the traditional post-illumination phosphorylation experiment (termed PIP^-) the light phase does not have ADP or Pi present, the missing component being added in the dark, after the light phase. Experiments of this sort will be described below. Both the onset lag flash number (estimated by two criteria, see Fig. 1 legend) and the PIP^+ post-illumination phosphorylation ATP yield can be easily measured.

Fig. 2 shows typical pyridine effects on the phosphorylation parameters at pH 8 [Fig. 2A, cf. (13) for details] and pH 7.0 [Fig. 2B, cf. (15)], for both the low and high KCl-stored thylakoids. At both pH values, the low salt samples showed phosphorylation parameters which were *insensitive* to 5 mM pyridine, showing close to the same onset flash number and similar post-illumination (PIP^+) ATP yields, consistent with energisation by localised proton gradients (6, 12). Table 1 (A:pH 8 and B: pH7) gives a compilation of data from several experiments. Chloroplasts stored in high salt, but assayed in the same medium as used for the low salt-stored sample, showed a pyridine-induced increase in the ATP formation onset

---

Fig. 1: Single-turnover flash initiated phosphorylation measured by luciferin-luciferase luminescence. Flashes were delivered at a rate of 5 Hz to thylakoids containing 10 μg of chlorophyll (Chl) suspended at 10°C in 1 ml of reaction mixture containing 50 mM Tricine-KOH (pH 8.0), 10 mM sorbitol, 3 mM $MgCl_2$, 2 mM KH_2PO_4, 0.1 mM ADP, 0.1 methylviologen, and 10 μM diadenosine pentaphosphate. Dithiothreotol (DTT) (5 mM) was included to protect critical luciferase sulphydryls. The vertical spike was the result of a light leak and served as a useful event marker. The flash lag (actual) for the onset of ATP formation was determined by the first detectable rise in luciferin-luciferase luminescence while the nmol ATP yield per flash was calculated from the linear rise in bioluminescence. The extrapolated lag for the onset of ATP formation was ascertained from where the linear rise in luminescence would intersect a baseline drawn through the initial, non-phosphorylating flashes [see Beard and Dilley (3)]. Valinomycin was omitted in A and 400 nM was included in B. The lag for the onset of ATP formation increased from about 3 flashes to 15 flashes after the addition of valinomycin, while the ATP yield per flash was about 0.75 nmol ATP/mg Chl, with or without valinomycin. After 50 flashes, phosphorylation continues to yield 2.7 nmol ATP/mg Chl (post-illumination phosphorylation, PIP^+) in the absence of valinomycin and 5.8 nmol ATP/mg Chl in its presence. PIP^+ was extrapolated to 200–300 ms after the last flash.

Inset: Comparison of the kinetics of luminescence due to addition of standard ATP (top trace), to the kinetics of the PIP^+ ATP yield (bottom trace). Just the last five flashes of the flash sequence are shown for the PIP^+ experiment. The data were taken from a different experiment than for Fig. 1A and B, but similar conditions to Fig. 1B were used. The indicated time scale was the same for both traces. Three separate experiments were performed, all of which gave very similar data.

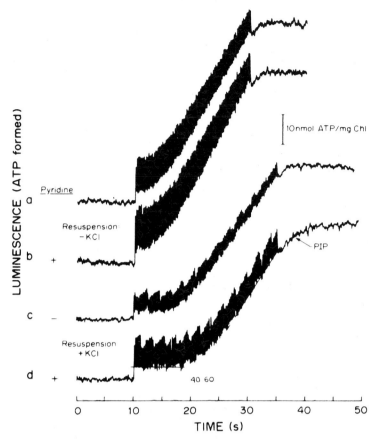

Fig. 2A: The effect of pyridine at pH 8.0 on single-turnover flash initiated phosphorylation with thylakoids stored in the absence or presence of 100 mM KCl. Thylakoids were washed and resuspended in 5 mM Hepes (pH 7.5), 2 mM $MgCl_2$, 0.5 mg/ml bovine serum albumin and 200 mM sucrose (a, b: −KCl or 100 mM KCl in place of the sucrose (c, d: +KCl). Flashes (100 in a,b; 125 in c,d) were delivered at a rate of 5 Hz to thylakoids containing 14 µM chlorophyll suspended at 10°C in 1 ml reaction mixture containing 50 mM Tricine-KOH (pH 8.0), 10 mM sorbitol, 3 mM $MgCl_2$, 1 mM KH_2PO_4, 5 mM DTT, 0.1 mM ADP, 0.1 mM methyl viologen, 400 nM valinomycin, and 5 µM diadenosine pentaphosphate. The reaction mixture was pH adjusted at 10°C with (b,d) or without (a,c) 5 mM pyridine. The vertical spike is the result of a light leak and served as a useful event marker. The different flash-spike heights are an artifactual result of the sampling time of the signal averager. The flash lag for the onset of ATP formation was determined with the aid of two criteria: (i) the first detectable rise in luminescence (45 in d) and (ii) the back extrapolation of the steady rise in the flash-induced luminescence increase to the x-axis (65 in d). The increase in luminescence after the last flash was due to post-illumination phosphorylation.

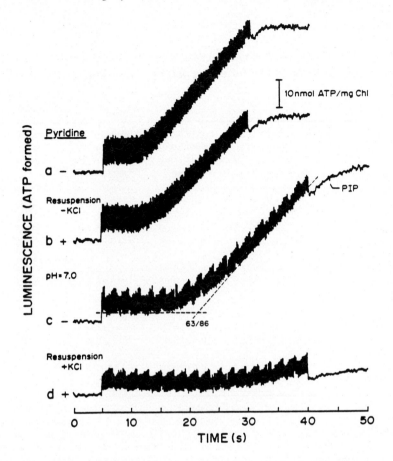

Fig. 2B: The effect of pyridine at pH 7.0 on single-turnover flash initiated phosphorylation with thylakoids resuspended in the absence or presence of 100 mM KCl. Flashes (125 in a, b; 175 in c, d) were delivered at a rate of 5 Hz to thylakoids with 15 μM chlorophyll in a reaction medium containing 50 mM MOPS-KOH (pH 7.0), and other additions as in Fig. 2A.

flash lag number of 10 flashes (Table 1A) at pH 8 and a much greater pyridine-induced lag at pH 7 (20 flashes, Table 1B). The pyridine-dependent increase in the energisation flash number is the predicted effect of a permeable buffer in the thylakoid lumen, interacting with a *de*localised proton gradient (val + K^+ present to keep $\Delta\psi$ suppressed). The high KCl storage medium in some way induced a change, expressed in the subsequent ATP formation assay, allowing the proton gradient to equilibrate with the lumen during the development of the energisation threshold ΔpH.

The much greater energisation lag increase at pH 7 importantly affirms that the effects of the pK 5.44 pyridine buffer are as predicted from theory for the delocalised coupling case for pH external of 7 compared to 8. At pH

TABLE 1

Effect of 100mM KCl or 200 mM sucrose in the thylakoid storage media and pyridine on the onset of ATP formation and PIP$^+$ post-illumination ATP yield

Conditions	Number of flashes to the onset of ATP formation	ATP yield flash [nmol ATP. (mg Chl flash)$^{-1}$]	Post-illumination ATP yield [nmol ATP (mg Chl)$^{-1}$]	Total ATP yield during flash train
A. pH 8				
Sucrose storage				
− pyridine	13±1/21±2	0.61±0.02	4.3±0.4	50±2
+ pyridine	14±1/25±1	0.64±0.08	4.3±0.4	51±6
KCl storage				
− pyridine	30±3/42±3	0.56±0.05	5.4±0.5	48±4
+ pyridine	40±3/65±5	0.53±0.05	9.5±0.4	33±5
B. pH 7				
Sucrose storage				
− pyridine	31±1/38±2	0.71±0.09	7.7±1.0	63±6
+ pyridine	36±3/45±1	0.62±0.06	7.9±0.8	51±5
KCl storage				
− pyridine	66±3/95± 6	0.41±0.03	11.5±0.9	34±4
+ pyridine	86±6/109±7	0.09±0.02	5.4±1.6	7±1

Conditions were as in Fig. 1. The lags for the onset of ATP formation were determined as described in Fig. 1 and represent the actual/extrapolated lags. The ATP yield per flash was determined from the linear rise in luminescence (see Fig. 1). The results are the mean of 4 observations ± standard error (SE). PIP (see text) ATP yield was determined from the increase in signal after the last flash in a flash sequence, while the remaining rise in signal from the onset of ATP formation to the beginning of PIP represents the total ATP yield from the flash train. The results are the means of four observations ± SE [see refs. (13, 15)].

7 outside, the lumenal pH of ≤ 4.7 is calculated to be required to overcome the thermodynamic driving force, ΔG_{ATP}, needed for ATP formation under these conditions, assuming 3 H$^+$ are needed per ATP. For a buffer with a pK of 5.4, about 85% of the buffer capacity must be overcome to reach pH 4.7, whereas at pH 8 outside, only 35% of the pyridine buffering is encountered by pH 5.7, the calculated energetic threshold for that situation. Thus, the data for the high salt-stored thylakoids are exactly as expected for delocalised gradient coupling, but the low salt-stored thylakoids respond as expected for localised coupling.

The PIP$^+$ post-illumination phosphorylation ATP yield data were also completely confirmatory of the above conclusion (Table 1). At pH 8, in the high salt-stored samples pyridine increased the PIP$^+$ ATP yield by 4 nmol (mg chl)$^{-1}$, nearly a doubling, and at pH 7 it *decreased* the PIP$^+$ yield by a factor of five. There was no effect of pyridine on the post-illumination ATP yield in the low salt-stored case (see ref. 12 for more discussion). References (10–15) give details on various experiments designed to check for any membrane perturbations caused by either storage treatment, such as might result in changes in thylakoid volume, H^+ ion permeability, electron or H^+ transort. No trivial effects were noticed which could cause the results.

Earlier work from two other laboratories had reported permeable buffer effects consistent with delocalised coupling, and it is noteworthy that 50 mM or more KCl was used in those experiments for storing the thylakoids (16,17). In the work of Ort et al. (6), which first reported permeable buffer-insensitive ATP formation onset lags, low salt media was used for thylakoid storage.

The question arises whether the high salt or the low salt treatments of the isolated thylakoids caused a type of state change of the membrane, shifting the phosphorylation responses irreversibly to different states. That this was not occurring was shown by washing and suspending high salt-stored thylakoids with the low salt medium. A subsequent phosphorylation assay showed no response to pyridine addition, by either phosphorylation criteria, thus a localised gradient coupling mode was reversibly restored due to the washing treatment (cf. Table IV of ref. 12). The reversibility of the localised-delocalised coupling responses argues in favour of there being a regulatable switch controlling the response, rather than the treatments causing irreversible membrane changes. This will be supported by reversible Ca^{++} ion effects described below. Before discussing the Ca^{++} effects, another aspect should be mentioned of the 'switching' from apparent localised proton gradient coupling to delocalised gradients, which can be observed with the low salt-stored thylakoids.

While the low salt-stored thylakoids showed localised coupling responses in the assays described above, they can also show delocalised coupling under different energisation conditions. That is, when the flash sequence was given in the absence of ADP and Pi and those components added immediately after the last flash (Fig. 3), the 'traditional' post-illumination ATP yield was affected by pyridine exactly as predicted for a delocalised coupling pattern (Table 2; see ref. 11 for details). Our interpretation is that the flash sequence builds up the proton accumulation in membrane-localised domains beyond the capacity reached when ADP and Pi are present, resulting in the 'excess' proton concentration opening a switch into the lumen with subsequent proton accumulation in the lumen. Pyridine present in the lumen would act as a buffer, giving increased proton accumulation during the flash train. We have directly measured the pyridine-dependent increased proton uptake with low salt-stored thylakoids under these conditions during a 5 Hz flash train at pH 8 outside to be 107 nmol H^+ (mg Chl)$^{-1}$ when the minus pyridine control gave 220 nmol H^+ (mg Chl)$^{+1}$ [Table III, ref. 11]. As shown in Table 2, in this

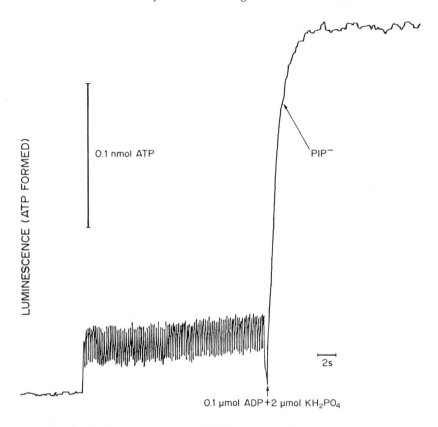

Fig. 3: Post-illumination phosphorylation (PIP⁻) as a result of 100 single-turnover flashes at pH 8.0. Conditions were the same as Fig. 1B except that ADP and P_i were added during the last flash. The Chl content was 13 μg and the final ATP yield was 17.4 nmol ATP/mg Chl. Due to residual ATP present in the ADP, 1.8 nmol ATP/mg Chl was subtracted from the observed PIP⁻ yield.

'traditional' post-illumination phosphorylation experiment at pH 8.0, the addition of ADP and Pi after the last flash gave 11.8 nmol ATP $(mg\ Chl)^{-1}$ for the control and 16.7 nmol ATP $(mg\ Chl)^{-1}$ for the +5 mM pyridine case. At pH 7, pyridine *inhibited* the PIP⁻ATP yield, as expected if the 125 flashes did not drive the internal pH to as low a value when pyridine was present. The traditional post-illumination phosphorylation showing bulk phase protons contributing to the ATP yield is contrasted, using the *same low salt-stored thylakoids*, with the PIP⁺ type post-illumination ATP yield, mentioned above, which was not at all influenced by pyridine (Table 1). As discussed in more detail below we interpret the development of a delocalised gradient, in the traditional post-illumination phosphorylation experiment, as due to excessive H^+ accumulation in the domains during the non-phosphorylating flash sequence causing protons to spill over into

TABLE 2

Effect of pyridine on the traditional post-illumination phosphorylation (PIP⁻) ATP yield[a] in low salt-stored thylakoids

pH	Number of Flashes	Pyridine	PIP^- ATP Yield (nmol ATP/mg Chl)
7	125	−	21±3
		+	14±0.7
8	100	−	11.8±0.4
		+	16.7±0.5

[a] The reaction medium consisted of 50 mM Tricine-KOH (pH 8.0), 10 mM sorbitol, 3 mM $MgCl_2$, 400 nM valinomycin, 0.1 mM methylviologen, 5 mM DTT, with or without 5 mM pyridine and 10 μM diadenosine pentaphosphate. After the last flash, ADP and KH_2PO_4 were added to a final concentration of 0.1 mM and 2 mM, respectively. For the assays at pH 7.50 mM MOPS-KOH replaced Tricine. The pH of the reaction mixture was adjusted at 10°C with and without 5 mM pyridine. The flash rate was 5 Hz. The values reported are the means ± S. E. of three determinations (cf. ref. 11 for details).

the lumen. With the low salt-stored chloroplasts, when ADP and Pi are present during the flash train, the utilisation of some of the H^+ gradient by the phosphorylation process would lead to less acidification in the domains than in the basal ($-$ADP, Pi) case, and the spillover of H^+s into the lumen would not occur.

The reversibility of the localised-delocalised coupling response by the washing treatment and the fact that the low salt-stored thylakoids showed delocalised gradient energy coupling properties in the traditional post-illumination phosphorylation experiment but not in the new, PIP^+, post-illumination mode, are consistent with the hypothesis that a regulatable switch must be present in thylakoids which controls the flux of protons. This set the stage for experiments designed to test for what factor(s) is responsible, in the high KCl-storage effect (and, in the PIP^- experiment with low salt-stored thylakoids, the excess proton accumulation effects) for the reversible switching between localised or delocalised proton gradient energy coupling modes.

Ca^{++} Ions Implicated in a Gating Function

Divalent cations, particularly Ca^{++}, were subsequently shown to be involved in an apparent switching function (14). This is shown in Fig. 4, where it is clear that 1 mM $CaCl_2$ added to the 100 mM KCl storage media (compare Figs. 4C and 4B), had the effect of blocking what would otherwise have been the delocalising action of the KCl and maintaining a localised

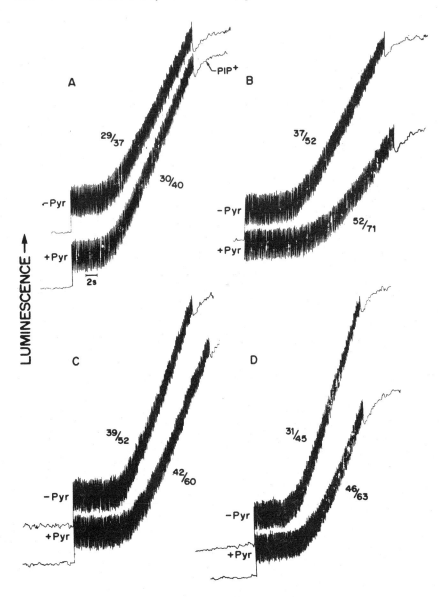

Fig. 4: Effect of ionic composition and EGTA, with and without pyridine, on the energisation lag for ATP formation and on post-illumination phosphorylation. The thylakoid storage treatment, phosphorylation medium and the luciferin-luciferase ATP assay were as described for Figs. 1 and 2. The energisation lag parameters are listed by each trace; i.e., for A, top trace, the top number, 29, is the number of flashes to the first detectable rise in the luminescence signal, and 37 is the flash number estimated by the intersection of the line given by the bottom of rising signal and the extension of the horizontal line defined by the bottom of the flashes occurring during the onset lag. In assays A–D, the top trace was from a sample without pyridine and the bottom trace was from a sample with 5 mM pyridine present, 3.5 min before beginning the flashes. Thylakoid samples were stored in the following media prior to

Table 3

Effect of Salts and Ca^{++} Chelators on Phosphorylation Parameters[a]

Treatments	Energisation Lag Number of Flashes		Lag Difference due to pyridine	PIP^+ ATP Yield nmol ATP $(mg\ Chl)^{-1}$	
	−Pyr	+Pyr	(+Pyr)−(−Pyr)	−Pyr	+Pyr
1. Control	29/37[b]	30/40	1/3	2.8±0.4	3.0±0.5
2. 100 mM KCl	37/52	52/71	15/19	3.3±0.2	6.7±0.2
3. 100 mM KCl 30 mM $CaCl_2$	39/52	42/60	3/8	3.4±0.1	4.7±0.1
4. 100 mM KCl 30 mM $MgCl_2$	32/47	35/53	3/6	2.4±0.2	2.6±0.2
5. 100 mM KCl 30 mM MgCl2 2 mM EGTA	31/45	46/63	15/18	3.7±0.3	7.8±0.7
6. 100 mM KCl 1 mM $CaCl_2$ 2 mM EGTA	37/54	49/71	12/17	4.1±0.2	7.0±0.5
7. 100 mM KCl 30 mM $MgCl_2$ 2 mM EGTA 4 mM $CaCl_2$	41/54	43/62	2/8	3.6±0.2	5.8±0.5
8. low salt 25μM TMB-8	33/43	45/54	12/11	3.9	6.2
9. low salt 25 μM TMB-8 5 mM $CaCl_2$	32/39	34/43	2/4	3.9	3.8

[a] Thylakoids were resuspended after isolation in either the low-salt medium (control), the 100 mM KCl containing medium (high salt) as defined in the text and in the legend of Figure 1, or medium with the addition of the various $CaCl_2$, $MgCl_2$, and EGTA combinations listed for treatments 3–7. The storage conditions and the subsequent dilution into the ATP formation assay medium are given in the text. Where present, pyridine was added at 5 mM 3.5 min prior to starting the 5-H_z flash sequence. The two lag parameters are defined briefly in Figure 1, and in more detail in Beard and Dilley (13), and are the average of at least three independent assays. The PIP^+ ATP yield refers to post-illumination phosphorylation that occurred after the last flash of a sequence during which ATP was formed (ADP and P_i were present from the beginning of the flash sequence).

[b] Average of three or more assays. Standard deviations were ±1 or ±2 in all cases except line 2 (-Pyr), where the data were 52 ± 3.

dilution (approx. 5 μl added to 800 μl of reaction medium) into the phosphorylation medium (identical for all samples): (A) Low salt (control)—200 mM sucrose, 5 mM Hepes-KOH pH 7.5, 2 mM $MgCl_2$ and 0.5 mg ml^{-1} bovine serum albumin. (B) High salt—100 mM KCl and 30 mM sucrose replaced the 200 mM sucrose used in part A. (C) High salt + 1 mM $CaCl_2$—the storage medium was as in B plus 1 mM $CaCl_2$. (D) High salt + 30 mM $MgCl_2$+ 2 mM EGTA—the medium was as in B plus 30 mM $MgCl_2$ and 2 mM EGTA. In (A), the bottom trace identifies the post-illumination phosphorylation ATP yield (PIP^+).

coupling pattern (little or no effect of pyridine on the ATP formation onset lag). 30 mM $MgCl_2$ added to the 100 mM KCl storage medium gave the same effect as addition of 1 mM $CaCl_2$ (Table 3). That Ca^{++} is the agent affected by the 30 mM $MgCl_2$ is suggested by the fact that 2 mM EGTA (a Ca^{++} chelator) blocked the much higher Mg^{++} concentration effect (Fig. 4D). Furthermore, a membrane-permeable Ca^{++} chelator, TMB-8 (3,4,5-trimethoxybenzoic acid 8-dimethylamino octyl ester) added to the *low salt* storage media caused the thylakoids to show pyridine effects on the ATP formation onset lags and post-illumination phosphorylation yield typical of the high salt-stored sample (Table 3, line 8). The proton gradient delocalising effect of TMB-8 was blocked when 5 mM $CaCl_2$ was included in the storage buffer (Table 3, line 9). Table 3 shows statistically significant differences induced by the $MgCl_2$, $CaCl_2$ and chelator influences which switched on or off the pyridine-dependent effects on ATP formation onset lag and on the post-illumination ATP yield parameters.

A model expressing the concept of a Ca^{++}-controlled gate for regulating localised or delocalised proton gradient coupling (14) is shown in Fig. 5. The working hypothesis embodied in the model is not meant to imply that we know specific proteins that interact with the putative 'gate' Ca^{++}; it is drawn as a speculative model, to stimulate the design of new experiments. Carboxyl groups associated with as-yet-unidentified membrane proteins are postulated as binding a Ca^{++} ion in the closed-gate configuration. Either K^+ ions in the 100 mM KCl storage medium, or H^+ ions in the case of high levels of H^+ accumulation from the redox reactions or from a sufficiently acidic storage medium (as in an acid-base phosphorylation protocol), could act to displace the Ca^{++}, allowing protons in the localised domains to equilibrate freely with the lumen (see ref. 14 for details). The model can explain all the effects we have thus far observed, and it resolves the long-standing controversy about whether (and when) *localised* proton coupling occurs in a chloroplast system that can, in certain circumstances, so clearly exhibit *delocalised* coupling. Our experiments show, and the model conveys this notion, that either coupling mode can occur, depending on conditions. For the model to account for the data, the pK of the putative Ca^{++}-binding carboxyl groups is predicted to be below the pH reached in the localised domains under efficient phosphorylating conditions. Blocking ATP formation by withholding ADP, for instance, is predicted to lead to a sufficiently low pH in the domains to protonate the gate carboxyls, thus allowing domain protons to equilibrate with the lumen. The lumen could provide additional carboxyl buffering groups located on lumen-exposed portions of membrane proteins, effectively acting as an additional reservoir for protons that can contribute to energy coupling upon emptying the reservoir. Evidence for there being additional buffering groups in the lumen that contribute to the energetically competent buffering range near pH 5.5 is: (a) the ATP formation onset lag, in the absence of pyridine, is more than doubled in the

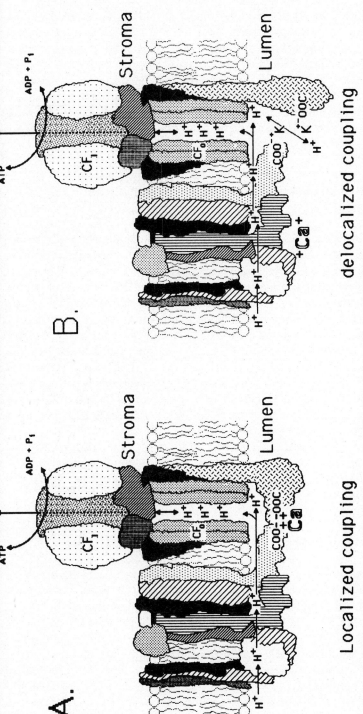

Localized coupling

delocalized coupling

Fig. 5: A model for a possible gating of proton fluxes between localised (A) or delocalised (B) energy coupling gradients. (A) A portion of a thylakoid membrane depicts, in a generalised and as yet speculative way, several intrinsic membrane proteins participating to form a localised proton diffusion domain from the proton releasing reactions in H_2O oxidation and plastoquinol oxidation into the CF_o channel (cf. ref. 2, Fig. 4 and accompanying discussion therein for details). Ca^{++} ions are hypothesised to form a cross-bridge between adjacent protein–COO^- groups to close a gated H^+ channel, although the cross-bridge could also form from tertiary structural parts of one polypeptide. (B) An open H^+ gate is shown owing to the putative Ca^{++} ligand being displaced by K^+ ion, producing an H^+ equilibration pathway between the localised domains and the lumen. The authors thank Dr F. C. T. Allnutt for crafting the model with computer graphics.

high salt-stored thylakoids, understandable if additional carboxyl buffers are exposed to the developing ΔpH (18); (b) with low salt-stored thylakoids the traditional post-illumination ATP yield (–pyridine) that occured after a flash train was also increased nearly twofold at pH 7.0 (outside) compared to pH 8 (Table 2). The latter effect, along with almost a two-fold increase in proton uptake (–pyridine) measured in a 5 Hz flash train at pH 7 compared to pH 8 (Table 3 ref. 11), is consistent with greater lumen carboxyl buffering as the internal pH is driven into the pH $\approx$ 5 range.

Localised-Delocalised H^+ Gradient Gating in Intact Chloroplasts

Intact chloroplast stored in either a 100 mM KCl-containing media or in a low salt sucrose-containing media also show Ca^{++} mediated switching between localised and delocalised responses (19). The intact chloroplasts were osmotically burst in the reaction cuvette immediately before running the ATP formation assay. Fig. 6 shows that the pyridine effects, diagnostic for delocalised proton gradient coupling, occurred in the high salt-stored sample (Fig. 6D, E, pyridine caused a 16 flash increase in the lag) but not in the low salt-stored case (Fig. 6A, B). Fig. 6C and F show that the chloroplasts had remained intact through the 30 min storage steps; i.e., they did not give very much ATP formation, because they were not given the osmotic bursting treatment. This is expected of intact chloroplasts which have been shown to have rather slow ADP/ATP exchange rates across the outer envelope (20).

Fig. 7D, E shows that addition of 5 mM $CaCl_2$ to the intact chloroplast high salt storage medium resulted in the pyridine-induced lag to be decreased to only 4 flashes, similar to the small pyridine effects observed in the low salt case (Fig. 5A, B).

Concluding Remarks

The question arises as to what predicted effect the normally-occurring mono- and di-valent cation concentrations found in intact chloroplasts would have on the coupling mode. Intact chloroplasts normally contain about 15 mM Ca^{++}, 30–35 mM Mg^{++} and 50 mM K^+ ion (21). With such high Mg^{++} and Ca^{++} levels, our results predict that the hypothesised Ca^{++} gate would normally be closed. Indeed, the data of Fig. 6 and 7 indicating localised coupling responses in the intact chloroplasts suspended in low salt media are consistent with this. Adding an additional 100 mM KCl to the intact chloroplast storage medium is apparently sufficient to induce opening of the gate in the phosphorylation assay, allowing pyridine

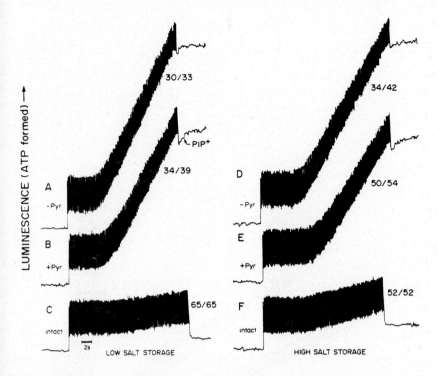

Fig. 6: Effect of pyridine on flash-initiated phosphorylation using intact chloroplasts stored in the absence and presence of 100 mM KCl, and osmotically burst prior to the assay. Chloroplasts were prepared as described in the text. Samples were osmotically burst immediately before the assay by at least a 10-fold dilution into H_2O to give a final concentration of $15\mu g/ml$ in 1 ml total reaction medium containing $10\mu l$ of reconstituted luciferin-luciferase ATP monitoring agent. Samples not burst were stirred in the same reaction medium except that 10 mM sorbitol was replaced by 0.33 M sorbitol. Saturating single-turnover flashes were delivered at a 5Hz repetition rate after 3.5 min dark incubation. The energisation lag parameters were determined as before by the first detectable rise in luminescence and from the back extrapolation of the steady-state rise in signal to the base line. Both parameters are given beside each trace. Top traces A and D were from osmotically burst samples without pyridine. Middle traces B and E were from osmotically burst samples with 5 mM pyridine. Bottom traces C and F refer to non-osmotically burst, intact chloroplasts. Storage media in A, B and C was Low Salt (control): 0.22 M sorbitol, 50 mM Hepes-NaOH (pH 7.5), 2 mM EDTA, 1 mM $MgCl_2$ and 1 mM $MnCl_2$. In D, E and F the storage media was High Salt: medium as above except that 0.33 M sorbitol was replaced by 0.21 M sorbitol and 0.1 M KCl.

buffering effects to be observed. Hence, *in situ*, conditions which would raise the stromal K^+ levels (especially if Mg^{++}, Ca^{++} were lowered) could lead to delocalised H^+ gradient formation. We would predict that electron–proton transport in the intact system occurring in the absence of full coupling to ATP formation, would also shift the proton flux to the lumen through the effect of protonating the carboxyl groups that bind the Ca^{++} ion. Thus, there seems to be a good basis to suggest that chloroplasts

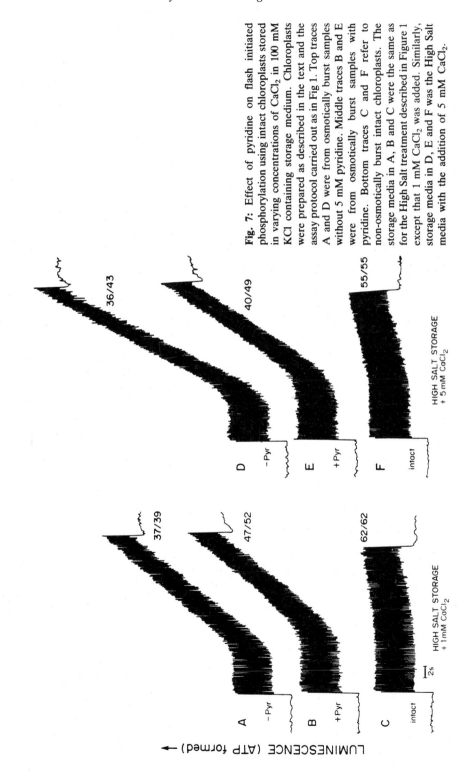

Fig. 7: Effect of pyridine on flash initiated phosphorylation using intact chloroplasts stored in varying concentrations of $CaCl_2$ in 100 mM KCl containing storage medium. Chloroplasts were prepared as described in the text and the assay protocol carried out as in Fig 1. Top traces A and D were from osmotically burst samples without 5 mM pyridine. Middle traces B and E were from osmotically burst samples with pyridine. Bottom traces C and F refer to non-osmotically burst intact chloroplasts. The storage media in A, B and C were the same as for the High Salt treatment described in Figure 1 except that 1 mM $CaCl_2$ was added. Similarly, storage media in D, E and F was the High Salt media with the addition of 5 mM $CaCl_2$.

in the leaf have the capacity to express either localised proton gradient energy coupling or delocalised gradient coupling.

In ref. (2) we discuss possible reasons for a physiological benefit that might accrue to the system by having a localised proton gradient coupling mode. It may be that energising ATP formation via localised gradients would reduce the tendency—that a more acidic lumen may promote—for $H_{in}^{+} \rightleftharpoons$ cation out (K^{+}, Na^{+}) exchanges to occur with subsequent Cl^{-} and water uptake. It is known that such exchange-driven salt uptake can occur in energised thylakoids (18,22,23), and the resulting high amplitude swelling can be deleterious. One way to modulate this sort of osmoregulatory stress would be to maintain proper thylakoid salt content and volume, partly through the ability of the system to keep the energy coupling proton gradients localised as much as possible.

Recent work suggests that one of the proteins involved in the Ca^{++} gating of H^{+} fluxer in the 8 kDa CF_O subunit (G. Chiang and R. A. Dilley, the French Sym. ref.).

Acknowledgements

This work was supported in part by grants from the National Science Foundation (USA) and the United States Department of Agriculture. The authors thank Janet Hollister for excellent assistance in preparation of the manuscript.

References

1. Ferguson, S. J. (1985) *Biochim. Biophys*. Acta 811, 47–95
2. Dilley, R. A., Theg, S. M., and Beard, W. A. (1987) *Annual Rev. Plant Physiol*. 38, 348–389
3. Nagle, J. and Tristram-Nagle, S. (1983) *J. Membr. Biol*. 74, 1–14
4. Cain, B. D. and Simoni, R. D. (1986) *J. Biol. Chem*. 261, 10043–10050
5. Cox, G. B., Fimmel, A. L., Gibson, F., and Hatch, L. (1986) *Biochim Biophys*. Acta 849, 62–69
6. Ort, D. R. Dilley, R. A., and Good, N. E. (1976) *Biochim. Biophys*. Acta 449, 108–124
7. Harold, F. (1977) *Curr. Top. Bioenerg*. 6, 85–149
8. Horner, R. and Moudrianakis, E. N. (1983) *J. Biol. Chem*. 258, 11643–11647
9. Sigalat, C., Haraux, F., de Kouchkovsky, F., Hung, S. P. N., and de Kouchkovsky, Y. (1985) *Biochim. Biophys*. Acta 809, 403–413
10. Beard, W. A. and Dilley, R. A. (1988) *J. Bioenerg, Biomemb*. 20, 85–106
11. Beard, W. A., Chiang, G., and Dilley, R. A. (1988) *J. Bioenerg. Biomemb*. 20, 107–128
12. Beard, W. A. and Dilley, R. A. (1988) *J. Bioenerg. Biomemb*. 20, 129–154
13. Beard, W. A. and Dilley, R. A. (1986) *FEBS Lett*. 201, 57–62
14. Chiang, G. and Dilley, R. A. (1987) *Biochemistry* 26, 4911–4916
15. Beard, W. A. and Dilley, R. A. (1987) *Prog. in Photosyn. Res*. Vol. III, 165–168
16. Vinkler, C., Avron, M. and Boyer, P. D. (1980) *J. Biol. Chem*. 255, 2263–2266
17. Davenport, J. W. and McCarthy, R. E. (1980) *Biochim. Biophys*. Acta 589, 353–357
18. Murakami, S. and Nobel, P. S. (1967) *Plant Cell. Physiol*. 8, 657–671

19. Chiang, G. and Dilley, R. A. (1988) *Plant Physiol.*, submitted
20. Heldt, H. W., Sauer, F., and Rapley, L. (1971) *2nd Intern. Congr. on Photosyn.*, Stresa, 1346–1355
21. Nakatani, H. Y., Barber, J., and Minski, M. J. (1979) *Biochim. Biophys.* Acta 545, 24–35
22. Nishida, K., Tamai, N., and Ryoyama, K. (1966) *Plant Cell Physiol.* 7, 415–428
23. Packer, L., Siegenthaler, P. A., and Nobel, P. S. (1965) *Biochem. Biophys. Res. Comm.* 18, 474–477

Plastoquinone Diffusion in Proteoliposomes Studied by Pyrene Fluorescence Quenching

MARY F. BLACKWELL AND JOHN WHITMARSH

U.S. Department of Agriculture/Agricultural Research Service
Department of Plant Biology
University of Illinois
289 Morrill Hall
505 S. Goodwin Ave.
Urbana IL 61801, USA

Summary

The rate limiting step of linear electron flow in chloroplasts appears to be the oxidation of plastoquinol by the cytochrome b/f complex. Thus, it is important to know whether long-range diffusion of plastoquinone from photosystem II to the cytochrome b/f complex can ever determine the rate of plastoquinol oxidation. Plastoquinone diffusion coefficients in phosphatidylcholine and thylakoid lipid vesicles are of the order $1-2 \times 10^{-7}$ cm^{2}s^{-1} and this would be unlikely to restrict the reaction rate. However, the effect that high (50% or greater) weight percentages of protein occurring in natural membranes would have on lateral diffusion remains unknown. The rate of plastoquinone diffusion was therefore studied in phosphatidylcholine proteoliposomes containing the following integral membrane proteins: gramicidin D, spinach cytochrome b/f complex, reaction centres from *Rhodobacter sphaeroides* and spinach cytochrome f. The rate of plastoquinone diffusion decreased by up to an order of magnitude when the protein composition of proteoliposomes exceeded 30–50 weight per cent. Based on the expectation of a protein composition of 60–65% in thylakoid membranes, we estimate that the plastoquinone diffusion coefficient in thylakoid membranes will be of the order 10^{-8} cm^{2}s^{-1}. In our view, the question remains open as to whether or not plastoquinone diffusion ever limits the rate of electron transport.

Introduction

As the oxidation of plastoquinol by the cytochrome b/f complex is thought to be the rate limiting step of linear electron flow in photosynthesis (1), it has been of interest to determine whether diffusion by plastoquinone from photosystem (PS) II to the cytochrome b/f complex ever determines the oxidation rate (for reviews of this topic, see refs. 2–5). To address this question, one needs to know both the distance that plastoquinone traverses and its rate of diffusion. The relative contributions of plastoquinone and plastocyanin to long-range electron transport and thus the required plastoquinone diffusion pathlength depend upon the distribution of the cytochrome b/f complex relative to those of PS I and II in the chloroplast thylakoid membrane. PS I and PS II appear to be largely segregated from one another, with PS II located primarily in appressed and PS I in non-appressed membrane regions (2–5). Several membrane fractionation and immunochemical studies indicate that cytochrome b/f complex is homogeneously distributed over the thylakoid membrane (5). However, in other fractionation studies the complex co-purifies with PS I and is depleted from appressed membrane fractions (5). Irrespective of the cytochrome b/f spatial distribution, lateral heterogeneity in the distributions of PS I and II raises questions about cytochrome b/f functional heterogeneity and the relative contributions of plastoquinone and plastocyanin to long-range electron transport.

Several publications have addressed the question of the lateral diffusion rate of plastoquinone and proposed diffusion coefficients ranging from 10^{-8} to 10^{-6} cm^{2}s^{-1} (2–4). The diffusion coefficients of both plastoquinone and plastoquinol have been measured in model membranes of soybean phosphatidylcholine (PC) and spinach thylakoid membrane lipids and found to be in the region 1–2×10^{-7} cm^{2}s^{-1} using the probe fluorescence quenching technique (6,7). Ubiquinone diffusion coefficients in model membranes and in mitochondria (8,9) measured by probe fluorescence quenching had similar values to those of plastoquinone after correction for transient dynamic quenching (6). A measurement of the lateral diffusion coefficient of a ubiquinone derivative in mitochondrial membranes using fluorescence photobleaching recovery was about two orders of magnitude lower than the values obtained by probe fluorescence quenching (10). However, the values measured by probe fluorescence quenching are in much closer agreement to values in the region 0.5–1×10^{-7} m^{2}s^{-1} measured for other lipids by fluorescence photobleaching recovery (11–13).

If the diffusion time from PS II to cytochrome b/f is assumed to be 2 ms (2), then plastoquinone can move as far as 300–400 nm with a diffusion coefficient of the order 1–2×10^{-7} cm^{2}s^{-1} according to the equation, $r = 2(Dt)^{1/2}$, where r is the two-dimensional root mean square displacement in time t and D is the diffusion coefficient (4). As this is roughly the magnitude of the longest estimated diffusion pathlength in thylakoids (2–5), i.e. for long-range diffusion from appressed to stromal regions, it would seem unlikely that plastoquinone diffusion normally poses any limitation on the rate of linear electron flow under normal conditions. However, the previous diffusion coefficient measurements were carried out in liposomes containing no integral membrane proteins and the high (60–65%) weight percentage of integral membrane protein estimated for thylakoid membranes (5) may significantly affect the plastoquinone lateral diffusion rate. Two studies of lipid or integral membrane protein diffusion in proteoliposomes have found that lateral diffusion coefficients are decreased by up to factor of ten at high protein concentration (11,12). Recent theoretical work describing diffusion in membranes (14–18) suggests a possible profound effect of high protein concentrations on diffusion rates. We have therefore measured plastoquinone diffusion coefficients by the probe fluorescence quenching technique in the region of 0–70 weight per cent protein for several integral membrane proteins in PC proteoliposomes.

Materials and Methods

Pyrene, soybean PC (type III-S), sodium cholate, β-octyl glucoside, gramicidin D and spinach cytochrome f were used as purchased from Sigma (St. Louis, MO). MEGA-9 detergent was synthesized by a published procedure (19). Plastoquinone-9 was a gift from Hoffman-La Roche (Basel, Switzerland). *Rhodobacter sphaeroides* R-26 reaction centers were a gift from Dr Colin Wraight. Cytochrome b/f complex was isolated from spinach by the method of Hauska (19).

Plastoquinone diffusion coefficients were estimated by the pyrene fluorescence quenching method (6), which involves measurement of steady-state pyrene fluorescence emission intensities at several plastoquinone concentrations. In preliminary experiments, we determined that plastoquinone-9 could not be incorporated into liposomes or proteoliposomes by the addition of aliquots of ethanolic stock solutions to aqueous suspensions of preformed membranes in the absence of detergent. The construction of proteoliposomes containing plastoquinone was therefore carried out either by addition of ethanolic plastoquinone to protein/lipid/ pyrene suspensions containing at least 1% detergent (method A) or by fusion of liposomes containing plastoquinone with preformed proteoliposomes (method B), as follows.

Method A. Detergent Dilution. For each weight percentage of protein, appropriate volumes of pyrene (0.124 mg/ml) and PC (100 mg/ml) in chloroform were combined in a test tube and dried to a film under a stream of nitrogen. The required amount of protein was added with buffer and sufficient detergent (usually 4–5% MEGA-9 or β-octyl glucoside, from 20% (w/v) stock solutions) to produce a clear suspension. The mixtures were dialysed against 500 volumes of 1% sodium cholate. Each dialysate was divided into 120μl aliquots in glass tubes and an appropriate volume of ethanolic plastoquinone was added. The samples were then diluted to 2 ml with water, resulting in a cholate concentration (1.4 mM) an order of magnitude lower than the critical micellar concentration (21). The final concentrations were 0.1 or 0.2 mg/ml lipid and 0.3–0.5 μM pyrene. The final ethanol concentration was less than 2%. At this detergent concentration, the membranes should be in the form of unilamellar bilayer vesicles (22) with partition of some detergent into the bilayer phase depending upon the lipid/water partition coefficient. A reasonable estimate for the combination of 1.4 mM cholate and 0.1–0.2 mg/ml PC is a lipid/detergent molar ratio in the bilayer phase of 5–10, assuming a lipid density of 1 g/ml and a lipid/water partition coefficient for cholate similar to that reported for glycocholate in egg PC, 100–200 mM/ml lipid bilayer per mM/ml of water (22). The plastoquinone diffusion coefficients were the same within experimental error as those measured in PC vesicles containing no detergent.

Method B. Detergent dialysis. Proteoliposomes were formed by dialysis of lipid/pyrene/protein mixtures, prepared as described in method A but with 20% less lipid, against 500 volumes of 10 mM Tris-HCl (pH 8, 5°C), with one change of dialysis buffer. Plastoquinone was incorporated with the remaining 20% lipid as follows. Plastoquinone-containing liposomes were prepared by drying appropriate amounts of lipid and plastoquinone in organic stock solutions under a stream of nitrogen followed by dispersal into distilled water by ultrasonic irradiation (10 s at 26 microns using a Gallenkamp Soniprep 150 probe sonicator). The plastoquinone containing liposomes were fused with the proteoliposomes by ultrasonic irradiation (10 s at 26 microns). That this technique resulted in fusion of plastoquinone liposomes with proteoliposomes was confirmed by the observation of pyrene fluorescence quenching to a similar degree as in the detergent dilution method or in control samples in which plastoquinone, lipid and pyrene were combined together from organic solutions prior to removal of solvent and dispersion into aqueous solution. The lipid concentration was 0.1 or 0.2 mg/ml at all protein concentrations. Weight percentages of protein were calculated as

$$f_p = \frac{m_p}{m_p + m_l} \times 100 \tag{1}$$

with m_p and m_l the protein and lipid mass concentrations in units of mg/ml, respectively. Protein concentrations were determined by the method of Lowry (23). Plastoquinone concentrations, determined in the organic stock solutions by spectrophotometric assay using Crane's extinction coefficients (24), were estimated in the membrane lipid phase as nanomoles per microgram lipid, assuming a lipid density of 1 g/ml as described previously (25).

Steady-state measurements of pyrene fluorescence were carried out on an SLM-Aminco SPF-500 spectrofluorometer at room temperature (22–24°C). Emission was measured at 390 nm (bandpass 10–20 nm) with excitation at 330 nm (bandpass 1–4 nm). A blank value of fluorescence from the measurement cuvette containing distilled water or buffer, typically less than 1–2%, was subtracted. Fluorescence kinetics were measured at 390 nm with excitation at 330 nm on a PRA single photon counting nsec fluorimeter in the laboratory of Dr L. Faulkner, at the

University of Illinois at Urbana, Illinois. The measurements and subsequent nonlinear least-squares kinetic analysis were otherwise as described previously (25).

Results and Discussion

In fluorescence quenching studies, steady-state probe fluorescence emission ratios, I^0/I, are obtained as a function of the concentration of a quencher, where I^0 and I are the respective intensities in the absence and presence of quencher. Figure 1 shows results obtained with pyrene as the probe and plastoquinone as the quencher in PC proteoliposomes containing *Rb. sphaeroides* R-26 reaction centres. The dependence of the fluorescence emission ratio I^0/I on the plastoquinone concentration is shown for control liposome samples containing no protein (Fig. 1a) and in proteoliposome samples containing 35 and 65 weight per cent protein

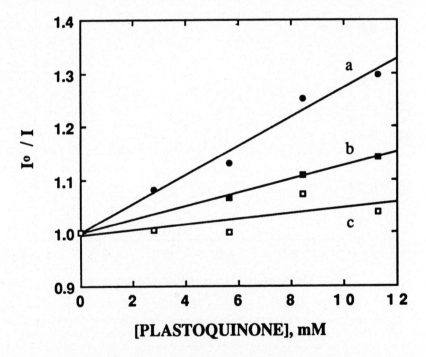

Fig 1: The effect of protein on the quenching of pyrene fluorescence by plastoquinone in soybean phophatidylcholine proteoliposomes containing *Rhodobacter sphaeroides* reaction centres. Steady-state pyrene fluorescence emission ratios, I^0/I, as a function of the plastoquinone concentration at the following protein weight percentages: 0 per cent (a), 35 per cent (b) and 65 per cent (c). I^0 and I are the pyrene emission intensities at 390 nm in the absence and presence of plastoquinone, respectively. The proteoliposomes were prepared by detergent dialysis (method B in the text).

(Figs. 1b and 1c, respectively). The slope of the plots is decreased to 44% of the control value at 35 weight per cent protein and 18% of the control value at 65 weight per cent protein, indicating a decrease in the efficiency of pyrene fluorescence quenching by plastoquinone in the presence of integral membrane proteins. This decrease most likely corresponds to a lower frequency of pyrene/plastoquinone collisions arising from a decrease in the rate of lateral diffusion of plastoquinone. The quenching studies were carried out at plastoquinone concentrations sufficiently low to avoid saturation effects observed in PC liposomes above about 10 mM (6).

The lateral diffusion coefficients can be estimated from the slopes of plots of I^0/I vs. the plastoquinone concentration if the probe fluorescence kinetic parameters are known (6). If m and m_0 are the respective slopes in the presence and absence of integral membrane proteins, the ratio m/m_0

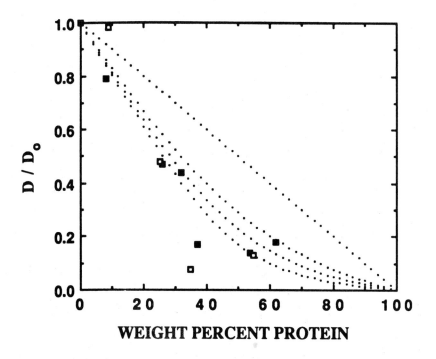

WEIGHT PERCENT PROTEIN

Fig. 2: The dependence of the rate of plastoquinone diffusion on the weight per cent protein in soybean phosphatidylcholine proteoliposome containing *Rhodobacter sphaeroides* reaction centres in: (open symbols) proteoliposomes prepared by the detergent dilution method (method A in the text); and (filled symbols) proteoliposomes prepared by the detergent dialysis method (method B in text). The solid curves are calculated from a theoretical model (Eq. 2) of diffusion in the presence of mobile proteins for the cases: (top curve) the diffusion coefficient at time zero and (lower three curves, top to bottom) the steady-state diffusion coefficient when the plastoquinone/protein diffusion coefficient ratio is 2, 4 or 8. D and D_0 are the diffusion coefficients in the presence and absence of protein, respectively. Other details are described in the text.

will differ from the corresponding ratio of diffusion coefficients, D/D_0, as a result of variability in the pyrene fluorescence kinetic parameters used to calculate D from m (6). However, we have determined that D/D_0 and m/m_0 agree to within 10% in the systems we have studied and thus for convenience we will assume that $D/D_0 = m/m_0$. Figures 2–5 show the dependence of the plastoquinone lateral diffusion rate on the protein concentration in PC proteoliposomes with several different integral membrane proteins: *Rb. sphaeroides* R-26 reaction centres (Fig. 2), spinach cytochrome b/f complex (Fig. 3), spinach cytochrome f (Fig. 4), and gramicidin D (Fig. 5). In all cases the lateral diffusion rate is decreased in the presence of more than 10–20 weight per cent protein. As indicated by the data in Figs. 2 and 5, we did not observe any obvious difference between the two protein incorporation procedures.

The plastoquinone diffusion rate in proteoliposomes containing the larger integral membrane proteins, R-26 reaction centre and cytochrome b/f complex, is decreased by a factor of 5–10 above about 50 weight per cent protein. This suggests that in thylakoid membranes, where the protein composition is estimated to be 60–65 weight per cent (5), the plasto-

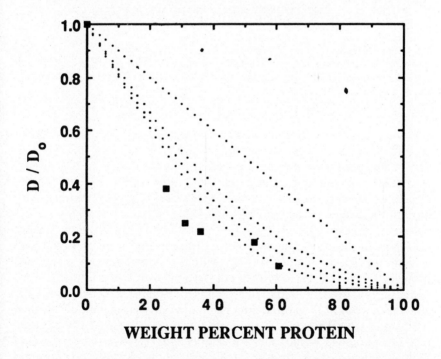

Fig. 3: The dependence of the rate of plastoquinone diffusion on the weight per cent protein in soybean phosphatidylcholine proteoliposomes containing cytochrome b/f complex isolated from spinach chloroplasts in proteoliposomes prepared by the detergent dialysis method (method B in text). Other details are as described for Fig. 2.

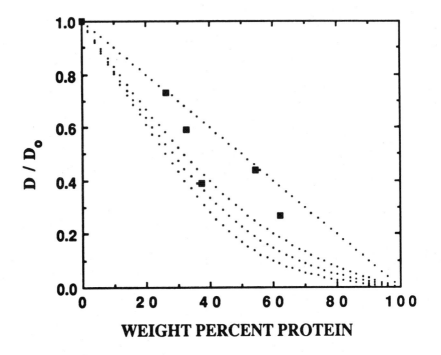

Fig. 4: The dependence of the rate of plastoquinone diffusion on the weight per cent protein in soybean phosphatidylcholine proteoliposomes containing spinach cytochrome f complex in proteoliposomes prepared by the detergent dialysis method (method B in text). Other details are as described for Fig. 2.

quinone diffusion coefficient could be decreased by as much as an order of magnitude below its value in pure lipid vesicles, i.e. to $1-2 \times 10^{-8} \, cm^2 s^{-1}$. The results presented here are in agreement with observations of similar large decreases in lipid lateral diffusion rates in fluorescence photo-bleaching recovery experiments on dimyristoyl-PC proteoliposomes with bacteriorhodopsin (11) or gramicidin (12).

As can be seen by comparing Figs. 1 and 2 to Figs. 3 and 4, smaller decreases in the plastoquinone lateral diffusion rate were observed with cytochrome f and gramicidin than with the larger integral membrane proteins. It is of interest to compare the latter two proteoliposome systems, in that cytochrome f has a single bilayer-spanning α-helix per 34 kD molecular weight (5) whereas gramicidin has one transmembrane channel per 3.4 kD (26). Thus, one would expect a larger decrease in the diffusion rate at a particular weight per cent of gramicidin relative to cytochrome f than is evident in Figs. 4 and 5. However, the gramicidin proteolipsomes were apparently not a uniform suspension of bilayer vesicles. Above about 30 weight per cent protein, the gramicidin proteoliposome suspensions contained a flocculent precipitate and the membranes did not appear to be

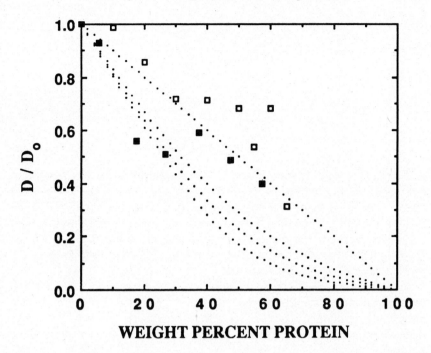

Fig. 5: The dependence of the rate of plastoquinone diffusion on the weight per cent protein in soybean phosphatidylcholine proteolipsomes containing Gramicidin D in: (open symbols) proteoliposomes prepared by the detergent dilution method (method A in the text); and (filled symbols) proteoliposomes prepared by the detergent dialysis method (method B in text). Other details are as described for Fig. 2.

uniformly dispersed as unilamellar bilayer vesicles in preliminary negative staining electron microscopy studies (data not shown), although X-ray diffraction studies (27) indicated a bilayer structure in dipalmitoyl PC proteoliposomes at as high as 50 weight per cent (33 mole per cent) gramicidin. Thus we believe that the weaker than expected concentration dependence of the diffusion rate in gramicidin proteoliposomes results from a lack of incorporation of gramicidin in excess of about 20–30 weight per cent.

Theoretical and/or numerical analyses (14–17) all predict a decrease in the lateral diffusion rates of lipids and proteins with increasing protein concentration in proteoliposomes (assuming no channelling (14)). When diffusion in proteoliposomes is modeled as a random walk process on a square point lattice (16,18), the dependence of D on the protein concentration is of the form

$$D^*(c,\gamma) = (1-c) \ F(c,\gamma) \tag{2}$$

with

$$F(c,\gamma) = \frac{\{[(1-\gamma)\ (1-c)\ f+c]^2+4\gamma\ (1-c)f^2\}^{1/2}-[(1-\gamma)\ (1-c)\ f+c]}{2\gamma(1-c)f}$$

where

$$f = \frac{[1-\alpha]}{[1+\ (2\gamma-1)\alpha]}$$

In Eq. 2, $D^* = D/D_0$ with D and D_0 the lateral diffusion coefficients of plastoquinone in the presence and absence of protein, c is the bilayer area fraction occupied by protein, α is a geometrical constant equal to 0.36338 and γ is the ratio of the lipid to the protein random walk jump frequencies (16). However, for the present analysis it will be assumed that γ can be represented as the diffusion coefficient ratio, $\gamma=D/D_p$, in which D_p is the protein lateral diffusion coefficient.

As pointed out by Peters and Cherry (11), lipid diffusion coefficients in proteoliposomes should be 1–2 times larger than those of typical membrane proteins. They based their argument on the weak size dependence for lateral diffusion coefficients in membranes derived by Saffman (29),

$$D = \frac{kT}{4\pi\eta h}\ (\ln \frac{\eta h}{\eta_w a} - \gamma_E) \tag{3}$$

with the diffusant represented as a cylinder of radius a spanning a membrane of thickness h and viscosity η; η_w is the viscosity of the suspending aqueous phase and γ_E is Euler's constant (0.5772). Peters and Cherry found that Eq. 3 is consistent with fluorescence photobleaching recovery measurements of both lipid and protein lateral diffusion in model systems and suggested that it should be valid in other membrane systems in which protein diffusion is not restricted by other factors. While the lateral diffusion coefficients of proteins measured by fluorescence photobleaching recovery *in vivo* in a number of biological membranes are 10–100 times lower than the values predicted by Eq.3, as discussed by Cherry (30), this is probably due to restriction of protein mobility by interactions with cytoskeletal structures. Vaz and co-workers (13) concluded from fluorescence photobleaching recovery studied of lipid probes as a function of fatty acid chain length that the dependence on h in Eq. 3 is not obeyed by lipids, but this should not effect the weak dependence on a in Eq. 3. Based on these considerations, we estimate that the ratio of protein/plastoquinone diffusion coefficients, γ in Eq. 2, will be of the order 1–8 for the proteolipsomes studied here if the diffusion coefficient of plastoquinone is 1–4 times larger than typical lipid diffusion coefficients. Eq. 2 is plotted in Figs. 2–5 for $\gamma = 0$ (or t=0 as discussed below) and for $\gamma=2,4$ and 8. For the purpose of plotting theoretical curves in Figs. 2–5, c in Eq. 2 is considered to be equal to the protein weight fraction. The results for *Rb. sphaeroides* reaction centre and cytochrome b/f complex (Figs. 2 and 3) appear to agree reasonably well with the behaviour predicted by Eq. 2 plotted on a protein weight per cent basis. However, the agreement with data for cytochrome f and gramicidin proteoliposomes appears to be less satisfactory. Saxton (16) found that the theoretical model failed to describe the concentration dependence of lipid diffusion coefficients in proteoliposomes of bacteriorhodopsin (11) and gramicidin (12). Thus it is possible that the apparent agreement of the data in Fig. 2 and 3 with Eq. 2 is fortuitous. There are two likely causes of disagreement with the theoretical model in Eq. 2. The first includes lipid/protein interactions that can influence lateral diffusion but are not included in Eq. 2 (14,16). The second is the crudeness of the approximation of c in Eq. 2 as the weight fraction of protein. The concentration c in Eq. 2 should properly be defined as the area fraction of membrane space occupied by protein,

$$c = \frac{a_p n_p}{a_L n_L + a_P n_P} \tag{4}$$

with a_P or a_L the areas of membrane occupied by individual protein or lipid molecules and n_P or n_L the molar concentration of protein or lipid, respectively. An earlier representation of c as the weight fraction of protein (14, 28) was convenient in that the experimentally measureable quantities are lipid and protein concentrations. The area fraction occupied by a particular membrane protein will depend on the number of lipid molecules it displaces and therefore on the percentage of protein that is hydrophobic and in the form of membrane-spanning α-helices. While protein radii can be estimated from structural data obtained by electron microscopy, X-ray crystallography or primary amino acid sequences, there is presently no direct experimental measurement of c and we are currently devising suitable approximations.

D* in Eq. 2 has been found to be time-dependent (14–16), taking 5–10 diffusive steps to fall from its value at time zero

$$D^*_{(t=0)} = 1 - c \tag{5}$$

to the steady-state value in Eq. 2(4). This could be an important factor in measurements of diffusion using the fluorescence quenching technique if the probe fluorescence lifetime is short enough to measure D* before it decays to its steady-state value. This may be the reason for plastoquinone and ubiquinone diffusion coefficients measured by fluorescence quenching techniques, $1-2 \times 10^{-7}$ cm^2 s^{-1} (6–9), being higher than those measured by fluorescence photobleaching recovery for lipids, $0.5-1 \times 10^{-8}$ cm^2/s (11–13). According to the equation

$$D = \omega \lambda^2 / 4 \tag{6}$$

with λ the lattice spacing and ω the jump frequency (16), plastoquionone can undergo 5–10 diffusive hops during the pyrene lifetime (approx. 100 ns (6,26)). This and the fact that the data points in Figs. 2–5 fall generally below the theoretical curve for D^* at $t=0$ suggest that pyrene fluorescence quenching measurements of diffusion coefficients are reasonably close to their steady-state values.

Conclusion

The results presented here suggest that the rate of plastoquinone lateral diffusion is strongly dependent on the protein composition of the membrane, decreasing from the value of $1-2 \times 10^{-7}$ cm^{2}s^{-1} in lipid vesicles up to an order of magnitude in the region of 40–60 weight per cent protein. As the thylakoid membrane is estimated to be roughly 60–65 weight per cent protein (5), we estimate the plastoquinone diffusion coefficient in thylakoid membranes to be of the order 10^{-8} cm^{2}s^{-1} or less. The time available for plastoquinone diffusion from PS II to the cytochrome b/f complex is thought to be about 2–3 ms (2–5). If so, then according to the Einstein-Smoluchowski equation, $r=2(Dt)^{1/2}$, the diffusion pathlength could well be lower than 150 nm. Thus plastoquinone diffusion could become rate-determining in photosynthesis if electron transport is required over comparable or longer distances.

Acknowledgements

This work was supported in part by the Photosynthesis Program of the Competitive Grants Office of the U.S. Department of Agriculture (Grant No. AG86-CRCR-1-1987).

References

1. Stiehl, H. H. and Witt, H. T. (1969) *Z. Naturforsch.* 246:1588–1598
2. Haehnel, W. (1984) *Ann. Rev. Plant Physiol.* 35:659–693
3. Millner, P.A. and Barber, J. (1984) *FEBS Lett.* 124:62–66
4. Whitmarsh, J. (1986) *Photosynthesis III*, Vol. 19, *Encyclopedia of Plant Physiology* (L. A. Staehelin and C. Arntzen, editors). Springer-Verlag GmbH & Co. Heidelberg. 505–525
5. Murphy, D. J. (1986) *Biochim. Biophys.* Acta 864:33–94
6. Blackwell, M. F., Gounaris, K., Zara, S. J. and Barber, J. (1987) *Biophys. J.* 51:735–744
7. Blackwell, M. F., Gounaris, K. and Barber, J. (1987) In: *Proceedings VII International Congress on Photosynthesis* (J. Biggins, editor). Martinus Nijhoff, The Netherlands. Vol. 2, 501–504
8. Fato, R., Battino, M., Parenti Castelli, G. and Lenaz, G. (1985) *FEBS Lett.* 179: 238–242

9. Fato, R., Battino, M., Degli Esposti, M., Parenti Castelli, G. and Lenaz, G. (1986) *Biochem.* 25:3378–3390
10. Gupte, S., E.-S., Hoechli, L., Hoechli, M., Jacobson., K., Sowers, A. E. and Hackenbrock, C. R. (1984) *Proc. Nat. Acad. Sci.* USA 81:2606–2610
11. Peters, R. and Cherry, R. J. (1982) *Proc. Nat. Acad. Sci.* USA 79:4317–4321
12. Tank, D. W., Wu., E.-S., Meers, P. R. and Webb, W. W. (1982) *Biophys. J.* 40:129–135
13. Vaz, W. L. C., Goodsaid-Zalduondo, F. and Jacobson, K. (1984) *FEBS Lett.*179:199–207
14. Saxton, M. J. (1982) *Biophys. J.* 39:165–173
15. Saxton, M. J. (1987) *Biophys. J.* 51:542a
16. Saxton, M. J. (1987) *Biophys. J.* 52:989–997
17. O'Leary, T.J. (1987) *Proc. Nat. Acad. Sci.* USA 84:429–433
18. Van Beijeren, H. and Kutner, R. (1985) *Phys. Rev. Lett.* 55:238–241
19. Hildreth, J. E. K. (1982) *Biochem. J.*207:363–366
20. Hauska, G. (1986) *Meth. Enzymol.* 126:271–285
21. Helenius, A. and Simon, K. (1975) *Biochim. Biophys.* Acta 415:29–79
22. Lichtenberg, D. (1985) *Biochim. Biophys.* Acta 821:470–478
23. Lowry, O. H., Rosebrough, N. J., Farr, A. L. and Randall, R. J. (1951) *J. Biol. Chem.* 193:265–275
24. Crane, F. L. (1959) *Plant Physiol.* 34:546–551
25. Blackwell, M. F., Gounaris, K. and Barber, J. (1986) *Biochim. Biophys.* Acta 858:221–234
26. Urry, D. W., Trapane, I. L. and Prasad, W. U. (1983) *Science* 221:1064–1067
27. Chapman, D., Cornell, B. A., Eliasz, A. W. and Perry, A. (1977) *J. Mol. Biol.* 113:517–538
28. Pink, D. A., Georgallas, A. and Chapman, D. (1981) *Biochem.* 20:7152–7157
29. Saffman, P. G. (1976) *J. Fluid Mech.* 73:593–602
30. Cherry, R. J. (1979) *Biochim. Biophys.* Acta 559:289–327

The Involvement of Sulphydryl Group of Coupling Factor 1 (CF$_1$) in Energy Transduction in Chloroplast

JAI PARKASH AND G. S. SINGHAL

School of Life Sciences
Jawaharlal Nehru University
New Delhi 110 067, INDIA

Summary

Both the control and chemically modified spinach thylakoid membranes were labelled with [^{14}C]-ADP to study the energy dependent release of adenine nucleotide from coupling factor 1 (CF$_1$). The photophosphorylation coupled to electron transport from water to methyl viologen was carried out under the same conditions as used for light-induced release of [^{14}C]-ADP, except that cold ADP was present. The alkylation of the -SH group of salt-washed thylakoid membrane by n-ethylmaleiimide (NEM) resulted in a 17 per cent decrease in the energy dependent release of [^{14}C]-ADP and a 53 per cent decrease in the rate of photophosphorylation. Dithiothreitol (DTT) could not reverse the inhibition of these activities caused by NEM. On treatment of NEM-modified chloroplasts with increasing concentrations of iodosobenzoate (IBZ), a further decline in both the energy-dependent release of [^{14}C]-ADP and ATP synthesis rate was observed. However, DTT could reverse the inhibition caused by IBZ. Our data show that thiol modulation of CF$_1$ alters the activation process and consequently the catalytic function of CF$_1$ in chloroplast.

Introduction

It was shown by McCarty and coworkers (1,2) that alkylation with n-ethylmaleiimide (NEM) of the -SH group of CF$_1$, exposed during illumination of chloroplast caused inhibition of photophosphorylation and Ca^{++}-dependent ATPase activity. The inhibition of these activities by NEM was sensitive to uncouplers and adenine nucleotides. ^{3}H-NEM incorporation into CF$_1$ labelled primarily the -SH group of γ-subunit of CF$_1$. However, the electron transport, light-induced H$^+$-uptake and electrochemical gradient of protons ($\Delta\mu_H+$) were not affected by NEM, indicating that inhibition of phosphorylation was a kind of energy transfer inhibition (3,4).

Similarly, the oxidation of the -SH group of CF$_1$, which gets exposed during illumination of chloroplast, by iodosobenzoate (IBZ) or by dithiobisnitropyridine (DTNP) caused inhibition of ATP synthesis and hydrolysis. However, dithiothreitol (DTT) could reverse the inhibition caused by IBZ or DTNP (5,6). Bifunctional maleiimides, e.g., O-phenylendimaleiimide have been shown to cross-link the -SH group within the γ-subunit of membrane-bound CF_1 resulting in inhibition of photophosphorylation (7,8). The inhibition of photophosphorylation by bis-maleiimides could be prevented by uncouplers or ATP and P_i. Based on these observations, it has been proposed that when the -SH group of the γ-subunit of CF_1 is cross-linked or oxidised to disulphide bond, the thylakoid membrane becomes leaky to H^+-ions, resulting in uncoupling of electron transport from photophosphorylation (9). However, when the -SH group (exposed during high-energy state) of the γ-subunit is blocked by a monofunctional reagent, an energy transfer inhibiton is observed (10).

IBZ is known to inhibit the dithiothreitol(DTT)-activated CF_1 (11). However, the inactivation of CF_1 by IBZ can be prevented if the DTT-treated CF_1 is alkylated with NEM prior to its treatment with IBZ. The activation of soluble CF_1 by DTT has also been studied by incorporation into the γ-subunit of CF_1 the fluorescent derivative of maleiimides (12). The Pi-ATP exchange reaction in chloroplasts was inhibited both by IBZ and uncouplers (13).

The reduced form of CF_1 in thylakoid membranes is the active form for photophosphorylation both in leaves and in intact chloroplasts (14,15). The endogenous reduction of disulphide linkage could be carried out by thioredoxin (16) or by other dithiols (17).

Thiol modulation of the CF_0-CF_1 complex by DTT lowers the threshold ΔpH at which ATP synthesis takes place and stimulates the overall yield of ATP synthesis at limiting ΔpH (14,18). It also results in energy-dependent release of ADP at significantly lower ΔpH. It seems that thiol modulation of the CF_0-CF_1 complex increases the fraction of active CF_0-CF_1 complex by decreasing the ΔpH for activation.

In this article, we report our results on the effect of thiol modulation of salt-washed thylakoid membranes on energy-dependent release of [^{14}C]-ADP and photophosphorylation. Our data show that oxidation of the -SH group of CF_1 exposed during the high-energy state with IBZ causes inhibition of both the light-induced release of [^{14}C]-ADP and photophosphorylation. Re-reduction of the oxidised -SH group by DTT resulted in reversal of inhibition. It is therefore suggested that thiol modulation of CF_1 affects both the activation and catalytic processes in CF_1.

Materials and Methods

Preparation of salt-washed thylakoid membranes: Salt-washed thylakoid membranes from spinach were prepared and suspended in 50 mM NaCl, 1 mM $MgCl_2$ and 2 mM Tricine, pH 7.8 to a chlorophyll (Ch1) concentration of 2 mg/ml (19).

Modification of chloroplasts with NEM in dark: The chloroplasts were resuspended in a medium containing 50 mM Tricine, pH 8.0, 5 mM $MgCl_2$, 50 mM NaCl and 1.23 mM NEM to a Chl concentration of 0.33 mg/ml in a final volume of 3 ml. The samples were placed in the dark at 25°C for 5 min. The samples were then centrifuged at 5000×g for 5 min in the SM-24 rotor of an RC-5 Sorvall Centrifuge (Sorvall Instruments, E. I. du Pont, USA). The pellet was washed twice with a medium containing 0.4 M sucrose, 2 mM Tricine, pH 8.00, 10 mM NaCl and 0.1 per cent (w/v) bovine serum albumin and suspended in the same medium to a Chl concentration of 2 mg/ml.

Modification of NEM-treated thylakoid membranes with IBZ: The control and NEM-treated thylakoid membranes were centrifuged at 10000×g for 10 min at 0°C and the pellets were resuspended in a reaction mixture containing 50 mM Tricine, pH 8.0, 50 mM NaCl, 5 mM $MgCl_2$, 50μM phenazine methosulphate (PMS) and various concentrations of IBZ (0–10 mM) in a final volume of 3 ml and Chl concentration of 0.33 mg/ml. The samples were immediately illuminated for 1.5 min with white actinic light provided by a 300-watt Kodak projector lamp. The light intensity at the centre of the reaction vessel was 290 Joules/m^2/sec. The samples were centrifuged at 5000×g for 5 min at 0°C in the SM-24 rotor of an RC-5 Sorvall Centrifuge. The pellet was washed twice with the ice-cold medium containing 0.4 M sucrose, 2 mM Tricine, pH 8.0 and 10 mM NaCl and resuspended in the same medium. The addition of DTT (final concentration = 12.66 mM) wherever required was carried out just before switching off the illumination.

Prelabelling with [^{14}C]-ADP and light-induced release of CF_1 bound [^{14}C]-ADP: The prelabelling of control and chemically modified thylakoid membranes was carried out as described in ref. 20. The Chl concentration was 0.5 mg/ml. Similarly, energy dependent release of [^{14}C]-ADP bound to CF_1 was carried out according to Shavit and Strotmann (20).

Photophosphorylation: The non-cyclic photophosphorylation coupled to electron transport from water to methyl viologen, both in control and in -SH-modified thylakoid membranes, was carried out according to ref. 21. However, in place of [^{14}C]-ADP cold ADP was used.

Sources of chemicals: PMS, IBZ and DTT were purchased from Sigma Chemical Co., USA. NEM was purchased from Calbiochem, California, USA. [^{14}C]-ADP was obtained from Amersham International Limited (UK).

Results

Treatment of salt-washed thylakoid membranes with 1.2 mM NEM in darkness resulted in 16 and 53 per cent inhibition in the rates of energy-dependent release of [^{14}C]-ADP and photophosphorylation respec-

TABLE 1

	Per cent Inhibition	
Preparation	$[^{14}C]$-ADP Release (1)	Photophosphorylation (2)
Control	–	–
,,,,, + 1.2 mM NEM (in dark)	16	53
,,,,, + ,,,,, + 12.66 mM DTT	29	59
Control + 4 mM IBZ (in light)	29	60
,,,,, + ,,,,, + 12.66 mM DTT	5	20

Control Activities
(1) 0.42 ± 0.05 nanomoles ADP released/ mg Chl
(2) $156.3 \pm 12 \mu moles$ ATP formed/mg Chl/hr

tively (Table 1). The addition of 12.66 mM DTT to the assay mixture did not reverse the inhibition caused by NEM (see Table 1, second row).

Similarly, treatment of salt-washed thylakoid membranes with 4 mM IBZ, in light, resulted in 29 and 60 per cent inhibition in the rates of energy-dependent release of $[^{14}C]$-ADP and photophosphorylation respectively (Table 1, third row). However, when 12.66 mM DTT was added just before putting off the illumination the maximum inhibitions of these activities were 5 and 20 per cent respectively (Table 1, fourth row).

A continuous decline in the rate of energy-dependent release of $[^{14}C]$-ADP was observed when salt-washed thylakoid membranes, pretreated in the dark with NEM, were allowed to react with IBZ in light (Fig. 1). The maximum inhibition of the activity was approximately 42 per cent at 10 mM IBZ. However, the addition of 12.66 mM DTT, in light, reversed the inhibition of energy-dependent release of $[^{14}C]$-ADP caused by treatment of NEM-modified thylakoid membranes with IBZ (Fig. 1). At 1 mM IBZ the addition of 12.66 mM DTT resulted in 17 per cent enhancement of the activity in addition to complete reversal of inhibition. Even at higher concentrations of IBZ, e.g., 6 or 10 mM, DTT was capable of reversing the inhibitory effect of IBZ.

In parallel to inhibition of energy-dependent release of $[^{14}C]$-ADP by IBZ, a continuous decline in the rate of non-cyclic photophosphorylation was observed when salt-washed thylakoid membranes alkylated with NEM in darkness were exposed to various concentrations of IBZ in light (Fig. 2). The maximum inhibition of the rate of non-cyclic photophosphorylation was approximately 48 per cent at 10 mM IBZ. In direct comparison to the reversal effect of DTT on inhibition of energy-dependent release of

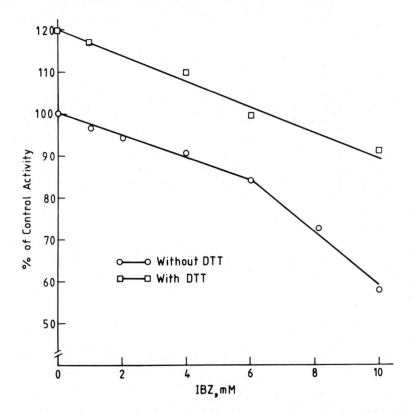

Fig. 1: The energy-dependent release of [^{14}C]-ADP as a function of thiol modulation. For details see 'Materials and Methods'. Control activities without any treatment with NEM or IBZ were 0.42 nanomoles [^{14}C]-ADP released/mg Chl in the absence of DTT and 0.52 nanomoles [^{14}C]-ADP released/mg Chl in the presence of DTT.

[^{14}C]-ADP by IBZ, the addition of 12.66 mM DTT to the IBZ-modified thylakoid membranes resulted in reversal of inhibition of photophosphorylation activity caused by IBZ (Fig. 2). At lower concentrations of IBZ, the addition of DTT not only completely reversed the inhibition but also increased the rate of non-cyclic photophosphorylation by 9 per cent (Fig. 2).

Discussion

Our observations on the alkylation of the -SH group of CF_1 in salt-washed thylakoid membranes with NEM in the dark suggests that the -SH group of CF_1 which is normally exposed in darkness and is thus available for alkylation with NEM is involved in the activation process of CF_1-ATPase

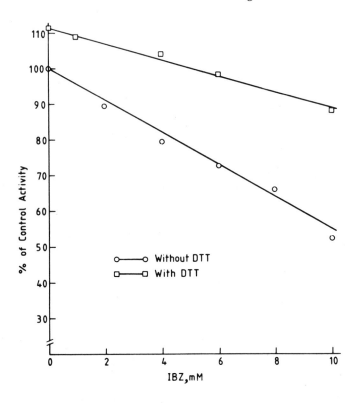

Fig. 2: Thiol modulation of salt-washed thylakoid membranes and its effect on non-cyclic photophosphorylation. For details see 'Materials and Methods'. The control activities were 156.3 μmoles ATP formed/mg Chl/hr in the absence of DTT and 178.5 μmoles ATP formed/mg Chl/hr in the presence of DTT.

(as shown by the energy-dependent release of ADP) and, consequently, in regulating the catalytic function of CF_1-ATPase, i.e., the process of photophosphorylation (Table 1). Since photophosphorylation was more sensitive than energy-dependent release of ADP (Table 1) towards alkylation with NEM, it is suggested that a small change in the fraction of activated CF_1 (defined as ∅) can result in significant changes in photophosphorylation activity.

The further inhibition in the rate of energy-dependent release of ADP and non-cyclic photophosphorylation on treatment of NEM-alkylated thylakoid membrane with DTT (Table 1) suggests that (1) initial inhibition of these activities by alkylating reagent is irreversible and, (2) DTT-modification of the -S-S-linkage during the high-energy state of salt-washed thylakoid membrane pre-treated with NEM is responsible for further inhibition of these activities, particularly the energy-dependent release of $[^{14}C]$-ADP. The latter may arise due to decrease in H^+-ion efflux required for the activation process or increase in $\triangle_{\mu H}+$ required for activation of CF_1.

Similarly, oxidation of the -SH group of salt-washed thylakoid membrane (exposed during the high-energy state) with IBZ resulted in inhibition of energy-dependent release of ADP and non-cyclic photophosphorylation (Table 1). This indicates that the -SH group of CF_1 which got exposed during the energy-dependent conformational rearrangements in CF_1 of salt-washed thylakoid membranes are involved in regulating both the activation process (i.e., changes in $\emptyset$) and consequently the catalytic function (ATP synthesis) of CF_1. It is interesting to note here that inhibition of photophosphorylation by IBZ was more than that of energy-dependent release of ADP. This observation, as seen with NEM modification also, suggests that small changes in the activation process of CF_1 results in significant changes in its catalytic function.

CF_1 is inactivated when ADP is tightly bound to the catalytic site at $0°$, $120°$ and $240°$ in a $360°$ catalytic configuration cycle and the removal of this tightly bound ADP requires a certain threshold value of protonmotive force. The modulation of -SH groups of the γ-subunit of CF_1 which participate directly in proton translocation (see 22) would cause changes in the threshold value of $\Delta_{\mu H}+$, resulting in changes in the activation and, consequently, photophosphorylation processes.

Our studies on IBZ modification of CF_1 are in agreement with such a suggestion. This is substantiated by our results on reversal of IBZ inhibitory effect by DTT which causes re-reduction of -SH groups previously oxidised by IBZ.

The data as presented in Figs. 1 and 2 show a parallel decrease of both the energy-dependent release of ADP and the rate of non-cyclic photophosphorylation when NEM-alkylated thylakoid membranes are treated with IBZ in light. These observations suggest that oxidation of the -SH group of the γ-subunit of CF_1 which gets exposed during energy dependent-conformational changes in CF_1 results in a decrease in the fraction $\emptyset$ of activated CF_1-ATPase. As pointed out above, the activation of CF_1-ATPase shown by removal of tightly bound ADP from the adenine nucleotide binding site requires a certain input of threshold protonmotive force. It has been shown previously by Mills and Mitchell (18) that the -SH group of the γ-subunit of CF_1 modulates this threshold pmf. Therefore, oxidation of such-SH groups which get exposed in light, by IBZ treatment would affect such modulation of the threshold pmf leading to decrease in $\emptyset$. It appears that oxidation of such -SH group by IBZ would increase the $\Delta\mu_{H+}$ required for the activation resulting in a decrease in $\emptyset$. The decrease in the fraction $\emptyset$ of the activated CF_1-ATPase would consequently result in a decrease in the rate of photophosphorylation (Fig. 2).

As shown in Figs. 1 and 2, DTT was found to fully restore the activities which were previously inhibited by IBZ. The increase in the rates of energy-dependent release of ADP by DTT suggests an increase in $\emptyset$

which may arise because of thiol modulation of threshold pmf leading to a decrease in $\Delta\mu_{H}+$ required for activation. This would result in increase in the rates of photophosphorylation (Fig. 2).

At the end, we would like to conclude that (1) modification of salt-washed thylakoid membranes with NEM in darkness leads to energy transfer inhibition, (2) oxidation of the -SH group of the γ-subunit of CF_1 exposed during illumination of thylakoid membranes results in decrease in $\emptyset$ as well as non-cyclic photophosphorylation and (3) re-reduction of oxidised -SH groups, i.e., the -S-S-linkage, by DTT reverses the inhibitory effect of IBZ.

References

1. McCarty, R. E., Pittman, P. R. and Tsuchiya, Y. (1972) *J. Biol. Chem.* 247, 3048–3052
2. McCarty, R. E. and Fagan, J. (1973) *Biochemistry* 12, 1503–1507
3. Magnusson, R. P. and McCarty, R. E. (1975) *J. Biol. Chem.* 250, 2593–2598
4. Portis, A. R., Jr., Magnusson, R. P. and McCarty, R. E. (1975) *Biochem. Biophys. Res. Commun.* 64, 877–884
5. Andreo, C. S. and Vallejos, R. H. (1976) *Biochim. Biophys.* Acta 423, 590–601
6. Vallejos, R. H. and Andreo, C. S. (1976) *FEBS Lett.* 61, 95–99
7. Moroney, J. V. and McCarty, R. E. (1979) *J. Biol. Chem.* 254, 8951–8955
8. Weiss, M. A. and McCarty, R. E. (1977) *J. Biol. Chem.* 252, 8007–8012
9. Moroney, J. V., Andreo, C. S., Vallejos, R. H. and McCarty, R. E. (1980) *J. Biol. Chem.* 255, 6670–6674
10. McCarty, R. E., Davenport, J. W., Ketcham, S. R., Moroney, J. V., Nalin, C. M., Patrie, W. J. and Warncke, K. (1984). In: *Adv. in Photosynth. Res.* Vol. II (Sybesma, C. ed.) pp. 371–378, Martinus Nijhoff/Dr W. Junk Publishers, The Hague, Netherlands
11. Arana, J. L. and Vallejos, R. H. (1982) *J. Biol. Chem.* 257, 1125–1127
12. Nalin, C. M., Beliveau, R. and McCarty, R. E. (1983) *J. Biol Chem.* 258, 3376–3381
13. Vallejos, R. H. and Ravizzini, R. A. (1984). In: *Adv. in Photosynth. Res.* Vol. II (Sybesma, C. ed.) pp. 519–522, Martinus Nijhoff/Dr W. Junk Publishers, The Hague, Netherlands
14. Mills, J. D. and Mitchell, P. (1982) *FEBS Lett.* 144, 63–67
15. Davenport, J. W. and McCarty, R. E. (1981) *J. Biol. Chem.* 256, 8947–8954
16. Mills, J. D. and Hind, G. (1979) *Biochim. Biophys.* Acta 547, 455–462
17. Anderson, L. E. (1975). In: *Proc. 3rd Int. Cong. Photosynth.* Vol. II (Avron, M. ed.) pp. 1393–1405, Elsevier, Amsterdam
18. Mills, J. D. and Mitchell, P. (1984). In: *Adv. in Photosynth. Res.* Vol. II (Sybesma, C. ed.) pp. 523–526, Martinus Nijhoff/Dr W. Junk Publishers, The Hague, Netherlands
19. Strotmann, H., Bickel, S. and Huchzermeyer, B. (1976) *FEBS Lett.* 61, 194–197
20. Shavit, N. and Strotmann, H. (1980). In: *Methods in Enzymology*, Vol. 69C (SanPietro, A. ed.) pp. 321–326, Academic Press, New York
21. Avron, M. (1960) *Biochim. Biophys.* Acta 40, 257–272
22. Mitchell, P. (1985) *FEBS Lett.* 182, 1–7

PART 3
Stress Effects on Photosynthesis
Organisms

The Bioenergetics of Stress Responses in Cyanobacteria

LESTER PACKER

Membrane Bioenergetics Group, Lawrence Berkeley Laboratory,
Department of Molecular and Cellular Biology,
University of California,
Berkeley, California 94720.

Our laboratory is trying to chart some new areas that offer attractive vistas for research in the response to stress in photosynthetic microorganisms. We use oxygenic photosynthetic cyanobacteria as a model system to examine stress responses. A useful approach has been to take a freshwater cyanobacterium suddenly exposed to seawater concentrations of sodium, and ask the question, 'What are the steps involved in the adaptation of the organism to grow in high salt?' In a similar way, we have investigated other kinds of stress, like temperature stress, toxic metal stress (change of the selenium to sulphur ratio), pH stress and photo-oxidative stress (Belkin *et al.*, 1987). The question that we seek answers to are: do universal mechanisms of response to stress exist, and what are the specific responses that the organisms show when they are exposed to one or another kind of stress? This is obviously a complex problem that involves changes in gene expression, bioenergetic changes and, in the case of salt tolerance, of course, osmoregulatory changes (Blumwald *et al.*, 1983a; Blumwald *et al.*, 1983b).

Salt Stress

The idea is to try and identify the steps involved by seeing what happens to the freshwater cyanobacterium *Synechococcus 6311* when exposed to 0.5M NaCl. Most experiments have been carried out in medium in which the cells grow. We emphasise noninvasive studies. If one examines the freshwater cyanobacterium after adding 0.5M salt medium, one finds a 50–80 per cent inhibition of photosynthesis and an inhibition of growth rate; after two days they recover photosynthetic activity and return to normal growth rates (Fry *et al.*, 1986). So the cells have to have some way of overcoming the energetic demand suddenly placed upon them by having to deal with pumping out sodium and also to meet the energetic demands needed to carry on biosynthesis.

We have tried to find out what the lesion is in the photosynthetic electron transport chain. We've looked at the reactions involved between water and Qa and Qb by fluorescence changes. As is known, if Qa is not working, and one adds DCMU, you get a large increase in fluorescence. We were interested to see if NaCl addition would uncouple this system. This is because photosynthesis is being driven to a large extent by the phycobiliprotein system attached to the membranes. Indeed, one way of isolating phycobiliproteins is to salt wash the membranes. Thus, we reasoned, maybe functional uncoupling of phycobiliproteins would occur. However, this was found to be only a small effect and actually only a

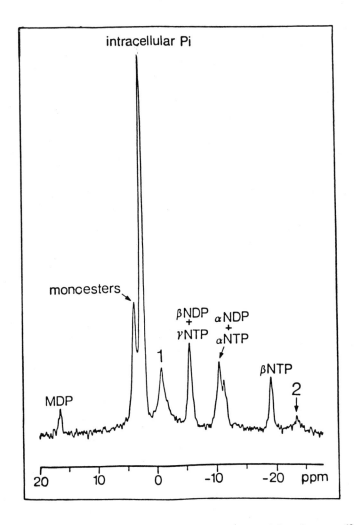

Fig. 1: NMR spectra of phosphorous metabolites of intact cells of *Synechococcus 6311* (after Packer *et al.*, 1987).

transient effect; it only occurs for a few minutes and then it is gone (G. Dubertret, M. Lefort-Tran, S. Spath collaboration). Thus, we know that this is not the site where photosynthetic electron transport is inhibited. Though we still don't exactly know where it is inhibited, we think that it is at the level of the ferredoxins, because ferredoxin-linked enzyme systems are usually very salt-sensitive, but we still haven't proven it.

Since photosynthesis is inhibited, one would expect that cell energetics would be affected. To study this, we use the 31–P NMR. We inserted nine fibre optical light pipes into an NMR sample tube for an AM 400 spectrometer with provision by capillary tubing for gas exchange (Packer *et al.*, 1987). In this way (Figure 1), we can get very good 31-P NMR spectra of *Synechococcus 6311* cell suspensions. One observes nicely resolved monoester phosphate and inorganic phosphate peaks, several nucleotide peaks and a pyrophosphate peak. After cells are exposed to 0.5M salt, a large increase in the inorganic phosphate peak occurs and one can't any longer clearly resolve the monoester phosphate peak. The ATP peak is almost gone, but after about an hour, the ATP peak returns, indicating the cells begin to recover their normal energetic profile of 31–P metabolites. To

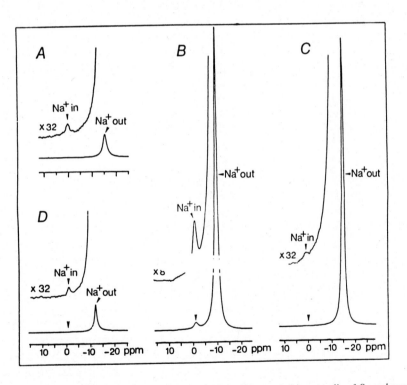

Fig. 2: Time sequence of Sodium NMR spectra of saline-treated intact cells of *Synechococcus 6311* (after Packer *et al.*, 1987) (A) Control Cells, (B) 0.3M NaCl-treated cells, 10 minutes, (C) 0.5M NaCl-grown cells, (D) Salt-grown cells plus 0.5 mannitol.

distinguish between the internal and external phosphate, we added an aliquot of pyrophosphate. Inorganic pyrophosphate doesn't enter the cells, but is hydrolysed by an extracellular phosphatase that appears allowing one to clearly observe it from the internal peak because of the different pH outside vs. inside the cells.

NMR is also a good method to observe the entry of sodium by 23 sodium NMR. One can separate the internal signal from the external signal by adding membrane impermeable shifts reagents like dysposium salts. This splits the signal, and one can see that initially there is very little sodium inside the cells (Figure 2), but after 0.5M sodium chloride addition, a lot of sodium accumulates inside. So there is very big intrusion of sodium into the cells and subsequent spectra reveal that after some time, the cells excrete most of the sodium. So a system for sodium exchange gets activated shortly after cells are exposed to NaCl.

Another extremely surprising observation that was made by thin section electron microscopy (Lefort-Tran *etal.*, 1988) is that one minute after adding sodium to the cell suspension, all signs of the intracellular accumulation products disappear. Cyanobacteria have internal inclusion products like glycogen, carboxysomes, cyanophycin, beta-hydroxybutyrate and poly-phosphate as granules. These are numerous different inclusion products that may occur under different conditions. All disappear one minute after salt exposure. Also, the DNA material which is initially coarse in appearance develops a fine, more threadlike appearance one minute after salt treatment. So there appears to be some large change in the intracellular physio-chemical environment. Figure 3 A and B summarise the ultrastructural changes. Maybe it is a pH change because when the Na enters a Na+/proton exchange system is activated. We're not exactly sure what brings this about, but it is a very striking finding.

Later, glycogen begins to be resynthesised. Cells grown on 30 per cent enriched C-13 bicarbonate show glycogen accumulation by 13C NMR. This is a good method to show accumulation and loss of glycogen under different environmental conditions (Tel-Or *et al.*, 1986). For example, in *Argamellum quadruplicatum* (PR6), a marine organism that grows between 1M and 1.5M salt, glycogen accumulation and loss can be followed dynamically with time. After salt growth, glycogen accumulation is massive in *Synechococcus 6311*, but the chemical and NMR studies do not reveal where the glycogen is located. Using the electron microscope (as described above) one minute after exposure to salt, loss of all glycogen occurs. After 4–5 hours of salt exposure glycogen is re-formed and about 24 hours after exposure, the glycogen is primarily located in the region between the cytoplasmic membrane and the thylakoid membranes. It's almost as if a pearl necklace has been placed around the edge of the cell (Figure 3). After 48 hours, when photosynthesis is completely recovered, a more random, normal appearance of glycogen is observed. However, the glycogen granules are still preferentially located between the photosynthe-

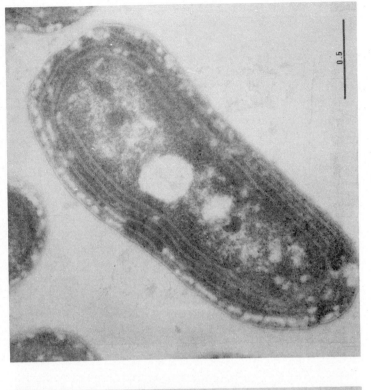

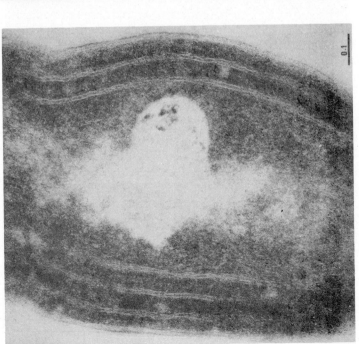

Fig. 3: Ultrastructure of *Synechococcus 6311* at various stages before and after exposure to 0.5M NaCl (after LeFort-Tran *et al.*, 1988) (A) one minute after exposure, (B) Salt-adapted cells, 48 hours.

tic membranes and the cytoplasmic membrane, but they are also found elsewhere in the cell. The distribution of the glycogen granules, when one examines a lot of different cells, shows that 4–5 hours seems to be a critical point. At this time, all of the cells are filled with glycogen again, whereas one minute after exposure, 71 per cent of the cells do not have any glycogen as detectable by electron microcopy. This striking glycogen accumulation suggests that respiratory activity may change because the respiratory activity, it is believed (Lefort-Tran *et al.*, 1988), is located in the cytoplasmic membrane. There is, however, a dispute of C. N. Murata Laboratory with regard to location of cytochrome oxidase and other respiration components in the cytoplasmic membrane.

Indeed, an accumulation of respiratory activity is seen in the salt-grown cells. Usually it is up to a ten-fold increase. Low temperature EPR, Cu^{+2} EPR signals of cytochrome oxidase are seen in *Synechococcus*. The signal is seen in the oxidised condition and it is not observed in the reduced sample. We have compared the oxidised Cu^{+2} signal to submitochondrial particles which showed a very similar signal to *Synechococcus*: not only with respect to the spectra, but also with respect to temperature-dependence and power saturation (Fry *et al.*, 1985).

The Na is pumped out and the question is, 'What kind of activity permits this?' We used several methods to show that a Na+/proton exchange mechanism, like that present in many bacteria, is present (Blumwald *et al.*, 1984). We used the fluorescence quenching of acridine dyes as one way to test for this activity. We can show that the increased respiratory activity was associated with increased Na+/proton exchange. Together with Huflejt *et al.*, (1988), we have used the dye, bis (carboxyethyl) -carboxy-fluorescein. This dye penetrated the cell in its uncharged form; after it enters the cell, groups are cleaved by nonspecific esterases to give rise to a charged molecule that is fluorescent and pH sensitive. We have been able to use this dye loading technique to show the activity of the Na^{+}/proton exchange system in *Synechococcus 6311*. So a new sensitive technique for noninvasively studying cytosolic changes in pH associated with Na+/proton exchange in intact cells is now available.

One cell structure that one would expect to be involved in any adaptation to stress would be the cytoplasmic membrane. The cytoplasmic membrane should contain pumps like Na+/proton exchange, and so we were interested in observing if the structure of the cytoplasmic membrane is altered in salt-stressed cells. Together with Lefort-Tran *et al.*, (1988), we looked at the cytoplasmic membranes in the EF and PF fracture faces. The EF fracture face has more particles. Analysis of these particles in the cytoplasmic membranes shows that there is a large shift in the particle size distribution in salt-adapted cells. In control cells 54 per cent of the particles are in the 7.5–9.5 nanometer diameter ragne, whereas after salt adaptation, only 17 per cent of the particles are in the small particle range. In salt-adapted cells 60 per cent of the particles are large, whereas in the control cells only 35 per cent of the particles are in the large size class (12–14 nanometer size range). These particle changes may represent a $Na^+/$ proton exchange system, or components of the respiration system or whatever else changes in the cytoplasmic membrane in response to stress adaptation.

Toxic Metal, Temperature Stress

Changing the growth temperature or introduing toxic species into the growth medium are examples of stress responses. In toxic metal stress, we have grown *Synechococcus6311* cells in the presence of different selenium to sulphur ratios (Fry *et al.*, unpublished results). If one increases the selenium to sulphur ratio, of 60 to 14, show a large increase in respiration activity (figure 4A). If one then adds sulphate to change the ratio of these two substances to the control ratio, then sulphate acts as a competitive inhibitor of the selenium, and the selenium-dependent increase in respiration is reversed. Also, if one adds sulphate, selenium which accumulates comes out (Figure 4B). Here then is another case of stress where one finds growth inhibition associated with an increased respiration, and glycogen accumlation. Temperature stress shows similar response. If cyanobacteria growing at $40°C$ have their temperature decreased to $20°C$, the growth rate slows, respiration increases, and glycogen accumulates. Thus a universal response to stress seems to be an increase in cell respiration and increased glycogen accumulation. So the cells which are normally obligately phototrophic, come to depend upon respiratory energy and presumably the proton electrochemical potential developed by a respiratory system. We think this respiratory system is probably in the cytopasmic membrane. Isolated cytplasmic membranes from salt-grown cells show increases in cytochrome C oxidation in the isolated cytoplasmic membranes and EPR methods also show increased CU^{+2} signals (Fry *et al.*, 1985). The lipid composition (Huflejt, Lang, Packer, unpublished resutls) and ESR lipid spin label determined order parameters (Khomatov and Packer, unpublished results) are also changed after salt adaptation. There is an increased amount and unsaturation of the fatty acids.

Concluding Remarks

There are two membrane systems in the cyanobacteria, the thylakoid and the cytoplasmic membranes. Both membranes contain proton ATPase and cytochrome oxidase. The cytoplasmic membrane systems certainly should additionally contain the ion transport systems like the Na^+/proton exchange. The thylakoid membrane contains the photosynthetic system. To what extent respiratory electron transport components in these two membranes is important in stress responses and in the energetics of these

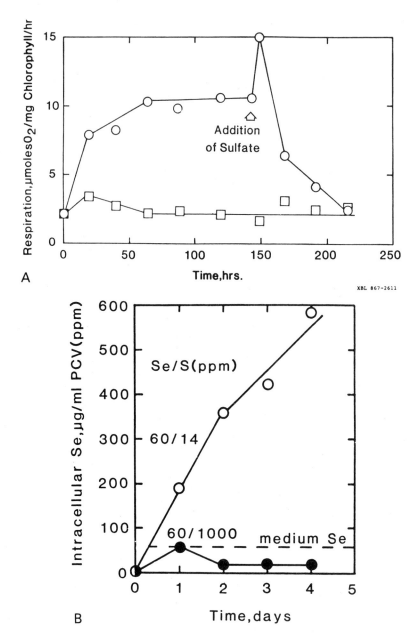

XBL 867-2611

Fig. 4: Effect of changing the selenium to sulphur rates on (A) respiration and (B) selenium accumulation in *Synechococcus 6311* (after Fry *et al.*, 1987).

cells, is still unresolved. We believe that stress adaptation may provide a tool to change the composition of these membranes perhaps selectively in the cytoplasmic membrane. This situation would allow for studies of the relative contribution to cell energetics of the photosynthetic and respiratory systems and to determine how they interact with each other through electron transport. This question is one of the unsolved problems in the oxygenic photosyntheic cyanobacteria.

Acknowledgements

Research described was supported by the NASA Closed Ecological Life Support System (CELSS) programme and the Office of Biological Energy Research, Department of Energy.

References

1. Belkin, S., Mehlhorn, R. J., and L. Packer, (1987) Determination of Dissolved Oxygen in Photosynthetic Systems by Nitroxide Spin-Probe Broadening, *Arch. Biochem. Biopys.*, 252 (2): 487–495
2. Blumwald, E., Mehlhorn, R. J., and Packer, L., (1983a) Studies of Osmoregulation in Salt Adaptation of Cyanobacteria Using ESR Spin Probe Techniques, *Proc. Natl. Acad. Sci.*, 80:2599–2602
3. Blumwald, E., Mehlhorn, R. J., and Packer, L., (1983b) Ionic Osmoregulation During Salt Adaptation of the Cyanobacterium *Synechococcus 6311, Plant Physiol.*, 73:377–380
4. Blumwald, E., Wolosin, J. M. and Packer, L., (1984) Na+/H+ Exchange in the Cyanobacterium Synechococcus 6311, *Biochem. Biophys. Res. Commun.* 122: 452–459
5. Fry, I. V., Hrabeta, J., Giauque, R. D. and L. Packer, Uptake and Cellular Distribution of Selenium by a Primary Producer, the Cyanobacterium *Synechococcus 6311*, (In Preparation)
6. Fry, I. V., Huflejt, M., Erber, W. W. A., Peschek, G. A., and Packer, L., (1986) The Role of Respiration During Adaptation of the Freshwater Cyanobacterium Synechococcus 6311 to Salinity, *Arch. Biochem. Biophys.*, 244: 686–691
7. Fry, I. V., Peschek, G. A., Huflejt, M., and Packer, L., (1985) EPR Signals of Redox Active Copper in EDTA Washed Membranes of the Cyanobacterium Synechococcus 6311, *Biochem. Biophys. Res. Commun.*, 129: 106–116
8. Huflejt, M., Negulescu, P., Machen, T. and L. Packer, (1988) Na$^+$ and Light-Dependent Regulation of Cytoplasmic pH in Cyanobacterium *Synechococcus 6311, Biophys. J.* 53(2): 616a
9. Lefort-Tran, M., Pouphile, M., Spath, S., and Packer, L., (1988) Cytoplasmic Membrane Changes During Adaptation of the Fresh Water Cyanobacterium *Synechococcus 6311* to Salinity, *Plant Physiology*
10. Packer, L. Spath, S., Martin, J., Roby, C., and Bligny, R., (1987) 23Na and 31P NMR Studies of the Effects of Salt Stress on the Fresh Water Cyanobacterium Synechococcus 6311, *Arch. Biochem. Biophys.* 256(1): 354–361
11. Tel-Or, E., Spath, S., Packer, L., and R. J. Mehlhorn, (1986) Carbon-13 NMR Studies of Salt Shock-Induced Carbohydrate Turnover in the Marine Cyanobacterium Agmenellum quadruplicatum, *Plant Physiol.* 82, 646–652

CAM Induction in *Mesembryanthemum crystallinum:* Protein Expression

J. M. Schmitt[1], R. Höfner[1], A. A. Abou-Mandour[1],
L. Vazquez-Moreno[1,2,3] and H. J. Bohnert[3]

[1]Institut für Botanik und Pharmazeutische Biologie
Universität
Mittlerer Dallenbergweg 64
D-8700 Würzburg. FRG

[2]Present address:
CIAD
Hermosillo, Sonora
Mexico

[3]Department of Biochemistry
University of Arizona
Tucson, AZ 85721, USA

Summary

When *Mesembryanthemum crystallinum* is irrigated with nutrient solutions containing high amounts of sodium chloride, the plants shift from C_3 to CAM photosynthesis. We have measured the amounts of three CAM enzymes, phosphoenolpyruvate carboxylase (PEP-Case), pyruvate, orthophosphate dikinase (PPDK) and NADP-malic enzyme (NADP-ME) over the induction period using quantitative immunoblot analysis. The antiserum against PPDK was raised using a fusion protein of PPDK with the carboxyterminal part of bacterial β galactosidase (expressed in *E. coli*) as antigen. The antiserum was specific for PPDK with no indication of cross reactivity with plant β galactosidases, showing that carboxyterminal fusions of protein coding regions with β-galactosidase may be of wider application for plant studies. PPDK and NADP-ME proteins accumulate during the first seven days of salt stress. This indicates that differential regulation of gene expression rather than enzyme activation accounts for the increase of apparent activity. Two subunit isoforms of PEPCase are present in C_3 leaves, roots and calli. They can be distinguished by their apparent molecular weights on SDS gels (M_r approximately 110,000 and 100,000). Only the 100-kDa subunit accumulates during CAM induction.

Introduction

Plants have developed numerous adaptations to cope with water deficits. One of the best studied metabolic responses occurring in some species of succulents is Crassulacean acid metabolism (CAM). CAM plants open their stomates at night and close them during the day when potential evaporative water loss is high. In the CAM mode of photosynthesis,

---

Abbreviations: PEPCase – phosphoenolpyruvate carboxylase; PPDK – pyruvate orthophosphate dikinase; NADP-ME – malic enzyme (NADP-dependent).

carbon dioxide is fixed by PEPCase into oxaloacetate at night. Oxaloacetate is reduced to malate and is stored in the vacuole. Malate is
decarboxylated during the day and the released carbon dioxide is utilised
by the Calvin cycle.

The Common ice plant (*Mesembryanthemum crystallinum*, L.) is a
facultative halophyte which has the capacity to fix carbon in both the C_3
and the CAM modes of photosynthesis [for a recent review see (1)]. Young
plants and well-irrigated old plants display the characteristics of C_3
photosynthesis. Under our experimental conditions, plants acquire the
ability to shift to CAM under salt stress (2,3) when they are approximately
six weeks of age (4). In the laboratory, CAM can best be induced by
irrigation with nutrient solution containing up to 0.5 M sodium chloride.
CAM is characterised by increased apparent activities of several enzymes
catalysing metabolic steps of the pathway (5).

The enzymic activity of PEPCase from CAM plants is increased
approximately 40-fold in comparison to C_3 plants (5). The increase in
apparent activity can be accounted for by accumulation of enzyme protein
(6). During the induction period, mRNA for PEPCase is present in higher
concentration than in C_3 plants as shown by *in vitro* translation (4) and
northern-type hybridisation (7). Upon relief from salt stress, PEPCase
mRNA and protein decrease rapidly (8).

Here we show that, in addition to PEPCase, amounts of PPDK and
NADP-ME also accumulate during CAM induction. Furthermore, we
present evidence that ice plant PEPCase is composed of two isoforms
which differ in appparent molecular weight on SDS gels (M_r 110,000 and
100,000). Only the 100-kDa form accumulates during salt stress.

Materials and Methods

Immunopurification of Antibodies

PEPCase was partially purified from crude extracts by precipitation with
15% polyethylene glycol (9;4). The protein was loaded on to preparative
SDS polyacrylamide gels and proteins were blotted to nitrocellulose filters.
After blocking (1% bovine serum albumin), PEPCase bands were
identified by a brief immunostaining and cut out. Antiserum (diluted 1:100
in 1% bovine serum albumin, 20 mM Tris, pH 8, 150 mM NaCl) was
incubated with the nitrocellulose-bound PEPCase overnight. The filters
were washed thoroughly with TBS (20 mM Tris, pH 8, 150 mM NaCl) and
the bound antibodies were eluted using 100 mM glycine-HCl, pH 2.5. The
solution containing the eluted antibodies was neutralised and used to probe
immunoblots.

Preparation of PPDK Fusion Protein for Immunisation

Protein was synthesised in *E. coli* from a cDNA clone (*lambda* McCAM 403) with an insert of 1.5 kbp in frame with β-galactosidase (10), as described previously (7). The crude bacterial extract was centrifuged for 5 min. in an Eppendorf centrifuge and the pellet was loaded on to an SDS gel. The β-galactosidase-PPDK fusion protein was localised by immuno-blotting as described previously (7) and cut out of coomassie-stained gels run in parallel. The gel bands were suspended in 150 mM NaCl, homogenised using an Ultra-Turrax and 2 ml aliquots were injected subcutaneously into a rabbit at two-week intervals.

Isolation of NADP Malic Enzyme

NADP malic enzyme was isolated from maize leaves using column chromatographic techniques. Details will be published elsewhere (Neubauer and Schmitt, in preparation).

Quantitative Immunoblot Analysis

Crude leaf extracts were separated on 8% SDS polyacrylamide gels. Proteins were transferred to nitrocellulose and immunolabelled using second antibody conjugated to alkaline phosphatase (Bio Rad). The protein bands were scanned with a CAMAG reflectance densitometer. The output signal from the densitometer was quantified using a Merck-Hitachi Integrator.

Peptide Mapping

Peptide mapping by partial proteolytic cleavage was performed as described by Cleveland et al. (11).

Tissue Culture

Conditions for callus culture of *M. crystallinum* will be described elsewhere (Abou-Mandour, in preparation).

Other Techniques

All other techniques have been described previously (6; 4; 7; 8).

Results

Antiserum against PPDK Using a Recombinant Antigen

PPDK activity is light-dependent and the enzyme is rapidly inactivated in the cold. Reliable measurement of PPDK is, therefore, difficult. We chose to use the PPDK-specific clone *lambda* McCAM 403 (7) as the antigen source to obtain an antibody probe for the protein. This clone was identified from a *Mesembryanthemum* cDNA library constructed in the vector *lambda gt11* (7) using antisera against maize PPDK (12; 13). *Lambda* gt11 allows the efficient expression of foreign sequences inserted in the unique EcoRI site located near the 3' end of the coding sequence for β-galactosidase (10). Gene fusions between β-galactosidase and another coding sequence in the proper reading frame lead to the expression of fusion proteins. The resulting chimeric proteins may retain their antigenic properties. The application of β-galactosidase fusions for generating antisera, used mainly in animal systems and with genes from microorganisms, is reviewed in (14).

When the antiserum raised against a β-galactosidase/PPDK fusion protein was used to probe protein blots from crude *Mesembryanthemum* extracts, a band migrating at approximately 95 kDa was detected (Fig. 1). This band migrated slightly faster than PEPCase but comigrated with the bands recognised by two anti-PPDK sera from independent sources kindly provided by K. Aoyagi (Fig. 1B) and G. Baer (data not shown; 12; 13). No reaction with PPDK was seen when the blot was probed with pre-immune serum. A second band of <48 kDa (which has been run off the gel in Fig. 1) reacted with both immune and pre-immune sera. As expected, the antiserum also reacted with β-galactosidase from *E. coli* (Fig. 1A).

Leaves from higher plants are known to contain β-galactosidases which

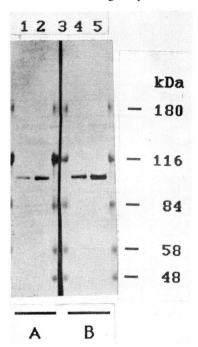

Fig 1: Immunoblot of crude leaf extracts probed with anti-PPDK/β-galactosidase (A) and anti-maize PPDK serum (B; 12). Lanes 1 and 4: C₃; lanes 2 and 5: CAM; lane 3 (divided): Pre-stained molecular weight markers, Sigma MW-SDS-BLUE. Approximate molecular weights are given. The marker at 116 is β-galactosidase from *E. coli*. Note the immunostaining of this band in panel A.

are localised in the vacuole [for a review see (15)]. Immunoblots of crude extracts from the ice plant failed to detect a band that could be assigned to β-galactosidase This failure can best be explained by a low concentration of β-galactosidase in the leaves, since vacuoles comprise only about 4% of the total cellular protein (16).

Vacuolar protein can be concentrated by preparing vacuoplasts (17; 48), plasmalemma-bound subcellular vesicles which contain the vacuole and some cytosol but no other organelles, i.e. nuclei, chloroplasts or mitochondria. When immunoblots of vacuoplast proteins were probed with anti-β-galactosidase/PPDK, again no specific reaction could be observed (M. Lang, personal communication), while anti-vacuole serum used as a positive control reacted strongly (18; data not shown).

Induction of Enzymes in Leaves

The increase in apparent activity of PEPCase during CAM induction is caused by a mass increase of enzyme protein (6). The experiment presented in Fig. 2 monitored the time course of induction for PEPCase, PPDK and NADP-ME after irrigation of the plants with 250 mM NaCl. The proteins accumulate roughly in parallel, although to different degrees, when the plants are stressed. We have included the lines of regression in the figures as a means of characterising the rate of accumulation.

In this experiment, PEPCase protein increased by a factor of 4–5 during seven days of stress. NADP-ME and PPDK accumulated about twofold. The build up of NADP-ME correlates better with time ($R=0.90$) than the accumulation of PPDK ($R=0.57$). We have repeatedly found widely varying concentrations of PPDK protein in unstressed and mildly stressed leaves. The time course of induction of PEPCase activity is included for comparison (Fig. 2a). It is similar to our previous results (4).

PEPCase Isoforms

Apart from its specific functions in C4 and CAM plants, PEPCase performs several roles in cellular metabolism. It is, for example, involved in nitrogen fixation, maintenance of electroneutrality and it interfaces different cellular pathways (19). PEPcase is composed of four subunits and is present in all plants. In the CAM mode of photosynthesis, PEPCase is inhibited by malate during the day, in order to avoid futile cycling of carbon [for reviews see (20; 21)]. We investigated whether PEPCase from C_3 and CAM leaves could be distinguished biochemically. Fig. 3 shows an immunoblot of crude soluble protein derived from C_3 leaves as compared to CAM leaves, roots, and callus tissue. All lanes were loaded with the same amount of PEPCase as determined by enzymic activity. The lanes loaded with extracts from tissues not performing CAM (C_3 leaves, root, callus) show two immunoreactive bands, approximately 100 kDa and 110 kDa in size. In the lane loaded with CAM extract, only the 100-kDa band

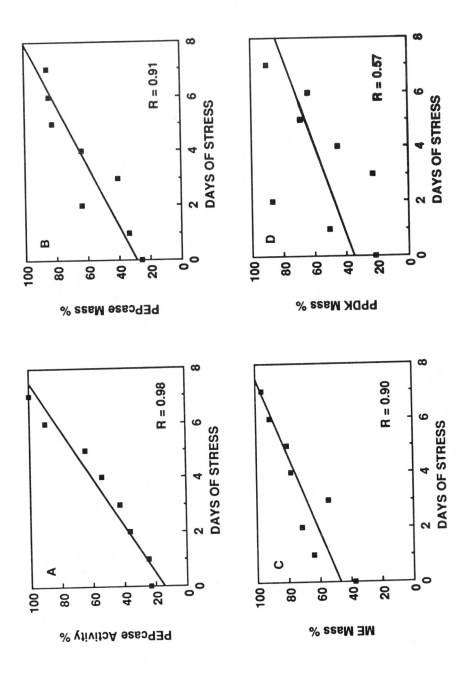

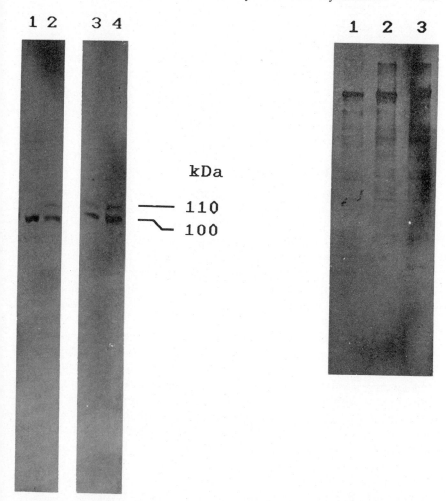

kDa

—— 110

⌐ 100

Fig. 3 *(left):* Immunoblot of crude extracts from *M. crystallinum* tissues. Samples were diluted to give comparable PEPCase activities per unit volume and separated on an 8% SDS gel. Immunopurified antibodies against PEPCase (6) were used to probe the blot. Lane 1: CAM leaf; land 2: C_3 leaf; lane 3: callus; lane 4: root.

Fig. 4 *(right)* Immunoblot of protease-digested PEPCase. Samples were electroeluted from a gel and digested with V8 protease as described in (11): The digests were separated on a 12.5% gel, immunoblotted and probed with affinity purified antibodies. Lane 1: CAM-PEPCase; lane 2: C_3 PEPCase 100-kDa species; lane 3: C_3 PEPCase, 110-kDa species.

Fig. 2: Induction of enzymes as measured by immunoblot analysis or photometrically. Plants were irrigated with 0.25 M NaCl for the time indicated. Day 0 represents the unstressed control. A: PEPCase activity (0.28 micromoles/mg protein/min at day 7). B: PEPCase mass. C: NADP malic enzyme (ME) mass. D: Pyruvate orthophosphate dikinase (PPDK) mass. Relative enzyme mass was determined using reflectance densitometry of immunoblots. Each data point represents the mean of leaves from two different plants. Values for a given day have been determined from the same leaf extracts. Regression (R) coefficients are given in the figures.

is observed. Although the 110-kDa form is no longer visible, it may still be expressed at its lower level.

The 110-kDa protein from leaves is homologous in sequence to the 100-kDa band from both C_3 and CAM leaves as can be seen from the peptide maps in Fig. 4.

Discussion

Anti-PPDK/β-galactosidase Serum

Raising antisera to proteins which are difficult to assay or which are of low abundance in plant tissues may be tedious. Here we demonstrate that plant sequences fused to β-galactosidase of *E. coli* near the carboxyterminus of the bacterial protein provide a convenient source of antigen. The large prokaryotic part (M_r114,000) of the fusion protein apparently did not interfere with the immunological recognition of the plant peptide. The lack of immunological cross-reaction between β galactosidase from *E. coli* and the corresponding enzyme from the ice plant reflects the large evolutionary divergence of the two species.

The initial identification of the PPDK-specific clone McCAM403 was based on the recognition of its fusion protein by two independent polyspecific antisera raised against PPDK from maize. Evolutionary divergence leading to weaker antibody binding and cross-reactivity of the antisera with other proteins are known sources of error, leading to false identification of fusion proteins by polyspecific antisera. We have, for example, obtained several clones whose fusion proteins reacted with anti-PEPCase. On further inspection, these putative PEPcase clones did not code for this enzyme (7): The identification of the DNA insert in McCAM403 as containing a large part of an ice plant PPDK gene is based on the observations that two independent antisera recognise PPDK/β-galactosidase fusion protein and that an antiserum against the fusion protein in turn recognises a band of the correct molecular weight (Fig 1).

PEPCase Isoforms

In *M. crystallinum*, there are at least two PEPCase isoforms, differing in molecular weight (Figs 3 and 4). In the non-induced (C_3) leaves, in root, and in undifferentiated callus, both isoforms can be seen.

Upon CAM induction in leaves, only the faster-migrating band (M_r 100,000) increases so that the slower-migrating band is no longer visible in lanes loaded with identical amounts of enzymic activity. Nimmo et al. (22) have found two closely migrating PEPCase bands in leaf extracts from the CAM plant *Bryophyllum fedtschenkoi*. The two forms co-purified through several chromatographic steps. Their *in vivo* relationship is unclear. Recently, Matsuoka and Hata (23) conducted a study in which PEPCases from C_3 and C_4 plants were compared by immunoblotting. Up to four subunit isoforms differing in apparent molecular weight were detected in some species. It was observed that distinct isoforms in rice and wheat accumulated differently during plant development and in response to light. No attempts were made to decide whether different genes were expressed or whether post-translational modifications produced the different banding patterns.

The identification of two protein isoforms in *M. crystallinum* is in accordance with the finding of more than one PEPCase gene as indicated by restriction length polymorphisms of genomic DNA (J. Cushman and G. Meyer, personal communication).

Accumulation of Enzyme Proteins

Holtum and Winter (5) could not detect PPDK in leaf extracts from C_3 plants by using activity measurements. The more sensitive immunological detection method never failed to detect PPDK protein in C_3 leaves (Figs. 1 and 2). Similarly, Aoyagi and Bassham (12) have found PPDK in leaves and seeds of C_3 plants. It may be that inactivation during isolation, or a physiological mechanism which regulates activity in C_3 ice plant explain the absence of measurable amounts of PPDK enzyme.

The transition from the C_3 to the CAM mode of photosynthesis involves the up-regulation of many enzymes. In the case of PEPCase (Fig. 2; 24; 6; 4), PPDK, and NADP-ME (Fig. 2) this up-regulation is achieved by increasing the amount of enzyme protein, although metabolic regulation of the enzymes may also occur. The genetic level of regulation of this process is as yet unclear. Concentrations of transcripts for PEPCase and PPDK are increased during CAM induction (4; 7; Michalowski et al., in preparation).

A simple way to control a battery of genes would be to put them under the control of a common regulatory element. We are in the process of comparing sequences of genomic clones specific for CAM and C_3 forms of PEPCase and PPDK. Characterisation of sequence motifs in regulatory regions will be helpful in future experiments designed to study how environmental stress leads to responses at the level of gene expression.

Acknowledgements

We are grateful to Drs Kazuko Aoyagi and Gianni Baer for providing antisera against PPDK, to Dr Catherine C. Wasmann for reading the manuscript, to Professor F.-C. Czygan for the use of his reflectance densitometer and to Mechtild Piepenbrock for skilled technical assistance. Our work is supported by USDA-CRGP-87-1-2475, by USDA-CRSR-86-2-2748, the Arizona Agricultural Experiment Station (ARZT#174441) and, in part, by NSF (PCM-8318166). JMS is supported by DFG (Schm 490/3). Travel support by NATO (RG84/230) is gratefully acknowledged.

References

1. Bohnert, H.J., J.A., Ostrem, J.C. Cushman, C.B. Michalowski, J. Rickers, G. Meyer, E.J. DeRocher, D.M. Vernon, L. Vazquez-Moreno, J. Velten, R. Höfner, and J.M. Schmitt. 1988, *Plant Mol. Biol. Reporter*, 6:10–28
2. Winter, K. 1973. *Planta* 109: 135–145
3. Winter, K. 1973. *Planta* 114: 75–85
4. Ostrem, J. A., S. W. Olson, J. M. Schmitt, and H. J. Bohnert. 1987. *Plant Physiol.* 84: 1270–1275
5. Holtum, J. A. M. and K. Winter. 1982. *Planta* 155:8–16
6. Höfner, R., L. Vazquez-Moreno, K. Winter, H. J. Bohnert, and J. M. Schmitt. 1987. *Plant Physiol.* 83: 915–919
7. Schmitt J. M. Michalowski, C. B., Bohnert, H. J. 1988. *Photosynth.* Res. in press
8. Vernon, D. M., J.A. Ostrem, J. M. Schmitt, and H. J. Bohnert. 1988. *Plant Physiol.* 86: in press
9. Hatch, M. D., and H. W. Heldt. 1985. *Anal. Biochem.* 145: 393–397
10. Huynh, T. V., R. A. Young, and R. W. Davis. 1985. In: Glover, D. M. ed. *DNA Cloning* Volume I, *A Practical Approach*, pp 49–78, Oxford, IRL Press
11. Cleveland, D. W., S. G. Fisher, M. W. Kirschner, and U. K. Laemmli. 1977. *J. Biol. Chem.* 252: 1102–1106
12. Aoyagi, K., and J. A. Bassham. 1984. *Plant Physiol.* 75: 387–392
13. Baer, G. R., and Schrader, L. E. 1985. *Plant Physiol.* 77: 608–611
14. Carroll, S. B., and Laughton, A. 1987. In: Glover, D. M. ed. *DNA Cloning* Volume III, *A Practical Approach*, pp 89–Oxford, IRL press
15. Matile, P. 1975. *The Lytic Compartment of Plant Cells*. (Cell Biology Monographs, Vol. 1) Springer, New York, Wien
16. Kaiser, G., E. Martinoia, J. M. Schmitt, D. K. Hincha, and U. Heber. 1986. *Planta* 169: 345–355
17. Lörz, H., C. T. Harms, and I. Potrykus. 1976. *Biochem. Physiol. Pflanzen* 169: 617–620
18 Schmitt, J. M., E. Martinoia, D. K. Hincha, and G. Kaiser. 1987. In: *Plant Vacuoles* (B. Marin, ed). Plenum Press, pp 43–50
19. Latzko, E. and G. J. Kelly. 1983. *Physiol. Veg.* 21: 805–815
20. Ting, I. P. 1985. *Ann. Rev. Plant Physiol.* 36: 595–622
21. Winter, K. 1985. In: J. Barber, N. R. Baker, eds., *Photosynthetic Mechanisms and the Environment*. Elsevier, Amsterdam, pp. 329–387
22. Nimmo, G. A., H. G. Nimmo, I. D. Hamilton, C. A. Fewson, and M. B. Wilkins. 1986. *Biochem. J.* 239: 213–220
23. Matsuoka, M., and Hata, S. 1987. *Plant Physiol.* 85: 947–951
24. Edwards, G. E., J. G. Foster, K. Winter. 1982. In: *Crassulacean Acid Metabolism*. (eds. Ting, I. P. and M. Gibbs) Am. Soc. Plant Physiol., Rockville, MD pp. 92–111

Sensitivity to Far-uv Light as a Parameter for the Study of *in vivo* Regulation of Development of PSII Particles in Cyanobacteria: An Analytical Model

S. K. BHATTACHARJEE

Molecular Biology and Agriculture Division
Bhabha Atomic Research Centre
Bombay 400 085, India

Summary

Assuming the chromosome and PSII particles in cyanobacteria as independent lethal targets of ultraviolet (UV) radiation, an analytical model for the development of the PSII particle has been proposed. The model assumes that the UV-sensitivity of PSII particles is regulated by functional association and dissociation of the high turnover herbicide binding B protein and the rest of the core particle. Preliminary data on mutants with altered UV-sensitivity are discussed in the light of this model.

Introduction

Visible light is known to regulate the UV-sensitivity of cyanobacteria (1, 2). Since there is a clear correlation between lethality and inactivation of PSII function, this seems to be mediated by the regulation of UV-sensitivity of PSII particles (3). The herbicide binding high turnover protein also appears to be directly involved in switching between the UV-sensitive and UV-resistant states of PSII (4).

We have earlier proposed a model in which the B protein plays a regulatory role in determining the UV-sensitive and -resistant states of PSII (4, 5). Here we present the dynamic aspect of the model to explain the kinetics of change in the UV-sensitivity when dark-survival is measured in different shift experiments with *A. nidulans*. We also discuss how mutants with altered UV-sensitivity may be used to investigate the regulation of development of PSII particles using the above model as a working hypothesis.

Abbreviations. CAP: Chloramphenicol; DCMU: 2/3–4, Dichloropheny 1/–1, 1–dimethylurea; Atrazine: 2–Chloro-4-ethylamino-6-isopropylamino-s-triazine; MNNG: N-Methyl-N'-Nitrosoguanidine.

Material and Methods

Strain: The strain used in these experiments was IU625, kindly supplied by Dr R. S. Safferman.

Medium for growth: Medium used was BG11 as described by Allen both for growth in liquid and on agar plates (6).

Growth Conditions: Light from tungsten lamps was used both for growth in liquid in a shaker at $35\pm1^\circ$C and on agar plates. Unless otherise mentioned, the intensity of light was 12 W/m^2 as measured by IL700 radiometer, International Light, INC, Newburyport, M. A. USA, with detector PT 171 C NO. 1027.

Dark-survival: Plates with irradiated and control cells were incubated in dark for 24h at room temperature (about 25°C) immediately after UV-irradiation. The plates were then transferred to an incubator for growth of colonies at 5 W/m^2 intensity of light from a bank of tungsten lamps. The temperature of growth was $35\pm1^\circ$C. (Other experimental details may be found in Ref. 3 and 17.)

The Assumptions of the Model

Assumption 1: UV damage to DNA and PSII which lead to lethality are independent events.

Consequently,

$$S = (1 - P_n)(1 - P_e) \qquad (i)$$

where S is the survival fraction; P_n and P_e are functions of UV-fluence, and represent probabilities of inactivation of a cell by damage only to DNA and PSII respectively (7).

Assumption 2: PSII particles exist in two states designated by $PSII_x$ and $PSII_y$ such that the former is UV-resistant, and the latter UV-sensitive. This assumption along with equation (i) above leads to

$$S = \alpha a + \beta X \qquad (ii)$$

where α, β are positive and X is the concentration of PSII particles in the UV-resistant state (see Appendix 1).

The above postulates do not define the multiplicity of targets and number of hits necessary for the lethal inactivation of a cell. The need for such a definition does not arise in the present formulation since α and β in equation (ii), which depend on probabilities of damage as well as repair are assumed constant for a given fluence of UV for both targets. Under such conditions, change in survival fraction, S, may conveniently be interpreted as solely due to change in X.

Assumption 3: The herbicide binding high turnover B protein and $PSII_x$

interact with 1:1 stoichiometry as shown below (8),

$$B_p + PSII_x \underset{K_{-1}}{\overset{K_1}{\rightleftarrows}} PSII_Y \overset{K_2}{\to} PSII_x + B_{p'}, \tag{iii}$$

where B_p represents B protein before functional association with $PSII_x$, and $B_{p'}$ the altered protein when dissociated from the complex $PSII_y$ by the light dependent step (9). $PSII_y$ simply represents the functional complex, $(PSII_x \cdot B_p)$ (10).

Assumption 4: In the 'steady state' (or equilibrium) under any given set of growth conditions like light quality and intensity, temperature, etc., the ratio of number of $PSII_y$ and $PSII_x$ particles per cell is regulated such that

$$Y/X = \phi \tag{iv}$$

where ϕ is a constant.

Using the rate equation for X and the above regulatory constraint for the case of steady state during the exponential phase of growth (see Appendix 2),

$$\dot{X} = -\lambda_e X \tag{v}$$

where $\lambda_e = K_1B - (K_{-1} + K_2)\phi_e - \rho$.

In the steady state the average number of PSII particles per cell is also assumed constant (See Appendix 1), and therefore $\dot{R} = O$. Since $R = X(1 + \phi_e)$ and ϕ_e is a positive constant, at equilibrium $\dot{X} = 0$. From (v), at equilibrium, we obtain the condition

$$\lambda_e = 0$$

or $K_1B_e - (K_{-1} + K_2)Y_e/X_e - \rho = 0$

Hence, the free pool of B protein at equilibrium during exponential growth of the culture with constant growth rate ρ, must satisfy the following:

$$B_e = \{(K_{-1} + K_2)\, Y_e/X_e + \rho\}/K_1 \tag{vi}$$

and the equilibrium ratio for the number of PSII parties in state $PSII_y$ and $PSII_x$ is then given by

$$\emptyset_e = Y_e/X_e = (K_1B_e - \rho)/(K_{-1} + K_2) \tag{vii}$$

It may be noted that the equilibrium values of B, X and Y will be constant for any set of constant values of the rate constants K_1, K_{-1}, K_2 and ρ. Thus a new equilibrium state will be reached in altered physiological state in which one or more of the four rate constants are changed. In the following we obtain the constraints on the system which would explain the transient behaviour of X, when the culture undergoes a physiological shift between two equilibrium states.

Results and Discussion

Case 1: L → L + CAP shift

Consider the shift experiment in which CAP is added to an exponentially growing culture in light. Assuming that CAP introduces physiological shift only by blocking protein synthesis, the rate equation for X at the instant when CAP is added,

$$\dot{X} = - \lambda_c X \qquad \qquad \text{(viii)}$$

where $\lambda_c = K_1 B_e - (K_{-1} + K_2) Y_e / X_e$, and B_e, X_e, Y_e are equilibrium values prior to addition of CAP. It may be noted that only growth rate

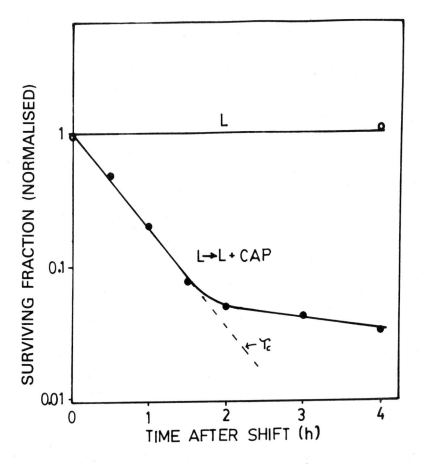

Fig. 1: The kinetics of change in UV-sensitivity when 260Jm^{-2} of far-UV was used to monitor the dark-survival following addition of CAP at 5 μg/ml final concentration. The experimental details were same as in (11). The survival fractions are normalised with respect to the value at 0 time.

constant ρ is altered and is assumed to be zero from the instant of shift to L + CAP. The other three rate constants are assumed to retain the constant values as in the exponential growth phase. The solution to (viii) is

$$X = X_e e^{-\lambda_c t} \tag{ix}$$

so long as λ_c is constant, i.e.

$$K_1 B_e - (K_{-1} + K_2) Y_e/X_e = \text{constant.}$$

Fig. 1 shows the first order decay kinetics, as predicted by (ix) substituted in (ii), for almost 90 minutes in our experimental conditions (11). The time constant of decay, τ_c, which is experimentally determined is given by

$$1/\tau_c = K_1 B_e - (K_{-1} - K_2) Y_e/X_e \tag{x}$$

Since the herbicide which specifically binds to B protein will titrate this

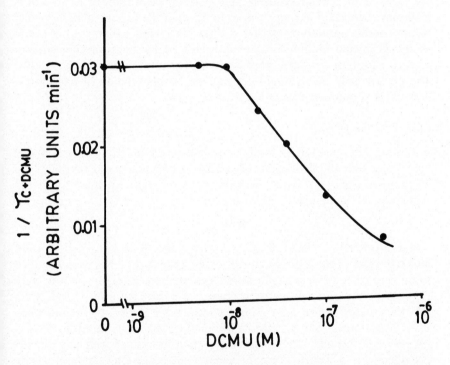

Fig. 2: *In vivo* titration of the herbicide binding B protein as observed by measurement of dark-survival following UV-irradiation in double inhibitor experiment. To an exponentially growing culture of the strain IU625 in BG11 medium CAP was added at 5 μg/ml final concentration. The culture with CAP was then divided equally (5 ml each) in 100 ml flasks, DCMU added to final concentrations as indicated and immediately transferred back to light. Dark-survival following 260 Jm^{-2} of UV was measured as in Fig. 1. From dark-survival levels measured within 70 min of addition of both the inhibitors, the decay time τ_{C+DCMU} was calculated.

protein from the free pool, the model predicts that $1/\tau_c$ should decrease with increasing concentration of DCMU in double-inhibitor experiments. The result of such an experiment in which CAP and DCMU are added together to titrate B protein with varying concentration of the herbicide shows that (Fig. 2),

(1) $\tau_{c + \text{DCMU}}$ is a function of DCMU concentration (for concentrations $> 10^{-8}$M).

and (2) with increasing concentration of DCMU, $\tau_{c+\text{DCMU}}$ decreases as expected from equation (x). These results, therefore, seem to clearly support the model.

If it is further assumed that there is no significant difference in the internal and external concentration of DCMU, the double-inhibitor experiment could be used to determine *in vivo* binding constant of the herbicide with the B protein. It is also evident from the results that even in the absence of protein synthesis free pool concentration of B protein is maintained constant at the steady state level of the exponentially growing culture, for a considerable period of time. This implies that free pool level of the herbicide binding protein is regulated by some mechanism not requiring protein synthesis, perhaps from a pool of its precursor (12, 13). The process may involve light-dependent phosphorylation or palmitolyation of B protein in the precursor pool (14).

Case 2: L → D shift

Assuming that one or more of the interaction constants K_1, K_{-1} and K_2 are light dependent and that magnitude of the corresponding rate constants are different in light and dark, we may write the rate equation for X(see Appendix 2) immediately following shift

$$\dot{X} = \lambda_d X \qquad (xi)$$

where $\lambda_d = (K'_{-1} + K'_2) \, Y_e/x_e - K'_1 \, B_e$ and K_1', K'_{-1} and K'_2 are altered values of the interaction constants in the dark. ρ has been assumed to be zero since protein synthesis is blocked immediately on shift to dark (15). Within the experimental accuracy, the dark-survival of the strain BD.1 increases exponentially for a considerable period (>3h) following L → D Shift (3). Similar results were obtained with the strain IU625 (unpublished results). The solution to (xi) is consistent with this kinetics provided λ_d is a positive constant. Hence, the condition that the new rate constants must satisfy is

$$(K'_{-1} + K'_2) \, Y_e/X_e > K'_1 \, B_e$$

Further, since it is known that B protein is not degraded in dark, $K'_2 = 0$, and hence the condition reduces to

$$K'_{-1} \, Y_e/X_e > K'_1 \, B_e$$

and $K'_{-1} \, Y_e/X_e - K'_1 \, B_e = \lambda_d$ = constant.
The solution for X is then,

$$X = X_e \, e^\lambda d^t$$

so long as $\dot{\lambda}_d \ll \dot{X}$, and $\dot{\lambda}_d \simeq 0$.

After prolonged incubation in dark, the new equilibrium is reached, and then the ratio of the numbers of PSII$_y$ to PSII$_x$ particles is given by

$$Y_d/X_d = K'_1/K'_{-1} \, B_d$$

This follows immediately from the fact that at equilibrium in dark, the association and dissociation rates are equal, i.e. $K'_{-1} \, Y_d = K'_1 \, X_d B_d$.

Case 3: D → L shift

This case may be looked upon as transition between two equilibrium states and the reverse of Case 2 (L → D) discussed above. During the transition the dark-survival shows damped oscillations until the level of the new equilibrium state, i.e. the level of the exponentially growing phase, is reached (3) (Fig. 3). In the following we discuss the conditions necessary for the solution of equation (iii) (Appendix 2), to explain the experimental results by the proposed developmental model.

On D → L shift, X has to satisfy the equation (xii) with coefficient λ altered because of new rate constants. Since cell division in *A. nidulans* is synchronised and not exponential, $\rho = 0$ for at least one generation time following shift (5,16)

$$\ddot{X} + \lambda_1 \dot{X} + \delta_1 X = 0 \qquad \text{(xii)}$$

where $\lambda_1 = K_1 B - (K_{-1} + K_2)\text{\o}$
$\qquad \delta_1 = K_1 \dot{B} - (K_{-1} + K_2) \, \dot{\text{\o}}$
and $\text{\o} = Y/X$

Further, since number of PSII particles per cell is considered constant,

$$R = X = Y = \text{constant}$$

so that $\dot{Y} = -\dot{X}$, which gives $\text{\o} = (-R/X^2) \, \dot{X}$. Since the model predicts relatively higher equilibrium value of X in dark than in light in the exponential phase, it may be assumed that R/X^2 is very small at least till the first synchronous division after D → L shift. Then ø may be treated as constant and hence $\dot{\text{\o}} \simeq 0$ transiently. The above equation will give damped oscillations for X(t) as solution so long as $K_1 B$ is a sufficiently slowly varying function of time to be effectively constant, and $K_1 B > \lambda^2_e /4$. The solution then is

$$X(t) = C_1 e^{-\lambda_1 t} \sin(\omega \, t + C_2)$$
$$\text{where } \omega = \sqrt{4K_1 B - \lambda_t^2}$$

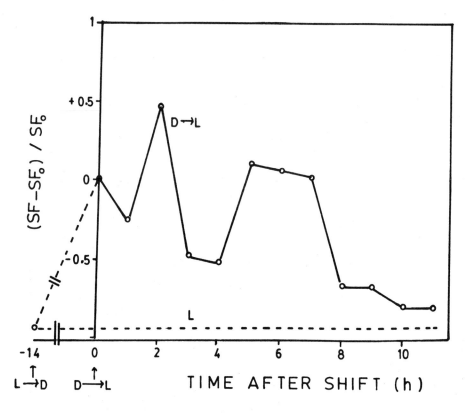

Fig. 3: The kinetics of change in UV-sensitivity when 260 Jm^{-2} of far-UV was used to monitor the dark-survival of an exponentially growing culture which was incubated in dark for 145h (L $\rightarrow$ D) before shift to light (D $\rightarrow$ L). The experimental details were the same as described earlier (3). The dashed line denotes the level of UV-sensitivity of the exponentially growing culture before transfer to dark.

C_1 and C_2 are constants of integration. It is obvious that the above analytical solution cannot be considered valid for the entire transit period since λ and δ are not constants indefinitely as the culture enters the exponential phase of growth to establish the new equilibrium state.

We have isolated, using MNNG, several UV-resistant and herbicide resistant mutants to identify genes which may be involved in the regulation of development of PSII particles (unpublished results) (Fig. 4). UV-resistant mutants are likely to be altered in the function ø or in the sensitivity (P_y) of the primary lethal target to far-UV radiation. A class of herbicide (DCMU and atrazine) resistant mutants are expected to alter the UV-sensitivity by either overproducing B_p compared to that in wild type or by altering the structure of B_p which affect both DCMU binding site and the UV-sensitivity of $PSII_y$ (i.e. by altering P_y). The herbicide resistant

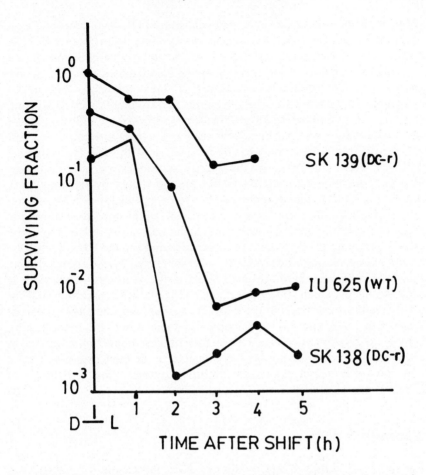

Fig. 4: The kinetics of change in UV-sensitivity of two mutants selected for resistance to the herbicide DCMU at $2X10^{-6}$ M (DC-r), following D → L shift. The experimental details were the same as described earlier that all the cultures were grown in relatively low light intensity (8 W/m) (3,4). UV-sensitivity was monitored by measuring the level of dark-survival following 390 Jm^{-2} of far-UV radiation.

mutants altered in P_y may be UVresistant or-sensitive compared to wild type depending on whether P_y of mutant is less or more than that of wild type under identical conditions of growth and UV-fluence. Two such DCMU resistant mutants (SK 138 and SK 139) altered in the kinetics of UV-sensitivity in D → L shift experiment in low light are shown in Fig. 4. However, we have not yet determined whether the alteration of UV-sensitivity of these mutants is due to change only in P_y. Mutations selected for altered UV-sensitivity, as the model suggest, should be useful in understanding the regulatory mechanisms in the development of functional PSII particles in cyanobacteria.

Cyanobacteria occupy a very significant status in the evolution of life on earth. It is thought that these organisms evolved when there was little oxygen and consequently no ozone in the atmosphere to absorb the harmful UV-radiation from the sun. Being solely dependent on light for energy, the photosynthetic apparatus of their ancestors had to be sufficiently adaptive to choose between photosynthetic efficiency and resistance to UV-radiation in order to survive the assault and still harvest the light energy for growth. Obviously this regulation could be best achieved through fine tuning of the proportion of the two types of PSII particles, $PSII_x$ and $PSII_y$, by the quality and intensity of the radiation. This proportion is represented in the present model by the regulatory function ø. Clearly the selective advantage of a mechanism of regulation not involving synthesis of the vital components of the apparatus starting at the gene expression level would be very high under such pressures since the cells could then tolerate a delay in repair of damage to the DNA itself (17). That such pressures did exist seem to be supported by the recent finding that in *A. nidulans* there are three allels of the gene psbA coding for B protein, all of which are transcribable (18). It will, therefore, not be surprising if it turns out that the dark-and perhaps also the photo-repair of damage to DNA and PSII are coupled in some way in cyanobacteria by complex regulatory processes. Unravelling of these regulatory mechanisms may help in understanding how cyanobacteria, or their primitive forms, may have ushered in eukaryotic evolution by tolerating high fluxes of UV(19).

Appendix 1

Let the probabilities of inactivation of PSII particles in state $PSII_x$ and $PSII_y$ be P_x and P_y respectively. Then

$$P_e = (P_x X + P_y Y)/R \tag{i}$$

where X, Y represent mean number of PSII particles in states $PSII_x$ and $PSII_y$ respectively;

$$R = X + Y \tag{ii}$$

and P_x and P_y <1;

R is the average number of PSII particles per cell and is considered constant under 'steady state' conditions.

From (i) and (ii)

$$P_e = [P_y R + (P_x - P_y)X]/R \tag{iii}$$

Since

$$S = (1 - P_n)(1 - P_e),$$

using equation (iii),

$$S = (1-P_n) \{(1-P_e)-(P_x-P_y) \, X/R\}$$

Since $PSII_x$ has been assumed to be UV-resistant state compared to $PSII_y$

$$P_y > P_x$$

Hence

$$S = (1-P_n) \, (1-P_y)+(1-P_n) \, (P_y-P_x)X/R$$
$$= \alpha + \beta X$$

where both α and β are functions of UV-fluence and positive.

Appendix 2

In order to obtain solution for X as a function of time during perturbed conditions induced by shift from steady state condition, consider the developmental equation for PSII,

$$B_p + PSII_x \; \underset{K_{-1}}{\overset{K_1}{\rightleftharpoons}} \; PSII_y \overset{K_2}{\longrightarrow} PSII_x + B_p$$

Rate equation for X at any time during exponential growth phase (in light) is

$$\dot{X} = -K_1B \, X + (K_2 + K_{-1}) \, Y + \rho X$$

where ρ is the rate of growth of X from its precursors.
Using the regulatory constraint, $Y = \phi X$,

$$\dot{X} = -K_1BX + (K_{-1} + K_2)\phi X + \delta X$$

On differentiating with respect to time, and rearranging,

$$\ddot{X} = \{K_1B- (K_{-1} + K_2) \, \phi \, -\rho\} \, \dot{X}+\{K_1\dot{B}- (K_{-1} + K_2) \, \dot{\phi}\}X=0$$

$$\text{or} \qquad \ddot{X} + \lambda\dot{X} +\delta \, X = O$$

$$\text{where} \qquad \lambda = K_1B-(K_{-1} + K_2)\phi-\rho$$

$$\text{and} \qquad \delta = K_1\dot{B}-(K_{-1} + K_2)\dot{\phi}$$

In cases of perturbations which stop growth, $\rho = 0$.

Acknowledgements

I am thankful to Dr S. K. Mahajan for critically going through the manuscript and for useful suggestions. I am also grateful to Dr A. J. Hoff for his valuable comments. The help of Dr K. A. V. David, Shri S. S.

Rane, Smt Manjula Mathur and Shri K. G. Khot is gratefully acknowledged. Shri N. D. Shirke very kindly supplied computer data on the
simulation studies of the model.

References

1. Bhattacharjee S. K. and K. A. V. David 1977. 'Unusual resistance in the dark phase of
   blue-green bacterium *Anacystis nidulans*'. *Nature* 265: 183–184
2. O'Brien, P. A. and J. A. Houghton 1982. 'Photoreactivation and excision repair of UV
   induced pyrimidine dimers in the unicellular cyanobacterium'. *Gleocapsa alpicola
   synechocystis* PCC 6308. *Photochem. Photobiol* 35: 359–364
3. David K. A. V., S. S. Rane, Manjula Mathur and S. K. Bhattacharjee 1988. 'Herbicide
   mediated UV-resistance in cyanobacteria: On the role of photosynthetic electron
   transport system rather than replicative DNA as lethal target determining dark-survival
   of *Anacystis nidulans*'. *Photochem. Photobiol.* 47: 107–113
4. Bhattacharjee S. K. and K. A. V. David 1987. 'UV-sensitivity of cyanobacteria
   Anacystis nidulans: II. A model inovolving PSII reaction centre as lethal target and the
   herbicide binding high turnover protein as regulator of dark-repair'. *Indian J. Exptl. Biol.*
   25: 837–842
5. Bhattacharjee S. K. 1987. 'On the role of herbicide binding B protein as regulator of
   UV-sensitivity of PSII: A model explaining dark-survival characteristics of cyanobacteria'. *Proc. Indian Natn. Sci Acad.* B.53 407–413
6. Allen, M. M. 1968, 'Simple conditions for growth of unicellular blue-green algae on
   plates'. *J. Phycol.* 4: 1–4
7. Zimmer K. G. 1961. *Studies on Quantitative Radiation Biology* (English translation by
   H. D. Griffith; Published by Oliver and Boyd Ltd., Edinburgh and London)
8. Koike M. and Inoue Y. 1985. 'Properties of a peripheral 34 kD protein in *Synechococcus
   vulcanus* Photosystem II particles. Its exchangeability with spinach 33 kD protein in
   reconstitution of O_2 evolution'. *Biochem. Biophys.* Acta 807: 64–73
9. Wettern M. 1986. 'Localization of 32000 Dalton chloroplast protein pools in thylakoids;
   significance in atrazine binding'. *Plant Sci.* 43: 173–177
10. Renger G. 1976. 'Studies on the structural and functional organization of system II of
    photosystem II of photosynthesis. The use of trypsin as a structurally selective inhibitor at
    the outer surface of the thylakoid membrane'. *Biochem. Biophys.* Acta 440: 287–300
11. Bhattacharjee S. K. 1977. 'Unstable protein mediated ultraviolet light resistance in
    Anacystis nidulans'. *Nature* 269: 82–83
12. Kyle D. J. and Ohad I. 1986. 'The mechanism of photoinhibition in Higher Plants and
    Green Algae'; in *Photosynthesis III* Eds. L. A. Staehelin and C. J. Arntzen. 468–475
    (Springer-Verlag)
13. Gaba V., Marder, J. B., Greenberg, B. M., Mattoo A. K. and Edelman M. 1987.
    'Degradation of the 32kD herbicide binding protein in far red light'. *Plant Physiol.* 84:
    348–352
14. Mattoo A. K. and M. Edelman 1987. 'Intramembrane translocation and posttranslational
    palmitoylation of the chloroplast 32-kDa herbicide binding protein'. *Proc. Natl. Acad.
    Sci.*USA 84: 1497–1501
15. Hayashi F, Ishida, M. R. and Kikuchi, T. 1969. 'Macromolecular synthesis in a
    blue-green alga *Anacystis nidulans*, in dark and light phases'. *A Rep. Res. Reactor Inst.* 2:
    56–66
16. Asato Y. 1983. 'Dark incubation causes reinitiation of cell cycle events in *Anacystis
    nidulans*'. *J Bacteriol.* 153: 1315–1321
17. S. K. Bhattacharjee, Manjula Mathur, S. S. Rane and David K. A.V. 1987. 'UV-

sensitivity of cyanobacteria *Anacystis nidulans*: I. Evidence for PSII as lethal target and constitutive nature of a dark-repair system against damge to PSII'. *Indian J. Exptl. Biol.* 25: 832–836

18. Haselkorn R. 1988. 'Molecular genetics of herbicide resistance in Cyanobacteria'. In the proceedings of the INDO-US workshop on application of Molecular Biology in Bioenergetics of Photosynthesis. KNY, New Delhi, India

19. Schopf J. W, Hayes, J. M. and Walter M. R. 1983. 'Evolution of earths earliest ecosystems: Recent progress and unsolved problems'. In *Earth's Earliest Biosphere.* 361–384. (Princeton University Press)

Effects of Light Chilling on Photophosphorylation in Cucumber

ROBERT R. WISE AND DONALD R. ORT

U. S. Department of Agriculture Agricultural Research Service &
Department of Plant Biology, University of Illinois,
289 Morrill Hall, 505 S. Goodwin Avenue,
Urbana, Illinois 61801, USA

Summary

The response of *in situ* photophosphorylation in attached cucumber (*Cucumis sativus* L.) leaves to chilling (5°C) under strong illumination (1000 $\mu E \cdot m^{-2} \cdot s^{-1}$) was investigated. A single-beam kinetic spectrophotometer fitted with a clamp-on, whole leaf cuvette was used to measure the flash-induced electrochromic absorbance change at 518–540 nm ($\Delta \Delta_{518-540}$) in attached leaves. The relaxation kinetics of the electric field-indicating $\Delta \Delta_{518-540}$ measure the rate of depolarization of the thylakoid membrane. Since this depolarization process is normally dominated by proton efflux through the coupling factor during ATP synthesis, this technique can be used, in conjunction with careful controls, as a monitor of *in situ* ATP formation competence.

Whole, attached leaves were chilled at 5°C and 1000 $\mu E \cdot m^{-2} \cdot s^{-1}$ for up to 6 h, then rewarmed in the dark and 100% R.H. at room temperature for 30 min. Leaf water potential, chlorophyll content, and the effective optical path length for the absorption measurements were not affected by the treatment. Light-and CO_2-saturated leaf disc oxygen evolution and the quantum efficiency of photosynthesis were inhibited by approx. 50% after 3 h of chilling and by approx. 75% after 6 h. Despite the large inhibition to net photosynthesis, the measurements of $\Delta \Delta_{518-540}$ relaxation kinetics showed photophosphorylation to be largely unaffected by the chilling and light exposure. Our measurements showed that the chilling of whole leaves in the light caused neither an uncoupling of photophosphorylation from photosynthetic electron transport nor any irreversible inhibition of the chloroplast coupling factor *in situ*. The sizeable inhibition observed after light chilling in net photosynthesis cannot, therefore, be attributed to a reduced capacity for photophosphorylation. The amplitude of the $\Delta \Delta_{518-540}$, however, was reduced by about half after 3 h of treatment. The cause of this decreased amplitude and its significance to the overall inhibition is under investigation.

Introduction

When held at temperatures lower than about 10°C, many species of crop plants show persistent and even irreversible reductions in light-and

---

Abbreviations: DCCD—dicyclohexylcarbodiimide; PAR—photosynthetically active radiation; ΔP-transmembrane protonmotive force.

CO_2-saturated photosynthesis after rewarming. This photosynthetic inhibition develops more rapidly and is more ruinous if the chilling occurs concurrent with illumination (1–4). Because the photosynthetic reduction is only partially restored by saturating CO_2 levels (4, 5), only a portion of the inhibition is stomatal-mediated while the remainder must lie at the level of the chloroplast itself. It is the non-stomatal, chloroplast biochemical dysfunction that has held the interest of our laboratory.

The chilling of whole plants in the light causes a reduction in photosynthetic activity of both attached leaves (4,6,7) and isolated thylakoids (8,9). However, Kee *et al.* (8) have shown that even though whole chain electron transport was substantially inhibited by light chilling, the remaining activity was in excess of that needed to support measured rates of light-and CO_2-saturated net photosynthesis. Thus, these reductions in the turnover capacity of electron transport carriers cannot be the direct cause of light-and chilling-induced photosynthetic inhibition.

The functioning of many chloroplast processes is vitally dependent upon the availability of ATP. Thus, before any potential direct effects of light chilling can be assigned to these 'ATP-dependent' processes, it is important to establish whether the ATP formation competence of these chill-sensitive plants is compromised by the stress. Photophorylation is a highly regulated and relatively labile process and thus a prime candidate for disruption by environmental stress. Indeed, there is strong evidence that ATP synthesis and ATP levels are adversely affected by water deficits (10–13). In the case of chilling adverse effects on ATP formation competence have been reported (9, 14–16) but the relevance of these *in vitro* studies to the *in situ* inhibition was not established. A special difficulty associated with studying stress-induced perturbations is to distinguish between inhibitions that are actually part of the *in vivo* process from inhibitions induced through manipulation of tissue made labile by the stress.

In this study we investigated the response of photophosphorylation to light chilling in attached leaves by monitoring the flash-induced electrochromic absorption bandshift measured as an absorption change at 518–540 nm (17). The basis for the measurement of *in situ* ATP formation is that the membrane depolarizing proton efflux through the coupling factor complex that drives the ADP phosphorylation reaction results in more rapid disappearance of the electric potential produced by the light flashes (18). From the analysis of the thylakoid membrane depolarization kinetics of light-chilled cucumber leaves we conclude that although photophosphorylation shows some marginal perturbation, it is not nearly severe enough to account for the magnitude of the inhibition of light-and CO_2-saturated photosynthesis.

Materials and Methods

Plant growth conditions. Cucumber plants (*Cucumis sativus* L. cv. Ashley) were raised from seed in a soil/peat/vermiculite mixture, watered daily and fertilised weekly. The plants were grown in a controlled environment chamber (600–800 $\mu E \cdot m^{-2} \cdot s^{-1}$, 14 hr photoperiod, day/night temperatures of 23°C/20°C) as datailed elsewhere (4). Attached leaves which had almost reached full expansion were used for the various *in vivo* or *in situ* measurements.

Chilling treatments. All chilling treatments were initiated at 10:00 a.m. (two hours into the photoperiod). Whole plants were wrapped in moistened Labmat (a two-ply, plastic/absorbant paper, disposable bench covering) with only a 15 cm^2 area of the leaf to be treated left exposed in order to prevent excessive water loss which could otherwise confound our studies (19). This leaf was positioned between two horizontal frames and the entire plant/frame assembly placed in a refrigerated cooler set at 2°C. Light (1000 $\mu E \cdot m^{-2} \cdot s^{-1}$) was provided by a 200 W spot light and passed through a dichroic filter ('cold' mirror) and 5 cm of water. Abaxial leaf temperature was measured with a thermocouple thermometer and fluctuated between 4–6°C.

Photosynthetic measurements. Oxygen evolution from 10 cm^2 leaf discs was measured at 23° C and 5% (v/v) CO_2 using a Hansatech LD1 polaragraphic oxygen electrode (Hansatech Ltd., Norfolk, England) as described in Delieu and Walker (20). Irradiance was provided by a slide projector lamp and attenuated with neutral density filters. Light saturation curves were constructed by starting at the lowest irradiance and increasing, stepwise, to full intensity (1550 $\mu E \cdot m^{-2} \cdot s^{-1}$ PAR). The quantum yields were calculated from the linear portion of the curve at irradiances below 200 $\mu E \cdot m^{-2} \cdot s^{-1}$. Absorbed irradiance was determined by subtracting reflected and transmitted light from incident light. Reflected light was measured 1 cm above the leaf surface at an angle of 45° and transmitted light was measured directly beneath the leaf, with a LiCor LI-185B quantum sensor (LiCor Inc., Lincoln, NE). Reflected light was 2.1±1% of incident (X±S.E., n = 15) and transmitted was 8.6 ± 3% of incident (n = 14) from 40 to 1400 $\mu E \cdot m^{-2} \cdot s^{-1}$. Therefore, absorbed light was calculated to be 89.3 ± 1.3% of incident irradiance.

Leaf water potential measurements. Leaf water potentials (ψ_L) were determined using a laboratory-built thermocouple psychrometer and the isopiestic technique of Boyer and Knipling (21).

Chlorophyll determinations. Chlorophyll concentrations in 80% acetone extracts were determined spectrophotometrically using the specific absorption coefficients for chlorophyll *a* and *b* of Ziegler and Egle (22, see also 23).

Spectrophotometric measurements of in vivo flash-induced absorbance changes. The flash-induced electrochromic absorption band shift was measured using a laboratory-built single beam spectrophotometer. An attached leaf was positioned in a foam-lined cuvette clamp. A shuttered, dim (6×10^{-2} J·m^{-2}·s^{-1}) measuring beam (4 nm band width) was provided by a diffraction monochromator (Model AH10, Instrument SA Inc. Metuchen, NJ), and delivered to a 1cm diameter spot on the adaxial leaf surface via a fiber optic light guide. The measuring beam passed through the leaf, was collected from the abaxial surface by another light guide and routed to a photomultiplier tube. The saturating actinic flash was generated by a xenon lamp (6μs duration at half peak height; model FX-200 Flash Tube, EG & G, Salem, MA), passed through a red blocking filter (Corning CS 2-59), and delivered to the abaxial surface by a third light guide. The brief duration of the very bright actinic flash caused a single turnover of >98% of all the photosystems in the illuminated portion of the leaf (24). The actinic flash was excluded from the photomultiplier tube by filters (a DT green wide band interference filter, 66.1055 Rolyn, Arcadia, Ca and a Corning 4-96) and by the positioning of the light guides. The photomultiplier output was digitised and stored in a Nicolet model 1174 signal averager (Nicolet Inst. Corp., Madison, WI). For the flash-induced absorption spectrum shown in Fig. 2, 8 to 64 flash events were summed (depending on signal amplitude) and normalised to one flash. For measurements of the flash-induced electrochromic effect, an equal number of events at 540 nm was subtracted from those at 518 nm to eliminate changes not due to electrochromism such as cytochrome changes and light scattering. The resulting trace of $\Delta A_{518-540}$ (Fig. 3, Trace A) can be seen to have a biphasic decay. Since only the initial, fast component of decay is associated with photophosphorylation (25) the slow component was subtracted. The remaining fast decay was expressed as the relaxation time constant, τ, according to the equation $\Delta A_{518-540} = \Delta A_{514-540 \, max} e^{-t/\tau}$ where both t (time) and τ are in ms. For the data given in Fig. 5, the differences in the various values before and after chilling were tested using an analysis of variance on a repeated measures design, testing τ against the within-plant residual error. Differences between the 3 chilling treatments were tested using a one-way analysis of variance at a 5% level of significance.

Results

Chilling (5°C) attached cucumber leaves in the light (1000 μE·m^{-}·s^{-1}) caused an inhibition of photosynthesis when measured at 23°C after the leaf had rewarmed in the dark (Fig. 1). This was seen as both a decrease in the initial slope of the light saturation curves (i.e. quantum yield) and a reduction in the light-and CO_2-saturated rate of leaf disc oxygen evolution (Fig. 1). Neither parameter was much affected by a 6 h treatment of 23°C and 1000 μE·m^{-2}·s^{-1}. The amount of inhibition of photosynthesis and

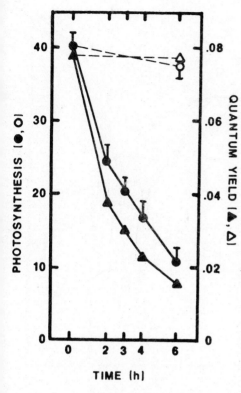

Fig. 1: Light-and CO_2-saturated photosynthesis (●) (in μ mol $O_2 \cdot m^{-2}$ leaf area $\cdot s^{-1}$) and quantum yield (▲) as a function of duration of chilling and light treatment (5°C, 1000 $\mu E \cdot m^{-1} \cdot s^{-1}$) of attached leaves. The open symbols (○,△) represent plants treated for 6 h at 23°C and 1000 $\mu E \cdot m^{-2} \cdot s^{-1}$. The plants rewarmed during a 30 min incubation in the dark at 100% RH and room temperature prior to measurements. Photosynthesis was measured as leaf disc oxygen evolution at 23°C, 5% (v/v) CO_2, and 1300 $\mu E \cdot m^{-2} \cdot s^{-1}$. The quantum yield ($\mu$mol O_2 evolved per μmol quanta absorbed) was calculated based on an absorbance of 89.3% of the incident irradiance by the leaf disc (see 'Materials and Methods') and for irradiances less than 200 $\mu E \cdot m^{-2} \cdot s^{-1}$. Mean ± s.e., n = 6.

quantum yield increased with increasing duration of light chilling. The observed inhibition to photosynthesis was not caused by leaf water deficit or by chlorophyll (Table 1).

Shown in Fig. 2 is the wavelength dependence of the absorption change measured 8 ms after the flash in a control, attached cucumber leaf. The flash-induced spectrum is indistinguishable from that of isolated spinach (x) thylakoids. Depolarization of the membrane by the presence of

TABLE 1

Leaf water potential (ψ_L) and chlorophyll content (Chl) after chilling (5°C) in the light (1000 $\mu E \cdot m^{-2} \cdot s^{-1}$). Mean ± S.E. are shown; Sample size is given in parentheses.

Duration of chilling (h)	ψ_L (MPa)	Chl (nmol.cm^{-2})
0	− 0.30 ± .01 (5)	43.3 ± 1.6 (17)
3	− 0.35 ± .01 (5)	42.0 ± 0.7 (5)
6	− 0.39 ± .02 (5)	40.8 ± 1.8 (8)

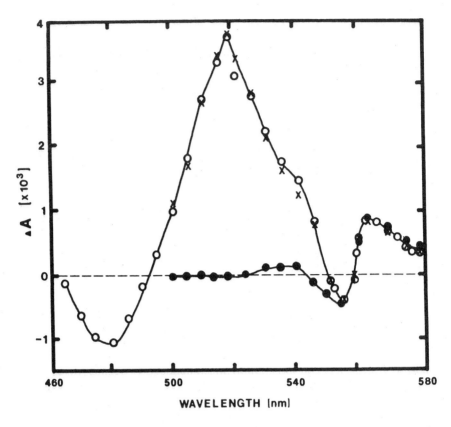

Fig. 2: Flash-induced absorbance spectra from 460–580 nm for attached cucumber leaves (0), isolated spinach thylakoids (X), and isolated spinach thylakoids treated with valinomycin (●). Amplitudes were taken 8 ms after the single-turnover, saturating, actinic flash.

valinomycin and K^+ (●) abolishes the electrochromic change and only the contribution by cytochrome turnover remains.

A typical signal averaged trace of the flash-induced electrochromic change ($\Delta A_{518-540}$) in attached cucumber leaf is shown in Fig. 3 A. We used a reference wavelength of 540 nm to isolate the electrochromic absorption change from any other contributory factors such as cytochrome redox changes (~545 to ~570 nm) and flash-induced scattering changes that exhibit a relatively broad and featureless spectrum. The 40 ms time constant for the relaxation of the electrochromic change seen in Fig. 3A is characteristic of active ATP synthesis. Trace B of Fig. 3 shows that DCCD, an inhibitor of ATP synthesis acting to prevent H^+ efflux through the coupling factor complex, dramatically slows the relaxation ($\tau \doteq 400$ ms).

Light chilling of attached cucumber leaves, while resulting in progressive loss in amplitude of the $\Delta A_{518-540}$ (Fig. 4; also see Fig. 6), caused relatively

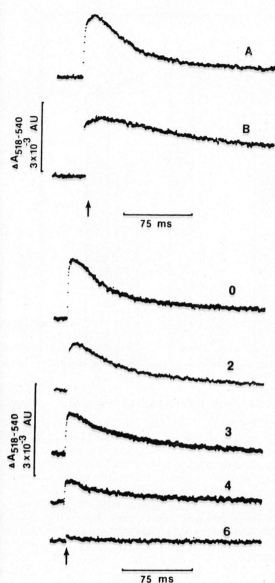

Fig. 3: Typical traces of the flash-induced absorbance change at 518 nm. A =Control cucumber leaf pretreated with 30 min of dark. B = Cucumber leaf given a 30 min surface application of 10 mM DCCD in 2% (v/v) methanol and 1% (v/v) Tween 20 in the dark.

Fig. 4: Typical flash-induced absorbance changes at 518–540 nm for attached cucumber leaves chilled (5°C) in the light (1000 $\mu E.m^{-2}.s^{-1}$) for the times (h) indicated, then incubated for 30 min at room temperature in the dark at 100% RH. The traces represent the normalised average of 16 actinic events at 540 nm subtracted from the normalised average of 8 actinic events at 518 nm. Single-turnover, saturating, actinic flashes were given at 0.1 Hz. The vertical arrow indicates the actinic flash.

little change in the relaxation kinetics (Fig. 4 and 5). Only at low flash frequency (e.g. 0.1 Hz) was detectably slower relaxation observed.

The possibility that the decrease in the amplitude of the flash-induced absorbance change at 518–540 nm (Figs. 4 and 6) was caused by a decrease in the optical pathlength through the leaf was investigated following the procedure of Rühle and Wild (26). Fig. 7 shows absorbance spectra of an intact cucumber leaf (trace a) and a suspension of cucumber thylakoids

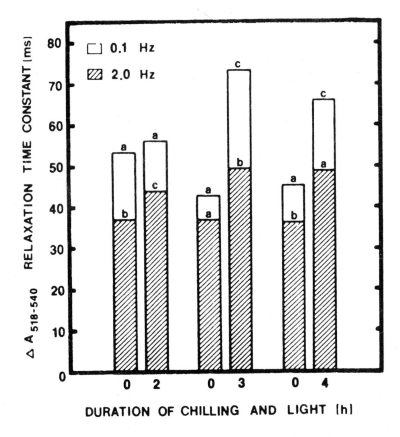

Fig. 5: Relaxation time constants for the decay of the fast phase of the flash-induced $\Delta A_{518-540}$ in attached cucumber leaves before and after a light chilling treatment of the indicated duration. Statistical differences are noted by lower case letters among the 4 bars within each of the 3 sets. Statistical comparisons among the 3 sets of bars are not presented. Bars with the same letter are *not* statistically different at the 5% level. Differences between the various time constant (τ) values before and after chilling were tested using an analysis of variance on a repeated measures design, testing τ aginst the within-plant residual error. Differences between the three chilling treatments were tested using a one-way analysis of variance at a 5% level of significance.

(trace C) of equal chlorophyll content (on a surface area basis). Trace D is the spectrum of a methanol-extracted, pigment-free leaf and is constant from 500 to 750 nm. By normalising traces A and C to trace D, the absorbance of the pigments themselves can be estimated. The intensification of absorbance due to increased pathlength caused by leaf architecture is then the ratio of the leaf absorbance (A_L) to that of the thylakoid suspension (A_T) at any discrete wavelength. Values for the intensification factor (β) are given for 4 wavelengths in Table 2 for control and chilled leaves. Chilling attached cucumber leaves for 3 h in the light did not significantly affect the

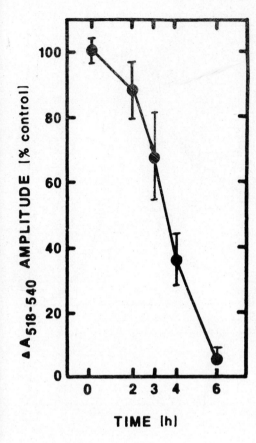

Fig. 6: Maximum amplitides of the fast phase of the $\Delta A_{518\text{-}540}$ immediately after the actinic flash for the plants in Fig. 5. The fast phase of the absorbance decay (see Fig. 3) was linearised on a log/linear plot and the ΔA at $t = 0$ was determined by extrapolation of the linearised plot back to the time of the actinic flash. Mean $\pm$ S.E., $n = 24$ for control and 6 for stressed plants. Control amplitude value was $2.1 \pm 0.1 \times 10^{-3}$ A. U.

TABLE 2

Effect of chilling ($5°C$) in the light (1000 $\mu E.m^{-2}.s^{-1}$) for 3 h on the intensification factor (β) of whole cucumber leaf absorbance at various wavelengths (λ). The intensification factor, β is the ratio of the leaf absorbance (A_L) to the thylakoid suspension absorbance (A_T) and represents the degree to which the leaf architecture increases the effective pathlength for absorbance on an equal chlorophyll per surface area basis.

	Before chilling			After Chilling			
λ	A_L	A_T	β_1	A_L	A_T	β_2	β_1/β_2
700	0.54	0.21	2.55	0.56	0.21	2.64	.96
677	2.00	1.42	1.41	2.04	1.42	1.44	.98
540	0.53	0.21	2.48	0.51	0.21	2.38	1.04
518	0.81	0.41	1.97	0.80	0.41	1.95	1.01

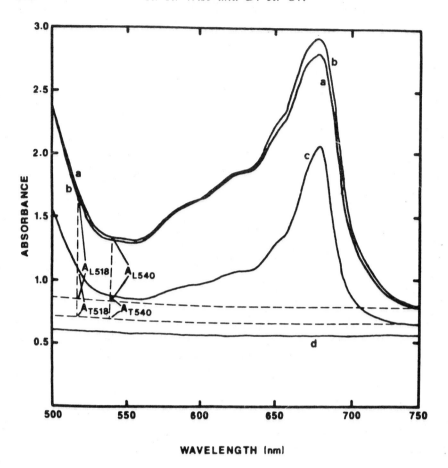

Fig. 7: Absorbance spectra for attrached cucumber leaf before (a) and after (b) a 3 h treatment of 5°C and 1000 $\mu E \cdot m^{-2} \cdot s^{-1}$. Trace C is a spectrum of a suspension of cucumber thylakoids at a chlorophyll content per cm^2 equal to that in (a). Trace D is of a pigment-free, methanol-extracted cucumber leaf. Dashed horizontal lines are base lines for the calculation of the absorbance of the leaf at 518 and 540 nm (A_{T518}, A_{L540}) and the thylakoid suspension (A_{T518}, A_{T540}). An SLM-Aminco 2C spectrophotometer was used for these measurements with the leaf tissue positioned within 5mm of the photomultiplier.

leaf absorbance at 518 and 540 nm. The values of β at 700 nm (2.55) and 677 nm (1.41) are in good agreement with those reported by Rühle and Wild (26) at 700 nm (2.44 ± 0.10) and 680 nm (1.70 ± .10) for *Sinapsis alba*.

Discussion

The inhibition of photosynthesis seen after a chilling and light treatment of a chilling sensitive plant (Fig. 1) has been well documented (1,3,6–8, 14,

27). Despite decreases in the quantum efficiency of CO_2-dependent leaf disc O_2 evolution (Fig. 4), previous work in our laboratory (8) has demonstrated that, after light chilling of tomato, the rate of electron transport, although impaired, was still in substantial excess of the requirements of CO_2 reduction. Therefore, the focus of this study was to investigate whether the capacity for photophosphorylation was being impaired and whether the supply of ATP was limiting to photosynthesis. For instance, a limiting ATP supply could underlie the inhibition of the Calvin cycle reported by Sassenrath *et al.* (27). An inhibition of the chloroplast coupling factor has been demonstrated as a mechanism in the photosynthetic inhibition accompanying a low leaf water potential (10,11,13) and impaired photophosphorylation has been implicated in chilling injury (9, 14–16).

Based primarily on studies performed with isolated thylakoids (28,29), it is anticipated that the coupling factor will be in a catalytically inactive state in dark-adapted leaves. The energetic threshold for the activation of the enzyme is similar to that of ATP synthesis itself (30). It appears likely that thioredoxin, located in the chloroplast stroma and photoreduced by photosystem I, mediates the *in situ* reduction of the γ subunit of the coupling factor under physiological conditions (28,29). Although reduction of γ is not required for activation, it lowers the energetic threshold from a Δp equivalent to about 2.9 pH units for the oxidised enzyme to about 2.6 pH units once the γ subunit is reduced (30). Furthermore, reduction stabilises the catalytically active state of the enzyme once activation has occurred. In the experiments reported in this paper, we used plants that had been incubated in the dark for 30 min prior to performing any measurements. Except when DCCD was present, relaxation of the electrochromic change indicated ATP synthesis was occurring and perforce that the coupling factor population was largely activated. At this time we cannot tell whether the active and/or reduced state of the enzyme was maintained throughout the 30 min dark adaptation or whether the subsequent flash series used in the electrochromic measurement was responsible for reactivating the enzyme. These alternatives are currently under investigation.

Chilling attached cucumber leaves in the light had only minor effects on the decay kinetics of the flash-induced electrochromic shift (Fig. 5). Only at a low flash frequency of 0.1 Hz was the relaxation time constant larger after chilling than before. Because of the involvement of coupling factor reduction in the enzyme's activation and because oxidants are produced in the cucumber chloroplast stroma during light chilling (31), it may be that the somewhat slower relaxation time constants measured at low flash frequencies represent an antagonistic effect of such oxidants on coupling factor activation. Flashing at 2.0 Hz restored the decay kinetics to close to those of the control and argues that the slow decay kinetics seen at the low flash frequency after lightchilling can be overcome by the larger Δp values produced when the individual contributions of closely spaced flashes to the Δp become additive. Since the reduction in photosynthesis caused by chilling in the light (Fig. 1) was measured at 23° C and in continuous illumination it is expected that the coupling factor population during the net photosynthetic measurements would be in the state more similar to that generated by the high rather than the low flash frequency. As a comparison, leaf water deficits have been shown to cause relaxation time constants in sunflower in excess of 400 ms that are not fully restored to control values by rapid flashing (13). This is also why we take care to prevent any accompanying changes in leaf water status (Table 1) in the light-chilled plants. McWilliam et al. (19) showed that low ψ_L can greatly exacerbate chilling injury to photosynthesis.

Peeler and Naylor (9) provided evidence for an uncoupling of photosynthetic electron transfer from photophosphorylation in cucumber thylakoids isolated from chilled and irradiated plants. Uncoupling would cause a marked decrease in the relaxation time constant (accelerate the decay of the $\Delta A_{518-540}$) and was not seen in the present study in over hundred experiments with plants chilled for 1 to 6 h. Perhaps the thylakoid uncoupling seen by Peeler and Naylor was differential artifact of the isolation procedures. Indeed, they showed in the same study that cucumber thylakoids are particularly sensitive to uncoupling during isolation.

Light chilling was observed to diminish $\Delta A_{518-540}$ to the point where it became difficult to detect after 6 h of exposure. The decrease in amplitude cannot be accounted for by a shortening of the effective optical pathlength through the leaf (Fig. 7). The smaller flash-induced amplitude after light chilling may reflect a smaller flash-generated membrane potential or a change in the sensitivity of the absorption band shift to the size of the electric field. These alternatives are under investigation.

Acknowledgements

The authors extend their thanks to A. Ortiz-Lopez for helpful discussions and technical assistance and to Dr. M. A. Topa for statistical analysis. This work was supported in part by a grant from USDA Competitive Grants Office.

References

1. Taylor, A. O. and Rowley, A. (1971) *Plant Physiol.* 47, 713–718
2. Van Hasselt, Ph.R. and Van Berlo, H. A. C. (1980) *Physiol. Plant.* 50, 52–56
3. Long, S. P., East, T. M. and Baker, N. R. (1983) *J. Expt. Bot.* 34, 177–188
4. Martin, B. and Ort, D. R. (1985) *Photosyn. Res.* 6, 121–132

5. Drake, B. and Raschke, K. (1974) *Plant Physiol.* 53, 808–812
6. Baker, N. R., East, T. M. and Long, S. P. (1983) *J. Expt. Bot.* 34, 189–197
7. Powles, S. B., Berry, J. A. and Björkman, O. (1983) *Plant Cell Environ.* 6, 117–123
8. Kee, S. C., Martin, B. and Ort, D. R. (1986) *Photosyn. Res.* 8, 41–51
9. Peeler, T. C. and Naylor, A. W. (1988) *Plant Physiol.* 86, 147–151
10. Keck, R. W. and Boyer, J. S. (1974) *Plant Physiol.* 67, 985–989
11. Younis, H. M., Boyer, J. S. and Govindjee (1979) *Biochim. Biophys.* Acta 548, 328–340
12. Kaiser, W. M., Kaiser, G., Prachaub, P. K., Weldamn, S. G. and Heber, U. (1981) *Planta* 153, 416–422
13. Ortiz-Lopez, A., Ort, D. R. and Boyer, J. S. (1987) In: *Progress in Photosynthesis Research. Proceedings of the 7th International Congress on Photosynthesis*, Vol. 4 (J Biggins, ed.), pp. 153–156, Martinus Nijhoff/Dr. W. Junk Publishers, Dordrecht
14. Kislyuk, I. M. and Vas'kovskii, M. D. (1972) *Sov. Plant Physiol.* 19, 688–692
15. Garber, M. P. (1977) *Plant Physiol* . 59, 981–985
16. Santarius, K. A. (1984) *Plant Physiol.* 61, 591–598
17. Witt, H. T. (1979) *Biochim. Biophys.* Acta 505, 355–427
18. Morita, S., Itoh, S. and Nishimura, M. (1982) *Biochim. Biophys.* Acta 679, 125–130
19. McWilliam, J. R. and Musser, R. L. (1982) *Aust. J. Plant Physio* 9, 343–352
20. Delieu, T. and Walker, D. A. (1981) *New Phytol.* 89, 165–178
21. Boyer, J. S. and Knipling, E. B. (1965) *Proc. Natl. Acad. Sci. USA* 54, 1044–1051
22. Ziegler, R. and Egle, K. (1965) *Beitr. Biol. Pflanzen.* 41, 11–37
23. Graan, T. and Ort, D. R. (1984) *J. Biol. Chem.* 259, 14003–14010
24. Graan, T. and Ort, D. R. (1986) *Biochim. Biophys.* Acta 852, 320–330
25. Petty, K. M. and Jackson, J. B. (1979) *Biochim. Biophys.* Acta 547, 463–473
26. Rühle, W. and Wild, A. (1979) *Planta* 146, 551–557
27. Sassenrath, G. F., Ort, D. R. and Portis, A. R., Jr. (1987) In: *Progress in Photosynthesis Research, Proceedings of the 7th International Congress on Photosynthesis*, vol 4 (Biggins, J. ed), pp. 103–108, Martinus Nijhoff/Dr. W. Junk Publishers, Dordrecht
28. Mills, J. D., Mitchell, P., and Schurmann, P. (1980) *FEBS Lett.* 112, 173–177
29. Shahak, Y. (1982) *Plant Physiol.* 70, 87–91
30. Hangarter, R. P., Grandoni, P. and Ort, D. R. (1987) *J. Biol. Chem.* 262, 13513–13519
31. Wise, R. R. and A. W. Naylor (1987) *Plant Physiol.* 83, 278–282

Effect of Heat Stress on Photosynthesis and Respiration in a Wheat (*Triticum aestivum* L.) Mutant.

JASJEET KAUR, INDER S. SHEORAN AND
HIMMAT S. NAINAWATEE

Department of Chemistry and Biochemistry
Haryana Agricultural University
Hisar-125004, INDIA

Summary

Wheat plants of *cv.* WH147 and its thermotolerant mutant WH147M were subjected to heat stress of 40 to 55°C for short periods and subsequently, membrane damage, photosynthesis and respiration were studied in the leaves. The tetrazolium reduction activity of leaves of heat-stressed plants of WH147M increased, whereas it decreased in the heat-stressed plants of WH147. Heat stress effected changes in ion-leakage, photosynthesis and oxygen uptake activity were also comparatively less pronounced in WH147M plants. It seems probable that these possess thermostable membrane system.

Introduction

Environmental stresses, such as those resulting from rapid changes in nutrition, water supply, anaerobiosis and temperature limit the productivity and adaptability of crops. Heat stress affects crop productivity by influencing almost all the physiological and biochemical processes from seed germination to grain development. If heat stress occurs in the life cycle of a crop plant, it is important that it should possess a certain degree of heat tolerance to survive the stress period. Adaptation to high temperature has been reported for many wild plants (1–3). Though evidently based on some physiological parameters, heat-tolerant genotypes of some commercial crops have been identified (4–6), the exact molecular mechanism of heat tolerance has not yet been understood. A thermoinducible system of heat-shock protection is known to be present in several organisms including plants (7, 8). In these cases, certain genes are induced which result in the synthesis of heat-shock polypeptides. None of these polypeptides have so far been identified and, moreover, protection is provided for a very short period. However, adaptation of crop plants to high temperature has been reported to be correlated with photosynthesis (9). High temperature causes membrane disruption, which alters solute movement, photosynthesis and respiration (10). Sheoran *et al.* (11) have developed a wheat mutant, WH147M, which has been reported to possess high thermal requirement (12). Seedlings of WH147M were found to have a relatively better heat-inducible thermoprotection mechanism (13). Results reported here show that the thermotolerant WH147M plants possess relatively more thermostable photosynthetic and respiratory activities.

Material and Methods

Plant Material

Seeds of wheat (*Triticum aestivum* L.) *cv.* WH147 and its thermotolerant mutant, WH147M, were obtained from the Department of Plant Breeding of this University. The crop was raised in pots in the *rabi* season, using standard agronomic practices.

Temperature Treatment

The plants, at the tillering stage, were subjected to heat stress by placing pots in a BOD incubator at the desired temperature. Heat stress, during ion-leakage determination, was effected by placing leaf discs on a temperature-controlled shaker water-bath.

Ion-leakage

Leaves were washed and cut into small discs, using a punch. Ten discs were placed in a test tube containing 10 ml deionised water. The test tube was kept for heat-stress in a shaker water-bath at the desired temperature. The sample was cooled to room temperature and the conductivity of the liquid was measured. The same sample was then placed in a boiling-water bath for 10 min. The water loss during boiling was replenished, the sample was shaken for 30 min, cooled and conductivity was measured. The values were expressed as percentage of the total conductivity.

Photosynthesis

The photosynthetic rate was measured with an Infra Red Gas Analyser, model ADC type 225/2K. The measurements were made at 11 a.m. at an average light intensity of 1000 μ E m^{-2} S^{-1}. The leaves were cut under water and placed in a 13 × 8 × 8 cm airtight Perspex chamber. Leaf-area was measured using a Leaf-Area Meter, model LI-3000, and the rate of photosynthesis expressed as mg CO_2/dm^2/h.

Oxygen Uptake Activity

Oxygen uptake activity was measured polarographically, using an oxygen electrode (Yellow Spring, Ohio, USA). Leaf discs were placed in the monitor cuvette containing 0.1 M phosphate buffer pH. 7.5. Oxygen uptake activity was expressed as μl O_2 consumed h/g fresh weight.

TTC Reduction Activity

Small discs punched from the leaves of temperature-stressed plants were weighed and kept in 0.05M phosphate buffer pH 7.5, containing 0.6 per cent TTC and 0.05 per cent Triton X-100. The discs were incubated at 30°C, 15 h, in the dark and then formazan was extracted in 95 per cent ethanol in a boiling-water bath for 5 min. Absorbance of the colour was measured at 530 nm and the results expressed as absorbance units/g leaf.

Results

Ion-leakage

Ion leakage from leaves was estimated by measuring the electrical conductance of leachates from leaf discs. The results, presented in Table 1 show that temperature treatment at 50° and 55° C, for varying periods, resulted in more ion-leakage from WH147 leaf discs as compared to that from WH147M discs. A 55°C treatment for 20 min resulted in 100 per cent leakage (total leakage determined by a 100°C treatment for 10 min) in WH147 leaf discs, while in WH147M it was 76 per cent.

TABLE 1

Temperature Stress-Effected Ion-leakage from WH147 and WH147M Leaves.

Treatment °C, min	Ion-leakage (per cent conductivity)	
	WH147M	WH147
50°C		
45 min	42.8	45.4
90 min	71.4	81.8
55°C		
5 min	19.0	31.3
10 min	38.0	62.5
20 min	76.1	100

NOTE: Each value is an average of ten observations.

Tetrazolium Reduction Activity

Results presented in Table 2 show that temperature stress of 45°C for 40

TABLE 2

Tetrazolium Reduction Activity of Temperature-stressed WH147 and WH147M Leaves.

Treatment °C, min	Tetrazolium reduction activity (OD units g^{-1} leaf)	
	WH147M	WH147
45°C		
20 min	8.9 (+42)	9.1 (+2)
40 min	10.9 (+73)	8.8 (−13)
50°C		
20 min	10.1 (+66)	5.3 (−39)
40 min	8.1 (+28)	4.2 (−52)

NOTE: Values in parenthesis indicate per cent deviation from control. Each value is an average of ten observations.

min did not appreciably affect the tetrazolium reduction activity of leaf discs of the parent, while an increase was observed in the mutant. When stress treatment of 50°C was given the decrease in tetrazolium reduction activity was appreciable in WH147, while it increased in WH147M leaf discs. The 40 min stress treatment resulted in nearly 50 per cent reduction in the WH147, whereas in WH147M there was an increase of 28 per cent.

Oxygen Uptake Activity

Temperature stress resulted in a higher reduction in oxygen uptake activity of WH147 leaves as compared to WH147M leaves (Table 3). Treatment at 50°C for 45 min resulted in 72 per cent reduction in oxygen uptake by

TABLE 3

Oxygen Uptake Activity of Temperature-stressed WH147 and WH147M Leaves.

Treatment °C, min	Oxygen uptake (μl O_2 min^{-1} g^{-1} leaf)	
	WH147M	WH147
50°C		
15 min	294 (−15)	238 (−60)
30 min	203 (−31)	207 (−66)
45 min	220 (−36)	167 (−72)

NOTE: Values in parentheses indicate per cent deviation from control. Each value is an average of four observations.

WH147 leaves, whereas the corresponding decrease in WH147M leaves was only 36 per cent.

Photosynthetic Activity

Table 4 shows the results of photosynthetic activity of WH147M and WH147 temperature-stressed wheat leaves. Reduction in photosynthetic activity due to temperature stress was relatively more marked in WH147 leaves as compared to WH147M leaves. Stress of 45°C for 60 min resulted in 82 and 33 per cent reduction in WH147 and WH147M respectively. At 50°C treatment for 1h, the leaves of WH147 completely lost photosynthetic activity whereas the loss was only 30 per cent in WH147M leaves.

TABLE 4

Photosynthetic Activity of Temperature-stressed WH147 and WH147M Plants

Treatment	Photosynthetic activity (mg CO_2 dm^{-2} h^{-1})	
	WH147M	WH147
45°C	21.4 (−63)	12.5 (−40)
30 min	22.8 (−33)	5.6 (−82)
60 min		
50°C	21.3 (−37)	1.7 (−95)
30 min	23.6 (−30)	
60 min		

NOTE: Values in parentheses indicate per cent deviation from control. Each value is an average of four observations.

Discussion

Thermo-insensitive wheat mutant WH147M is a product of gamma irradiated parent WH147 (11). Grain yield of the mutant is highest when it is sown in the high temperature days of September (mean maximum temperature, 35.5°C, minimum temperature 21.0°C), as compared to the normal sowing time of November (mean maximum temperature 29.1°C, minimum temperature 10.6°C) for the parent. Thus, the mutant possesses some thermotolerant developmental mechanism which facilitates normal crop production even under adverse climatic conditions. Under different stress conditions, alterations in gene expression takes place in crop plants (14). Heat-shock polypeptides are known to be synthesised in response to heat stress in soyabean (15) maize (16, 17) and wheat (18). The mechanism by which heat-shock polypeptides provide protection against heat stress in plants is not understood. However, their intracellular localisation, particularly of hsp 18, with plasma membrane suggests that they protect membrane destruction (17). In WH147M the membrane damage caused by simulated temperature stress, as indicated by ion-leakage, was less as compared to that in WH147. Electrolyte leakage has been reported to be a good index of heat-stress tolerance in pear (19), tomato, barley and onion bulbs (20), soyabean (21) and sorghum (22). The maintenance of membrane integrity under temperature stress is also indicated by an active endoplasmic reticulum. Belanger *et al.* (23) reported the destabilisation of mRNA and disruption of rough endoplasmic reticulum at high temperature in barley aleurone. In the leaves of WH147M, incorporation of ^{3}H-leucine increased at high temperature, while in parent WH147 there was a reduction in ^{3}H-leucine incorporation (24). The respiratory activity of WH147M, as indicated by tetrazolium dye reduction and oxygen uptake activity also showed a thermotolerant character. Tetrazolium reduction activity is a suitable selection test for screening of heat-tolerant genotypes. In WH147M leaves, heat stress increased the tetrazolium reduction activity, while the oxygen consumption decreased. This perhaps was due to a reduction in respiratory activity without any change in the energy charge, as has been reported in pea seedlings subjected to high temperature (26). The photosynthetic apparatus of WH147M was also relatively more thermotolerant than that of WH147. Heat stress inhibits photosynthesis by affecting chlorophyll accumulation, protein synthesis and photosystem activities (27, 28). Synthesis of chloroplast proteins, particularly ribulose 1, 5 bisphosphate carboxylase large subunit, is drastically reduced by heat shock (29). Thermotolerant photosynthetic activities are known in plants (9, 30, 31). As an adaptational change, high-temperature-induced conformational change in ribulose 1, 5 bisphosphate carboxylase of wheat is already known (32). Though the molecular mechanism of heat tolerance in WH147M is not understood, this mutant has opened up a possibility of wheat cultivation under wider agroclimatic conditions, particularly under early sown and rainfed conditions, which, though desired, was hitherto not feasible (33).

References

1. Alexandrov, V. Y. (1964) *Q. Rev. Biol.* 39, 35–77
2. Raison, J. K. and Berry, J. A. (1979) *Carnegie Inst. Ann. Report.* pp. 149–152
3. Bjorkman, O. and Badger, M. (1979) *Carnegie Inst. Ann. Report.* pp. 145–148
4. Stevens, M. A. and Rudich, J. (1978) *Hort. Sci.* 13, 673–678
5. Mendoza, H. A. and Estrada, R. N. (1979) In: *Stress Physiology in Crop Plants* (Mussel H. and Staples, R. C. eds.) pp 227–262, John Wiley, New York
6. Halterlein, A. J., Clayberg, C. D. and Teare, D. (1980) *J. Amer. Soc. Hort. Sci.* 105, 12–14
7. Atkinson, B. G. and Walden, D. B. (1985) *Changes in Eukaryotic Gene Expression in Response to Environmental Stress.* Academic Press, London

8. Nainawatee, H. S., Gupta, M. and Kaur, J. (1986). In *Proc. Natl. Symp. Physiol. Biochem. Genetic aspects of crop plants in relation to environmental stress.* (Singh, R., Sheoran I. S. and Saharan, M. R. eds.) pp. 98–100, HAU, Hisar

9. Berry, J. A. (1975). *Science* 188, 644–650

10. Christiansen, M. N. (1978) In: *Crop Tolerance to Suboptimal Land Conditions.* (Jung, G. A., ed.) pp 173–191, Am. Soc. Agron. Wisconsin

11. Sheoran, I. S., Kuhad, M. S., Behl, R. K., Nandwal, A. S. and Singh D. (1983) *Indian J. Agric. Sci.* 53, 1076–1078

12. Behl, R. K., Sheoran, I. S., Singh, G. and Kuhad, M. S. (1986) *Intern. Sem. Water Management in Arid and Semi Arid Zones.* pp. 360–373, HAU, Hisar

13. Gupta, M., Behl, R. K. and Nainawatee, H. S. (1987) *Ann. Biol.* 3, 11–13

14. Sachs, M M. and Ho, T.-H. D. (1986) *Ann. Rev. Plant Physiol.* 37, 363–376

15. Lin, C. Y., Roberts, J. K. and Key, J. L. (1984) *Plant Physiol.* 74, 152–160

16. Baszczynski, C. L., Walden, D. B. and Atkinson, B. G. (1982) *Can. J. Biochem.* 60, 569–579

17. Cooper, P. and Ho, T.-H. D. (1983) *Plant Physiol.* 71, 215–222

18. Gupta, M. and Nainawatee, H S. (1985). In *Natl. Sem. Pl. Molec. Biol. Nitrogen Met*, p. 9 IARI, New Delhi

19. Wu, M. T. and Stephen, J. W. (1983) *Plant Physiol.* 72, 817–820

20. Onwueme, I. C. (1979) *J. Agric. Sci.* (Camb.) 92, 527–531

21. Martineaue, J. R., Specht, J. E., Williams, J. H. and Sullivan, C. Y. (1979) *Crop Sci.* 19, 75–78.

22. Towill, L. E. and Mazur, P. (1974) *Can. J. Bot.* 53, 1097–1102

23. Belanger, F. C., Brodl, M. R. Ho, T.-H. D. (1986) *Proc. Natl Acad. Sci. USA* 83, 1354–1358

24. Kaur, J. (1986) 'Studies on the Effect of Temperature Stress on Wheat Plants'. M.Sc. thesis, Haryana Agricultural University, Hisar

25. Chen, H. H., Shen, Z. Y. and Li, P. H. (1982) *Crop Sci.* 22, 719–725

26. Nikulina, G. N. (1985) *Fiziol. Biochem. Kult Rast* 17, 131–134

27. Onwueme, I. C. and Lawanson, A. O. (1973) *Planta* 110, 81–84

28. Thebud, R. and Sanatarius, K. A. (1982) *Plant Physiol.* 70, 200–205

29. Vierling, E. and Key, J. L. (1985) *Plant Physiol.* 78, 155–162

30. Pearcy, R. W. (1977) *Plant Physiol.* 59, 873–878

31. Burke, J. J., Hatfield, J. L., Klein, R. R. and Mullet, J. E. (1985) *Plant Physiol.* 78, 394–398

32. Weidner, M. and Fehling, E. (1985) *Planta* 166, 117–127

33. Asana, R. D. (1974) *Indian J. Genet.* 34, 190–96

Inhibition of Energy Transfer Reactions in Cyanobacteria by Different Ultraviolet Radiation

G. Kulandaivelu, V. Gheetha and S. Periyanan

School of Biological Sciences
Madurai Kamaraj University
Madurai 625 021, INDIA

Summary

When *Anacystis nidulans* cells were irradiated with ultraviolet-C (UV-C, 254 nm), UV-B (285–325 nm) or UV-A (320–400 nm) radiation, the overall photosynthetic rate decreased progressively with time of treatment. Since a decrease in the rate of photosynthesis could be either due to a loss of excitation energy transfer from phycobilisomes (PBS) to chlorophyll and/or inhibition of various electron transfer reactions, both the fluorescence spectra and electron transfer reactions (rate of O_2 evolution) were measured in UV irradiated cells. UV irradiation brings about drastic changes in the excitation and emission spectra of *Anacystis* cells. A comparative investigation on the nature of action of these different UV wavelengths indicates strong action by UV-C radiation on both PBS to Chl *a* energy transfer and Chl *a*-mediated electron transport reactions than that of UV-B and UV-A. UV-C radiation brings about drastic loss of phycobilins, as evidenced from absorption and fluorescence excitation spectra. In contrast to this, UV-B and UV-A treatment resulted in loss of energy transfer from allophycocyanin (APC) to Chl *a*. This was indicated by increased fluorescence emission from phycobilins with concomitant decrease in chlorophyll fluorescence.

Introduction

In the past three decades, several workers have made a number of investigations on the mechanism of inhibition of photosynthetic electron transport by ultraviolet (UV) radiation. Many (Bishop 1961; Yamashita and Butler 1968; Erixon and Butler 1971; Katoh and Kimimura 1974) have investigated in great detail the action of short wavelength UV radiation, namely UV-C (< 280 nm). Attention has also been focused on long wavelength UV, particularly UV-B (285–325 nm), as this radiation constitutes a part of solar radiation, and the level of which at the earth's surface is regulated by atmospheric ozone density (Berner 1972; Green *et al*. 1974). The action of UV-C and UV-B on photosynthetic electron transport has been compared and shown to be quite distinct (Kulandaivelu and Noorudeen 1983). Recently we have found that long wavelength blue-UV radiation (UV-A, 320–400 nm), which forms a major part of UV radiation in sunlight, inhibits the photosynthetic reaction specifically at the water oxidation site.

Among the lower algae, cyanobacteria are used extensively for studies on photochemical energy transfer reactions as they have phycobilisomes (PBS) as the major light harvesting complex. Besides, the unicellular forms are easy to manage, unlike the red and brown algae. The PBS, being protein chromophores, are likely to undergo denaturation on absorption of UV radiation. Hence, the present investigation has been undertaken to study in detail the nature of structural and photochemical changes occurring in PBS and energy transfer from PBS to chlorophyll upon treatment with different UV radiations in a typical unicellular cyanobacterium, *Anacystis nidulans*.

Materials and Methods

Algal cultures: Cultures of *Anacystis nidulans* were developed photoauto-trophically in 500 ml culture tubes at 25°C. The cultures were aerated with filtered air. For all experiments, cells were harvested at the mid-log phase by centrifugation, washed once, and suspended in fresh culture medium.

UV treatment: Cell suspension at a chlorophyll concentration of 0.5 mg/ml was transferred to thermostated irradiation vessel (3 cm diameter $\times$ 3 mm depth) as a thin layer (approx. 1 mm thick) for UV treatment. UV radiation of different wavebands was obtained using either a Philips fluorescent black light lamp, type 05 (UV-A, λ emission max. 365 nm) or Philips sun lamp, type 12 (UV B, λ emission max. 315 nm) or Philips germicidal lamp (UV-C, λ emission max. 254 nm). Temperature during the treatment was maintained at 20°C. To avoid sedimentation of cells, the irradiation vessel was vibrated, using a motor at a speed of 5 Hz. Fluence rate was 5 W.m^2. Control samples were covered with a plastic filter to remove all radiation below 400 nm. UV radiance was measured using a IL 700A radiometer (International Light, Inc. USA).

Photosynthetic measurement: The rate of photosynthetic O_2 evolution was measured in a Hansatech O_2 electrode under saturating white light at 25°C. White light at a radiance of 100 W.m^{-2} was provided by a slide projector. Radiance level was measured using a Li-cor 188 quantum/radiometer (Li-cor, Inc., USA). Chl-*a* content was determined by the method of Myers and Kratz (1955).

Absorption and fluorescence spectra: Room temperature absorption spectra of cells were measured using a Hitachi 557 spectrophotometer. Fluorescence excitation and emission spectra were measured in a Hitachi MPF 4 spectrofluorimeter. Spectra presented here are not corrected for the differences in emission characteristics of the monochromator and photomultiplier sensitivity (S20 response). The chlorophyll content of the cell suspension was maintained below 1 μg/ml to minimise self-absorption.

Results and Discussion

Figure 1 shows the changes in the rate of photosynthetic O_2 evolution in *Anacystis* cells upon treatment with equal doses of UV radiation of different wavelengths. Among the three different UV wavelengths used, UV-C radiation showed a rapid inactivation of photosynthetic O_2 evolution. A 30 min UV-C irradiation completely abolished the photo-

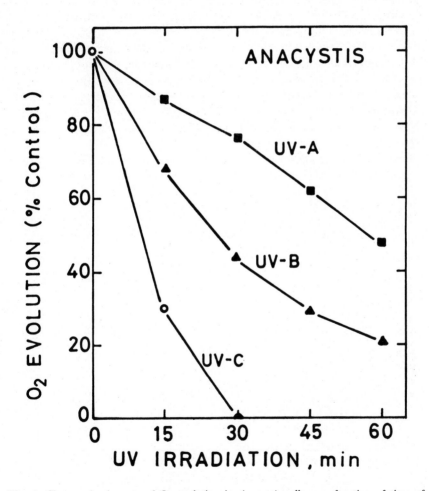

Fig. 1: Changes in the rate of O_2 evolution in *Anacystis* cells as a function of time of irradiation with different UV wavelengths. Conditions for UV treatment and measurement are as under 'Materials and Methods'. The 100% value was 126 μ moles O_2. mg $Chl^{-1}.h^{-1}$.

synthetic O_2 evolution of intact algal cells. However, UV-B and UV-A radiations brought about, respectively, only 50 and 25% inactivation after 30 min of treatment. Loss of O_2 evolution capacity in *Anacystis* could be either due to an inhibition of photosynthetic electron transport by the inactivation or alteration of the electron transport intermediates or by a block in the energy transfer between the primary light-harvesting complexes (PBS) and the reaction centres. The possible sites of inactivation of the photosynthetic electron transport by UV-C and UV-B have already been studied (Bishop 1961; Yamashita and Butler 1968; Erixon and Butler 1971; Katoh and Kimimura 1974; Kulandaivelu and Noorudeen 1983).

Anacystis cells possess a complex light-harvesting system. In fully developed cells, 80% of the available light is captured by PBS. Hence, it is of interest to follow how the energy transfer from PBS complex to chlorophyll is altered on UV treatment. To study the changes in the chromophore and thylakoid organisation, room temperature absorption spectra in control and UV treated *Anacystis* cells were first measured.

The changes in absorption spectral characteristics of *Anacystis* cells after 30 and 60 min of treatment with different UV radiations are shown in Fig. 2. No measurable change in both the PBS (A 625) and chlorophyll (A 682) absorption bands was noticed even after 60 min of UV-A irradiation, whereas UV-B irradiation caused a 10% decrease in the PBS absorption band. In contrast, UV-C irradiation caused a much larger decrease (70% at

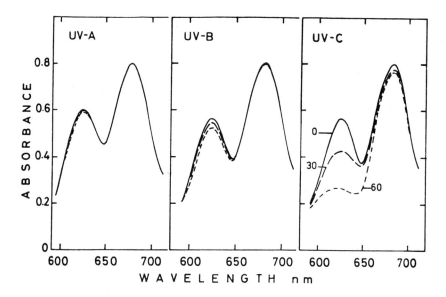

Fig. 2: Effect of 30 and 60 min UV irradiation on the room temperature absorption spectra of *Anacystis* cells. Numbers along trace indicate duration of UV treatment.

60 min) in the PBS absorption band. UV-A radiation, 320 to 400 nm, is strongly absorbed by the phycobilin chromophores, Chl *a* and some electron-transport components such as cytochromes. Since these wavelengths are not absorbed by the aromatic amino acids, the protein moiety of the phycobiliproteins remains intact. Unlike UV-A, UV-B radiation is strongly absorbed by the aromatic amino acids of the phycobiliproteins and hence caused some absorption changes. UV-C radiation appears to have multiple and unspecific action on biological membranes. Although proteins have low absorption at these wavelengths, they showed extensive damage, possibly because the unsaturated double bonds of the chromophore pyrrole chain have strong absorption in the region of these wavelengths. The decrease in chlorophyll absorption could be due to the interaction of UV-C radiation with the membrane and destruction of chlorophyll pigments (pheophytinisation). The decrease of the phycobilin absorption band was shown to be due to conformational changes in the PBS proteins and bleaching of the chromophore (Scheer and Kufer 1977; Laczko and Barabas 1981).

Since treatment of *Anacystis* cells with different UV irradiation brought about characteristic absorption changes, detailed investigations on the energy transfer between PBS and chlorophyll have been carried out. For this, fluorescence excitation and emission spectra were analysed in control and un-treated *Anacystis* cells. At room temperature, *Anacystis* cells showed two major fluorescence emission peaks, at 652 nm from phycobilins and at 682 nm from Chl *a*. UV-A irradiation caused a gradual decrease in Chl *a* fluorescence with concomitant increase in phycobilin fluorescence (Fig. 3). The fluorescence emission spectrum was measured under 370 nm light, which excites PBS preferentially. The increased fluorescence emission at 652 nm over the 682 nm upon UV-A irradiation could be either due to an inhibition of the PBS to Chl *a* energy transfer and/or uncoupling of the PBS from the thylakoid membranes. Detachment of PBS from the thylakoid membranes by cold treatment was shown to increase the PBS fluorescence with concomitant decrease in Chl *a* fluorescence (Schreiber 1980). Such uncoupling of PBS from the thylakoids could be due to physical separation that may include loss of the colourless 95 kD linker polypeptide (Ruskowski and Zilinskas, 1982; Zilinskas and Howell 1983). Though there was no specific change in absorption behaviour of these pigments, their fluorescence characteristics were altered by UV-A radiation.

The fluorescence emission spectra of UV-B-irradiated cells showed a decrease in Chl *a* emission and the APC fluorescence level either remained unchanged or marginally increased. This indicates that UV-B radiation reduced the PBS to chlorophyll energy transfer. Inhibition of energy transfer from PBS to chlorophyll should increase in the fluorescence yield at 652 nm. However, such an increase of 652 nm fluorescence was inhibited, possibly because of the changes in the PBS. Changes in the PBS

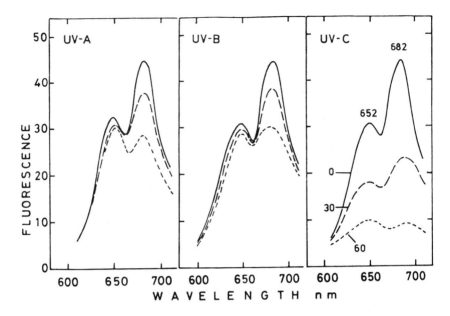

Fig. 3: Effect of 30 and 60 min UV treatment on the room temperature fluorescence emission spectra of *Anacystis* cells. Fluorescence was excited by 370 nm. Excitation energy was adjusted for changes in absorption at 370 nm in UV-treated samples. Numbers along the trace indicate duration of UV treatment and peak emission wavelengths.

could be at the protein level and/or at chromophore level. This is evidenced by the decrease in absorption at 625 nm (Fig. 2). UV-C radiation showed a larger decrease in phycobiliprotein that in Chl *a* absorption (see Fig. 2) The drastic decrease seen in both chlorophyll and PBS fluorescence (Fig. 3) supports the conclusion of possible structrual denaturation of the PBS itself by UV-C radiation.

Figure 4 shows changes in excitation spectra of *Anacystis* cells irradiated for 60 min with different UV wavelengths. The fluorescence emission of Chl *a* at 682 nm arose by the excitation of both PBS and Chl *a*. The PBS showed three prominent peaks of excitation, viz., At 370, 382 and 418 nm. The chlorophyll excitation peaked at 436 nm. Both UV-A and UV-B irradiation caused no change in the level of 370 and 382 nm PBS excitation peaks, but reduced the level of the 436 nm Chl *a* peak. The level of the 418 nm excitation band remains unchanged in 60 min UV-A irradiated cells. In contrast to this, the excitation spectrum of UV-C treated cells showed very low level of excitation in the 320–382 nm region. This indicates a reduced energy transfer from PBS to Chl *a* which is partially due to chromophore bleaching and also conformational changes in the phycobilin tetraphyrrole. Absorption spectral analysis indicates that chromophores in UV-C treated cells undergo degradation.

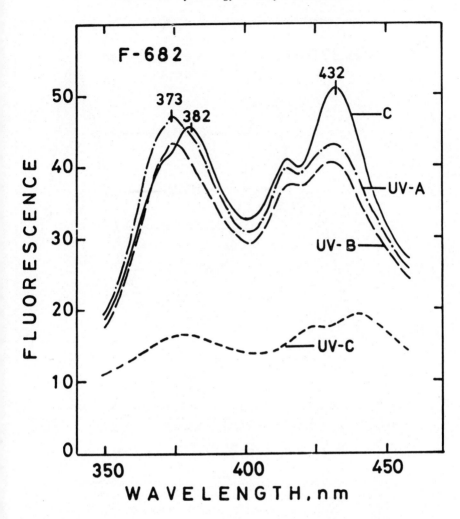

Fig. 4: Changes in the excitation spectra for F-682 in *Anacystis* cells upon 60 min treatment with different UV radiations.

To check UV-induced changes in the efficiency of the PBS to Chl *a* energy transfer, and also the PS II mediated electron transfer reaction, fluorescence emission spectra were followed with respectively, 370 and 430 nm excitation. The emission spectra with preferential excitation of chlorophyll showed only one prominent peak at 682 nm (Fig. 5). Both UV-A and UV-B irradiation showed no change in the spectrum indicating the stability of the pigment complex and absence of membrane conformational changes. In contrast, UV-C radiation, reduced the level of 682 nm emission, which could be mainly due to membrane conformational changes. It has been shown that a 30 min UV-C irradiation of isolated leaf

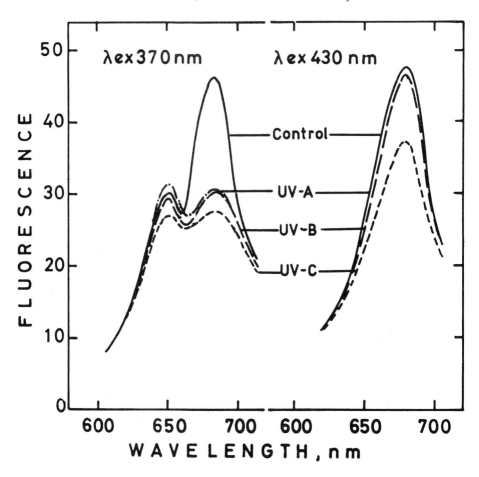

Fig. 5: Changes in the fluorescence emission spectra of *Anacystis* cells irradiated for 60 min with different wavelength UV radiation. Fluorescence was excited either with 370 or 430 nm. Excitation energy was adjusted for changes in absorption at both 370 and 430 nm in UV-treated samples.

cells brings about extensive swelling of thylakoid and granal membranes (Kulandaivelu and Noorudeen 1986). A comparison of the emission spectra with 370 nm as excitation wavelength reveals a drastic reduction in chlorophyll emission in UV-C irradiated cells, but much less decrease in both UV-A and UV-B irradiated cells. These observations support the earlier conclusion that UV-B and UV-A bring about no major structural damage but induce dissociation of PBS and also loss of certain phycobiliproteins, thereby affecting the overall photochemical efficiency, while UV-C radiation causes extensive damage to both PBS and thylakoid membranes. Additional evidence from direct protein analysis is, however, necessary to substantiate this conclusion.

Acknowledgements

We thank Professor Govindjee for critical reading of the manuscript and his valuable comments. Financial supports from the Department of Science and Technology and Council of Scientific & Industrial Research are gratefully acknowledged.

References

1. Berner, P. 1972. Appropriate values of intensity of natural ultraviolet radiation for different amounts of atmospheric ozone. *Final Tech. Report.* Eur. Res. Office. U.S. Army, London, Contract No. DAJA 37–68–C–1017, p. 55
2. Bishop, N. I. 1961. The possible role of plastoquinone (Q–254) in the electron transport system of photosynthesis. In *Ciba Foundation Symposium on Quinone in Electron Transport,* pp. 385–404 (Wolstenholme, C. E. W. and O' Connor, C. M., eds., Churchill Ltd).
3. Erixon, K. and Butler, W. L. 1971. Destruction of C–550 by UV radiation. *Biochim. Biophys.* Acta 253: 483–486
4. Green, A. E. S., Sawada, T. and Shettle, E. P. 1974. The middle ultraviolet reaching the ground. *Photochem. Photobiol.* 19: 251–259
5. Katoh, S. and Kimimura, M. 1974. Light-induced changes of C-550 and fluorescence yield in ultraviolet-irradiated chloroplasts at room temperature. *Biochim. Biophys.* Acta 333: 71–74
6. Kulandaivelu, G. and Noorudeen, A. M. 1983. Comparative study of the action of ultraviolet–C and ultraviolet–B radiation on photosynthetic electron transport. *Physiol. Plant.* 58: 389–394
7. Kulandaivelu, G. and Noorudeen, A M. 1986. Ultraviolet–C (UV–C) radiation induced changes in chloroplast ultrastructure. *Pl. Physiol. Biochem.* 13: 90–98
8. Laczko, I. and Barabas, K. 1981. Hydrogen evolution by Photobleached *Anabaena cylindrica.* Planta 153: 312–316.
9. Myers, J. and Kratz, W. A. 1955. Relation between pigment content and photosynthesis in a blue-green alga. *J. Gen. Physiol.* 39: 11–22
10. Rusckowski, M. and Zilinskas, B. A. 1982. Allophycocyanin I and the 95 kilodalton polypeptide. The bridge between phycobilisomes and membranes. Plant Physiol. 70: 1055–1059
11. Scheer, H. and Kufer, W. 1977. Studies on plant bile pigments. IV. Conforamtional studies on C–phycocyanin from *Spirulina platensis.* Z. *Naturforsch.* 32C: 513–519
12. Schreiber, V. 1979. Cold-induced uncoupling of energy transfer between phycobilins and chlorophyll in *Anacystis nidulans.* FEBS Lett. 107: 4–9
13. Zilinskas, B. A. and Howell, D. A. 1983. Role of the colourless polypeptides in phycobilisome assembly in *Nostoc* sp. Plant Physiol. 71: 379–387

Carbon Dioxide Fixation by PEP Carboxylase in Pod-walls of Chickpea (*Cicer arietinum* L.)

RANDHIR SINGH

Plant Biochemistry Laboratory
Department of Chemistry and Biochemistry
Haryana Agricultural University
Hisar 125 004, INDIA

Summary

Activities of some key enzymes of the PCR cycle and C_4 metabolism, rates of $^{14}CO_2$ fixation in light and dark, and initial products of photosynthetic $^{14}CO_2$ fixation were determined in the pod-wall and seed-coat (fruiting structures) and their subtending leaf in chickpea. Compared to the activities of RuBP carboxylase and other PCR cycle enzymes, the levels of PEP carboxylase and other enzymes of C_4 metabolism were generally much higher in pod-wall and seed-coat than in the leaf. Pod-wall and seed-coat fixed $^{14}CO_2$ in light and dark at much higher rates than the leaf. Short-term assimilation of $^{14}CO_2$ by illuminated fruiting structures produced malate as the major labelled product with less labelling in 3-phosphoglycerate, whereas the leaf showed a major incorporation into 3-phosphoglycerate. PEP carboxylase purified to homogeneity from immature pod-wall and having a molecular weight of about 200,000 daltons was a tetramer of four identical sub-units. Mg^{2+} ions were specifically required for the enzyme activity. The enzyme showed typical hyperbolic kinetics with PEP with a km of 0.74 mM, whereas sigmoidal response was observed with increasing concentrations of HCO_3^- with S0.5 value of 7.6 mM. The enzyme was activated by inorganic phosphate (Pi) and phosphate esters like glucose 6-phosphate, α-glycerophosphate, 3-phosphoglyceric acid, and fructose 1, 6-bisphosphate, and inhibited by nucleotide triphosphates, organic acids and divalent cations Ca^{2+} and Mn^{2+}. Non-competitive inhibition caused by oxaloacetate and malate was reversed by glucose 6-phosphate, indicating that glucose 6-phosphate and oxaloacetate and malate through their opposing effects might be important in controlling PEP carboxylase activity *in vivo*.

Introduction

Developing pods of legumes are known to be photosynthetically active and are generally considered to perform a major role in carrying out the refixation of CO_2 released during either respiration or photorespiration (1). However, the structural and carboxylation characteristics of pods vary widely, depending upon the legume species. In *Pisum arvense* L., the pod-wall supplies 66% of the carbon required by the developing seed during the period of maximum growth (2). Compared to the leaf, more translocation of photosynthates has been observed to occur from the pod-wall during later stages of pod development in pea (3). Similarly, pod-wall of chickpea has also been reported to contribute significant amounts of photosynthates to the developing fruit, and leaf photosynthesis could be complemented by pod-wall photosynthesis during pod development (4, 5). However, compared to leaf, detailed studies on the photosynthetic characteristics of pod-wall and seed-coat of legume crops have not yet been carried out during the entire period of seed development (6–9). Here we report the detailed photosynthetic characteristics of fruiting structures of chickpea along with the physical and kinetic characteristics of pod-wall PEP carboxylase; the enzyme responsible for carrying out the refixation of respired or photorespired CO_2.

Materials and Methods

Chemicals

All the biochemicals and enzymes were purchased from Sigma (St. Louis, Mo, USA). $NaH^{14}CO_3$ (specific Activity 1.85 TB_q mol^{-1}) was purchased from Bhaba Atomic Research Centre (India). All other chemicals used were of analytical grade.

Plant material

Chickpea crop (cv H 75–35) was raised in the fields of the Pulse Section of Haryana Agricultural University, Hisar, following recommended agronomic practices. Fully opened flowers were tagged on the day of anthesis.

Twenty pods sampled at random at three-day intervals, starting from day 3 after anthesis and going on up to full maturity, were brought to the laboratory and their fresh and dry weights recorded after separating the pods into pod-walls and seeds. Carbon dioxide exchange studies were carried out at 3-day intervals following the procedure described elsewhere (10). During the period of rapid seed growth (20 days after anthesis), $^{14}CO_2$ feeding was done both from outside and inside of the pod-wall as described earlier (10). After 20 sec incubation in the presence of labelled CO_2, the pods were sampled immediately after feeding and at 60,120, and 300 sec and 24 and 48 h after the start of feeding. The pods were separated into pod-walls and seeds and killed in boiling 80% ethanol. ^{14}C distribution in the ethanol-soluble fraction was determined by a liquid scintillation counter as done earlier (11). Internal CO_2 concentration of pods was measured with an Infra-red Gas Analyser operating in the differential mode. These measurements were made in full sunlight.

Products of $^{14}CO_2$ Assimilation

Immediately after separation, the tissues were incubated with $^{14}CO_2$ (500 μl L^{-1}) in a perspex chamber for 20, 40, 60, 120 and 300 sec, using the technique described previously (11). After incubation, the tissues were immediately killed and extracted in boiling 80% ethanol. The ethanol extract was evaporated to dryness. Chlorophyll was extracted from the solids by washing twice with chloroform. After evaporating excess chloroform, the solids were taken up in water. A suitable aliquot of the water extract was then subjected to paper chromatography, using n-butanol–propionic acid–water (10:5:7, v/v) as the solvent, according to the method of Benson *et al.* (12). Radioactive products were identified by Co-chromatography and autoradiography. Areas containing ^{14}C were punched from the paper, extracted in 80% ethanol and radioactivity determined by liquid scintillation counting (Beckman LS 100 C) with an efficiency of 80%.

Enzyme Extraction

Five hundred mg of each tissue were used and enzyme extracts prepared as described previously (11).

Enzyme Assays

Enzyme activities were determined spectrophotometrically at 340 nm by following the oxidation of NAD(P)H or reduction of NAD(P)$^+$. All assays were carried out at 30°C. Preliminary assays were done for all the enzymes to determine optimum conditions where linear reaction rates with respect to time and enzyme concentrations were obtained. The activities of various key enzymes of the Calvin cycle, viz. RuBP carboxylase (EC 4.1.1.39), NADP$^+$-glyceraldehyde 3-phosphate dehydrogenase (EC 1.2.1.13), and ribulose 5-phosphate kinase (EC 2.7. 1.19) and C$_4$ metabolism, viz. PEP carboxylase (EC 4.1.1.31), NAD$^+$ malate dehydrogenase (EC 1.1.1.37), NADP$^+$-malate dehydrogenase (EC 1.1.1.82), NAD$^+$-malic enzyme (EC 1.1.1.39), NADP$^+$-malic enzyme (EC 1.1.1.40), glutamate oxaloacetate transaminase (EC 2.6.1.1) and glutamate pyruvate transaminase (EC 2.6.1.2), were determined by following the standard assay methods, as reported earlier (11).

Purification of PEP Carboxylase

The enzyme PEP carboxylase from immature pod-walls of chickpea was purified as per the details contained in an earlier report (13).

Determination of Purity, Molecular Weight and Molecular Weight of Subunits of the Enzyme

The purity of the enzyme preparation obtained from gel filtration through Sephadex G-200 was judged by PAGE at 4°C in 7.5% gel, using tris-glycine buffer (pH 8.3) following the method of Davis (14). The molecular weight of the purified enzyme was estimated by passing it through a Sephadex G-200 column which had previously been calibrated with catalase (Mr 210,000), aldolase (Mr 158,000), alcohol dehydrogenase (Mr 125,000), and bovine serum albumin (Mr 67,000). The sub-unit molecular weight was determined by SDS-PAGE carried out in 10% gel at 4°C according to the method of Weber and Osborne (15). Bovine serum albumin (Mr 67,000), egg albumin (Mr 45,000), trypsinogen (Mr 24,000), and β-lactoglobulin (Mr 18,400) were used as reference proteins. The protein bands were detected by staining with Coomassie brilliant blue.

Estimation of Protein and Chlorophyll

Protein in the enzyme extracts was measured following the method of Lowry *et al.* (16) after precipitation with trichloroacetic acid. Chlorophyll was estimated according to Strain *et al.* (17).

Results

Pod development in chickpea can be divided into two phases; the first phase comprising of pod-wall development and continues up to day 15

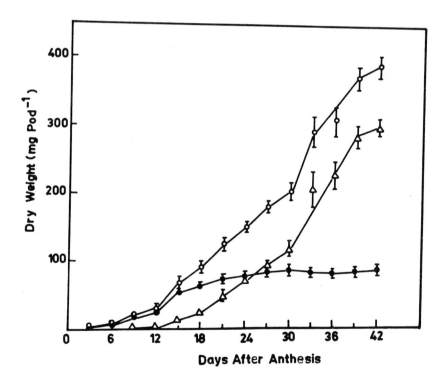

Fig.1: Dry weight of pod (○—○), seed (△—△) and pod-wall (●—●) at different stages after anthesis. Each value in the figure is an average of 20 replicates.

after anthesis; the second phase is that of seed development starting from day 15 after anthesis and ending on day 42 after anthesis.

Dry weight of pod-wall increased up to day 15 after anthesis and seed growth starting after this period was maximum during the period 30 to 39 days after anthesis (Fig. 1).

Pod fixed net CO_2 in light up to day 21 after anthesis, the maximum fixation being at day 18 after anthesis (Fig. 2). In the dark, there was a net loss of CO_2, which increased continuously up to day 21 after anthesis. At maturity, CO_2 loss during light was more than the loss in the dark.

CO_2 concentration in the pod cavity increased up to day 18 after anthesis. Thereafter, it was constant up to day 33 and then decreased. Expressed as a percentage, the CO_2 concentration increased up to day 18 and then remained unchanged up to day 39, the period of active seed

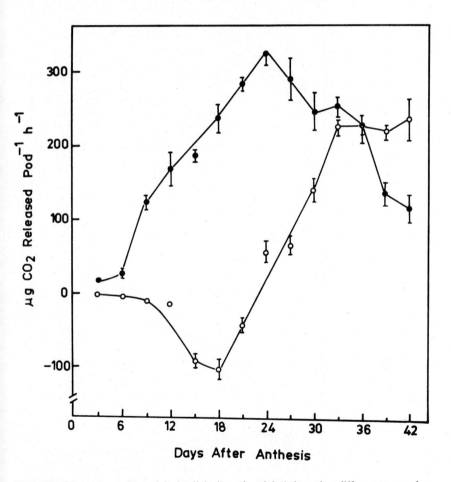

Fig.2: Net CO_2 exchange by pod during light (o—o) and dark (•—•) at different stages after anthesis. Each value in the figure is an average of five replicates.

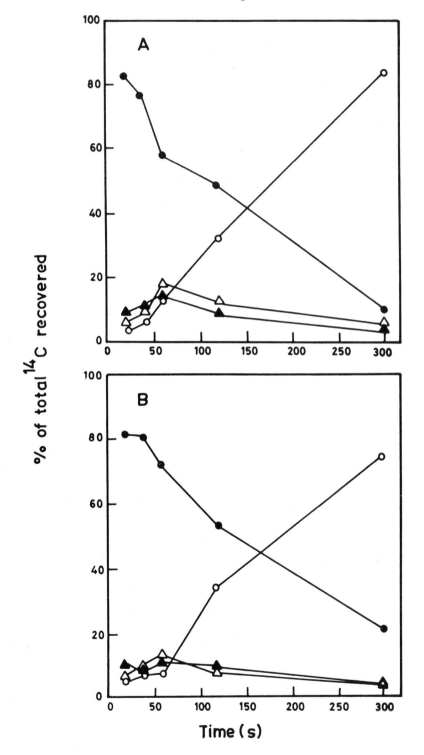

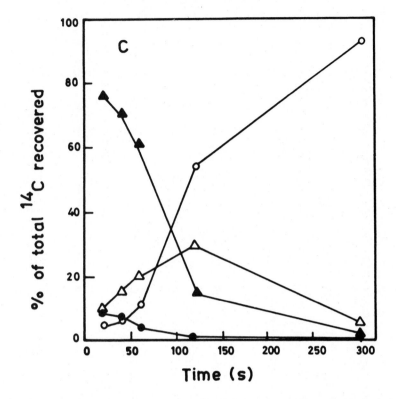

Fig.3: Time course distribution (%) of ^{14}C between malate (●), 3-phosphoglycerate (▲), sucrose (○) and glucose 6-phosphate (△) in A, pod-wall; B, seedcoat and C, leaf of chickpea.

growth. Seed yield was reduced by 20% when the pod was covered with aluminium foil on day 3 after anthesis. However, covering the pod on day 18 had no effect on the final seed yield. Similarly, the removal of the subtending leaf did not affect the seed yield.

No labelling was observed in the seeds up to 5 min, when $^{14}CO_2$ was fed from outside. Only at 48 hr, seeds contained about 50% of the label. On the contrary, seeds fixed only a small amount of CO_2 (10%) when the pod was fed through an internal cavity. In the latter case, the fixation occurred mainly in the pod-wall.

The distribution of radioactivity between malate, 3-phosphoglycerate, sucrose and glucose 6-phosphate obtained as products of $^{14}CO_2$ assimilation, was determined after 20, 40, 60, 120 and 300 sec of incubation. Of the $^{14}CO_2$ fixed after 20 sec of photosynthesis in pod-wall, seed coat and leaf, about 82, 80 and 8% was in malate, whereas 9, 10 and 77%, respectively, was in 3-phosphoglycerate (Fig. 3). The remaining label was in sucrose and glucose 6-phosphate. The ratio of malate to 3-phosphoglycerate was 9. 1, 8. 0 and 0.1 in pod-wall, seed coat and leaf, respectively. After 120 sec,

TABLE 1

Activities of Key Enzymes of Calvin Cycle and C_4 Metabolism in Leaf and Pod-wall of Chickpea at Different Days after Flowering (n mol min^{-1} mg^{-1} protein)

Enzymes	10 days		20 days		30 days		40 days	
	Leaf	Pod-wall	Leaf	Pod-wall	Leaf	Pod-wall	Leaf	Pod-wall
RuBP Carbokylase	403 ± 5.1	73 ± 0.8	141 ± 1.6	93 ± 1.3	90 ± 0.8	55 ± 0.4	9 ± 0.1	2 ± 0.0
NADP-glyceraldehyde-3-phosphate dehydrogenase	1780 ± 31.9	257 ± 3.9	900 ± 13.5	299 ± 4.7	642 ± 9.2	293 ± 4.5	103 ± 1.1	14 ± 0.3
NAD-glyceraldehyde-3-phosphate dehydrogenase	1081 ± 14.7	335 ± 5.3	893 ± 6.3	401 ± 5.1	382 ± 4.7	216 ± 2.9	124 ± 1.5	40 ± 0.6
Ribulose-5-phosphate kinase	4563 ± 83.9	915 ± 13.4	3042 ± 44.8	1638 ± 24.9	1901 ± 27.9	1025 ± 14.9	332 ± 4.6	75 ± 0.6
PEP Carboxylase	25 ± 0.2	74 ± 0.7	11 ± 0.1	99 ± 0.8	5 ± 0.0	159 ± 2.7	0 ± 0.0	27 ± 0.2
NADP malate dehydrogenase	20 ± 0.3	30 ± 0.2	17 ± 0.1	40 ± 0.6	14 ± 0.2	122 ± 1.4	10 ± 0.2	19 ± 0.2
NAD malate dehydrogenase	9811 ± 149.9	17818 ± 239.4	6832 ± 131.7	20667 ± 398.7	6356 ± 95.7	29981 ± 587.6	5279 ± 73.9	8201 ± 121.6
NADP malic enzyme	30 ± 0.3	37 ± 0.4	27 ± 0.3	46 ± 0.4	23 ± 0.3	94 ± 1.1	12 ± 0.3	22 ± 0.5
NAD malic enzyme	356 ± 4.6	729 ± 11.5	318 ± 4.4	870 ± 10.4	292 ± 3.6	1339 ± 21.8	114 ± 1.2	211 ± 3.4
Glutamate oxaloacetate transaminase	125 ± 2.2	234 ± 3.7	112 ± 1.6	386 ± 5.2	97 ± 0.9	433 ± 5.9	67 ± 0.7	103 ± 1.1
Glutamate pyruvate transaminase	89 ± 0.9	139 ± 1.8	79 ± 0.7	153 ± 1.9	67 ± 0.5	224 ± 3.1	45 ± 0.5	62 ± 1.0

while sucrose was the major labelled product in leaf, the level of labelled malate still remained higher than that of sucrose, in spite of an increase in the level of labelled sucrose and a decrease in labelled malate. At 300 sec, sucrose accounted for 70 to 90% of the total ^{14}C recovered.

The activity of RuBP carboxylase, a key enzyme of the Calvin cycle, was higher in leaf then in pod-wall and seed-coat at all stages of seed development (Table 1). The enzyme activity was highest in the youngest tissues and declined as maturity advanced. However, in pod-wall, the enzyme activity declined only after 20 days of flowering. The other Calvin cycle enzymes examined, viz. $NADP^+$-glyceraldehyde 3-phosphate dehydrogenase, NAD^+-glyceraldehyde 3-phosphate dehydrogenase and ribulose 5-phosphate kinase showed patterns qualitatively similar to that of RuBP carboxylase. Ribulose 5-phosphate kinase activity, however, was very high compared to that of the other enzymes of the Calvin cycle at all stages.

Contrary to RuBP carboxylase, PEP carboxylase was more active in pod-wall and seed-coat than in leaf at all stages of seed development. Enzyme activity in leaf and seed-coat decreased as the maturity advanced, whereas in the pod-wall it first increased up to 30 days after flowering and decreased thereafter. The other enzymes of C_4 metabolism investigated, NADP-malate dehydrogenase, NAD^+-malate dehydrogenase, $NADP^+$-malic enzyme, NAD^+-malic enzyme, glutamate oxaloacetate transaminase, and glutamate pyruvate transaminase followed a pattern similar to that of PEP carboxylase.

PEP carboxylase was purified to homogeneity (Fig. 4) as judged by PAGE with 47-fold purification and about 29% recovery using $(NH_4)_2SO_4$ fractionation, ion exchange chromatography on DEAE-cellulose and gel filtration through Sephadex G-200 (Table 2). The molecular weight of purified enzyme, as determined from gel filtration through Sephadex

TABLE 2

Purification of PEP Carboxylase from Immature Pods of Chickpea

Fraction	Total Activity	Protein	Specific Activity	Recovery	Overall Purification
	nmol/min	mg	nmol/min.mg	%	-fold
Crude extract	4793.8	1621.4	2.9	100	
40–55% $(NH_4)_2SO_4$ precipitate dialysed	5562.5	113.9	48.8	116.0	16.8
DEAE-cellulose	2825.0	34.4	82.1	58.9	28.3
60% $(NH_4)_2SO_4$ precipitate dilaysed	2562.5	30.4	84.3	53.4	29.0
Sephadex G-200	1381.3	10.1	136.7	28.8	47.1

Fig. 4   Fig. 5

Fig. 4: Polyacrylamide gel electrophoresis of purified PEP carboxylase obtained after gel filtration through Sephadex G-200.

Fig. 5: SDS-polyacrylamide gel electrophoresis of purified PEP carboxylase obtained after Sephadex G-200 chromatography.

G-200 was found to be about 200,000. SDS-PAGE of the enzyme preparation yielded a single protein band (Fig. 5) indicating that the enzyme is composed of identical sub-units. The estimated molecular weight of the subunit was about 50,000, indicating that the enzyme from immature chickpea pods is a tetramer composed of four identical subunits.

The enzyme, exhibiting optimum activity at pH 8.1, showed typical hyperbolic kinetics with PEP as the substrate with Km of 0.74 mM. The enzyme however showed sigmoidal response to increasing concentrations of HCO_3^- (Fig. 6) indicating cooperative binding of HCO_3^- to the enzyme. The SO.5 value for HCO_3^- was 7.6 mM. Again with Mg^{2+}, the enzyme showed hyperbolic kinetics with Km of 0.56 mM.

Inorganic phosphate (Pi) and a number of phosphate esters activated the purified enzyme (Fig. 7). At 8 mM, glucose 6-phosphate caused maximum activation of the enzyme, followed by α-glycerophosphate, 3-phosphoglyceric acid, fructose 1, 6-bisphosphate and Pi. However, at 2 mM concentration, Pi caused maximum activation of the enzyme.

None of the amino acids tested at 5mM concentration had any significant effect on the enzyme activity. Similarly, nucleoside monophosphates and diphosphates did not exert any effect on the enzyme activity. However, the enzyme was inhibited to the extent of about 28, 28, 14 and 10%, respectively by ATP, UTP, CTP and GTP.

The purified enzyme required Mg^2 ions specifically for its activity. The enzyme showed no activity in the absence of Mg^{2+} and also when Mn^{2+} or

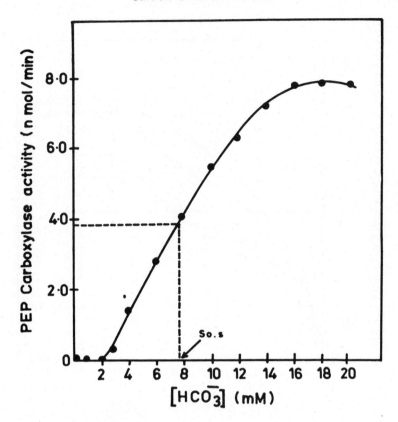

Fig. 6: Effect of varying concentrations of HCO⁻₃ on the activity of purified PEP carboxylase with fixed concentration of PEP.

Ca^{2+} were present alone. Mn^{2+} and Ca^{2+}, when present at 5 mM concentration in the presence of 5 mM Mg^{2+}, inhibited the enzyme by 72 and 83% respectively.

Oxaloacetate (OAA), the product of the enzyme reaction, inhibited the enzyme activity substantially, 50% inhibition occurring at 0.2 mM concentration. Kinetic analysis indicated OAA to be a non-competitive inhibitor of the enzyme with respect to PEP ($Ki = 0.115mM$). However, inhibition by malate was not that profound. At 12mM concentration, malate inhibited the enzyme activity by 33%. The inhibition again was of a non-competitive type with ki of 13.8 mM. Addition of glucose 6-phosphate relieved the inhibition of the enzyme by OAA and malate. Other organic acids like succinate, fumarate, glycolate, and malonate, when tested at 5mM concentrations, did not produce any significant effect on the enzyme activity.

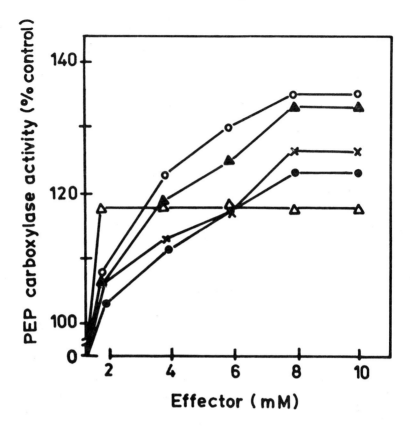

Fig. 7: Effect of varying concentrations of glucose 6-phosphate (—○—), α-glycerophosphate (—▲—), 3-phosphoglyceric acid (—×—), fructose 1, 6-bisphosphate (—●—), and P_i (—△—) on the activity of purified PEP carboxylase.

Discussion

Developing pods of legumes are known to be photosynthetically active (1). However, the extent of their contribution towards seed growth is still not clear. Though a number of studies conducted earlier have indicated that legume pods fix very little net CO_2 in light (6, 7); here, in the present case, there was net fixation of CO_2 in light, which increased up to day 18 after anthesis (Fig. 2), indicating that the pod-wall of chickpea fixes atmospheric CO_2 photosynthetically during the early phase of pod development. This was further confirmed by a labelling experiment wherein pod-wall fixed $^{14}CO_2$ when it was fed externally. The CO_2 fixed by the pod was transported to seeds with a lag of about 5 min. About 50% of the label was found in seeds within 48 hr. In pea, 60% of total $^{14}CO_2$ fixed by carpel was exported to seeds after 24 hr of feeding (18).

In the present case, pod-wall, through mobilisation (pod-wall dry weight decreased from 79 mg to 63 mg when covered on day 3) and net CO_2 fixation, contributed about 20% of photosynthates towards seed dry weight. This is contrary to the results reported by Singh and Pandey (4) where they concluded that pod-wall photosynthesis was not a significant source of assimilate for seed development. In pea also, pod photosynthesis was shown to improve the economy of carbon usage by about 16 to 20% (19).

Chickpea pod-wall was far more efficient (90%) in $^{14}CO_2$ fixation than the seeds (10%) when fed through internal cavity. Since no stomata was observed on the inner layer of the pod-wall, the source of CO_2 for this could only be the respired CO_2. In this way, pod-wall acts as an impermeable barrier to internal CO_2 accumulating in great amounts in the pod cavity, especially during the rapid phase of seed growth. Interestingly, in this crop, seeds did not develop when the pod-wall was made permeable by puncturing and allowing CO_2 to escape. In pea also, the inner layer of pod-wall has been shown to be the most active site for refixation of respired CO_2 (18).

Two experimental approaches, namely kinetic studies with $^{14}CO_2$ and enzyme profiles, were followed here to determine further the pathway of CO_2 assimilation in fruiting structures of chickpea. Interestingly, in pod-wall and seed-coat, the initial product of $^{14}CO_2$ assimilation after 20 sec of photosynthesis was malate. About 80% of the total radioactivity appeared in malate compared to about 10% in 3-phosphoglycerate (Fig. 3). However, the leaf showed 77% of radioactivity in 3-phosphoglycerate and only 8% in malate. The rapid labelling and relatively high proportion of ^{14}C in malate seems to reflect the synthesis of OAA catalysed by PEP carboxylase. Since 3-phosphoglycerate was also an early product of photosynthetic assimilation and also sugar phosphates were labelled, CO_2 fixation by RuBP carboxylase and operation of the PCR cycle is suggested. However, it could not be ascertained from these experiments whether or not the phosphorylated compounds were derived directly from the activity of RuBP carboxylase or subsequently from malate, as occurs in the leaves of C_4 plants.

The results of ^{14}C labelling were supported by enzymic studies. The activities of RuBP carboxylase and other PCR cycle enzymes were high in the leaf compared to the pod-wall and seed-coat (Table 1), which had higher levels of PEP carboxylase and other enzymes of C_4 metabolism at all stages of seed development. The observed increase in the level of pod-wall PEP carboxylase correlated very well with the corresponding increase in the activities of respiratory enzymes and rates of respiration in developing chickpea seeds (20), and with the level of dark fixation of CO_2 in the pod-wall. Thus it could be concluded that fruiting structures of chickpea utilise PEP carboxylase to recapture the respired CO_2. Hedley *et al.* (21) have suggested that in C_3 plants, high levels of PEP carboxylase are induced during the developmental phase, when the rate of respiration exceeds the rate of photosynthesis, resulting in net loss of CO_2. Such respiratory losses amount to 29–71% of the gross CO_2 fixed during photosynthesis (22). Thus PEP carboxylase, by minimising respiratory carbon losses, may help in improving considerably the carbon economy of the developing pods and enhancing plant productivity (23).

The enzyme PEP carboxylase exhibited some properties resembling those from C_4 and C_3 plants (24). However, in some other respects, its properties did not conform to either of the two groups of these plants. The high S0.5 value for HCO^-_3 is commensurate with the physiological conditions prevalent in pods. Since extracellular CO_2 cannot readily diffuse out of pod-wall and the CO_2 concentration inside the pod is very high as compared to its concentration in the atmosphere (10), the high S0.5 value for HCO^-_3 and sigmoidal response to increasing concentrations of this substrate thus indicate that the functioning of this enzyme is favoured under conditions when the respiratory CO_2 losses are high, which essentially occur during pod development. PEP carboxylase thus plays an important role in recycling the CO_2 released during either dark respiration or photorespiration. However, the extent of CO_2 recycled in relation to the total respiratory losses, the origin of PEP, and the fate of recycled CO_2 remain to be conclusively proved.

Legume seeds being rich in proteins require a large supply of amino acids, for the synthesis of which carbon skeletons are derived from the tricarboxylic acid cycle. PEP carboxylase, by playing an anapleurotic role in replenishing the intermediates of the above cycle, might also be helping the synthesis of amino acids in developing seeds.

Oxaloacetate and malate, the sequential products of CO_2 fixation by PEP carboxylase, inhibit the enzyme activity. *In vivo*, the concentration of oxaloacetate would be considerably less than that of malate (25) and also, in view of the fact that oxaloacetate rarely seems to accumulate in plant cells, it is likely that L-malate would be relatively more important as a possible allosteric regulator of PEP carboxylase, thus regulating the CO_2 fixation by feed-back or end-product inhibition of the enzyme. The reversal of inhibitory effects of oxaloacetate and malate by glucose 6-phosphate is interesting. In C_3 plants, PEP for anapleurotic reactions is derived from glycolytic reactions. When a sufficient amount of glucose 6-phosphate is available in the cell to ensure continued production of PEP, the inhibitory effects of oxaloacetate and malate are suppressed. It is thus likely that glucose 6-phosphate and oxaloacetate/malate, through their opposing effects, might be important in controlling PEP carboxylase activity *in vivo*.

Acknowledgements

This research was financed in part by a grant made by the USDA under the Co-operative Agricultural Research Grant Program (PL-480). Drs I. S. Sheoran and H. R. Singal are thanked for carrying out the experiments reported here.

References

1. Singh, Randhir (1987). *In : Recent Advances in Frontier Areas of Plant Biochemistry* (Singh, Randhir and Sawhney, S. K., eds.) pp. 102–126, Prentice Hall of India
2. Flinn, A. M. and Pate, J. S. (1970) *J. Exp.Bot.* 21, 71–82
3. Khanna-Chopra, R. and Sinha, S. K. (1976). *Indian J.Exp.Biol.* 14, 159–162.
4. Singh, B. K. and Pandey, R. K. (1980). *Aust. J. Plant Physiol.* 7, 727–735
5. Khanna-Chopra, R. and Sinha, S. K. (1982). *Photosynthetica* 16, 509–513
6. Crookston, R. K., O'Toole, J. and Ojbun, J. L. (1974). *Crop Sci.* 14, 706–712
7. Quebedeaux, B. and Chollet, R. (1975). *Plant Physiol.* 55, 745–748
8. Koundal, K. R. and Sinha, S. K. (1981). *Phytochemistry* 20, 1251–1252
9. Luthra, Y. P., Sheoran, I. S. and Singh, Randhir (1983). *Photosynthetica* 17, 210–215
10. Sheoran, I. S., Singal, H. R. and Singh, Randhir (1987), *Indian J. Exp. Biol.* 25, 843–847
11. Singal, H. R., Sheoran, I. S. and Singh, Randhir (1986), *Physiol. Plant* 66, 457–462
12. Benson, A. A., Bassham, J. A., Calvin, M., Goodale, T. C., Mass, V. A. and Stepka, W. (1950). *J.Am.Chem.Soc.*72, 1710–1718
13. Singal, H. R. and Singh, Randhir (1986). *Plant Physiol.* 80, 369–373
14. Davis, B. J. (1964). *Ann. NY Acad. Sci.* 121, 404–427
15. Weber, K. and Osborne, M. (1969). *J. Biol.Chem.* 244, 4406–4412
16. Lowry, O. H., Rosebrough, N. J., Farr, A. L. and Randall, R. L. (1951). *J. Biol.Chem.* 193, 265–275
17. Strain, H. H., Cope, B. T. and Svec, A. A. (1971). *Methods Enzymol.* 23, 452–476
18. Lovell, P. H. and Lovell, P. J. (1970). *Physiol. Plant.* 23, 316–322
19. Flinn, A. M., Atkins, C. A. and Pate, J. S. (1977). *Plant Physiol.* 60, 412–418
20. Sangwan, R. S. Popli, S. and Singh, Randhir (1983). *Indian J. Exp. Biol.* 21, 37–39
21. Hedley, C. L., Harvey, D. M. and Kelly, R. J. (1975). *Nature* 258, 352–354
22. Zelitch, I. (1975). *Science* 188, 626–631
23. Rao, A. S. and Singh, Randhir (1983). *J. Theor.Biol.* 104, 113–120
24. O'Leary, M. H. (1982). *Annu.Rev.Plant Physiol.* 33, 297–315
25. Ting, I. P. and Dugger, W. M. (1967). *Plant Physiol.* 42, 712–718

Genotypic Response to Assimilate Demand in a Variegated Leaf Mutant of *Nerium Oleander*

C. C. Subbaiah and A. Gnanam

Department of Plant Sciences,
Madurai Kamaraj University,
Madurai 625 021. INDIA

Summary

A naturally occurring variegated leaf biotype of *Nerium oleander* L. was tentatively identified to be a plastome coded mutant. The yellow sectors leaves in the mutant had less than 1–2% of the total quantity of chlorophyll, and 5–8% of the carotenoids, normally found in the wild type; and were totally lacking in light-dependent O_2 evolution as well as CO_2 fixation. However, the green sectors of the variegated leaves in the mutant compensated for the thinning effect imposed by the heterotrophic yellow tissue, since they had high rates of CO_2 assimilation. We looked at the basis of the superior photosynthetic efficiency of the green sectors of the mutant leaves, which was two times greater than that in the wild type leaves. The chloroplasts from the green sectors showed greater capacity for light utilisation with increased amounts of electron transport complexes (PQ and Cyt t) and ATP synthase per unit chlorophyll in their thylakoid membranes as compared to those in the wild type, thus resembling the chloroplasts from plants adapted to high irradiance. However, at the gross morphological level the variegated leaves mimicked shade leaves with greater surface areas and greater intercellular air space volume than the wild type leaves. We propose that these contrasting strategies at the organelle and organ levels help the green sectors to achieve greater photosynthetic rates and overcome the heterotrophic stress imposed by the non-green tissue. Also, these responses appeared to involve a co-ordinated gene expression in nucleo-cytoplasmic and plastid compartments.

Introduction

One of the fundamental problems in chloroplast molecular biology deals with the mechanisms by which the photosynthetic apparatus adapts to changes in environmental, genetic and developmental factors. One approach to this problem is to analyse photosynthetic mutants with lesions in specific reactions and study the consequences on the overall process of photosynthesis as well as the contribution of a particular reaction to the regulation of the process.

We located a naturally occurring variegated leaf mutant in *Nerium oleander* in our University campus, and this was transferred to the Botanic

Garden for detailed observation. In order to localise the genetic lesion, we began to characterise this variant, photosynthetically. To our interest, the green sectors of the variegated leaves showed very high rates of CO_2 fixation, well above the average values in wild type plants. In the present communication, we have analysed the structural and functional modifications of the chloroplasts in these sectors that would explain why their photosynthetic rate is greater than that of the wild type leaves. The intention is to understand the genetic regulation of the process.

Material and Methods

Leaves of similar plastochron index from field grown plants were used for all the studies. The plants were free from any pest and physiological stress.

Leaf morphological parameters:

Surface areas were measured using an automatic area meter (Hayashi Denko, Japan). Dry weights were taken after drying to constant weight at 65°C for 48 h. Leaf and mesophyll thickness were measured in a light microscope from hand sections of leaf tissue. Intercellular air space volume was determined by an infiltration technique (1).

Pigment analysis: Leaf pigments were extracted in 80% acetone and estimated spectrophotometrically (2).

TABLE 1

Pigment composition ($\mu g/cm^2$) in wild and mutant biotypes of nerium

Type of sample	Total chl	a/b	Total carot	a+b/carot
Wild type	53.0	2.76	11.13	5.13
Shaded wild	27.3	2.67	4.33	6.16
Green sector	54.2	3.10	13.50	4.50
Yellow sector	0.8	1.92	0.55	1.80
Yellow leaf	0.2	7 to 10	0.90	0.38

Field Photosynthesis: The net photosynthetic rates were measured using Li-Cor field photosynthesis system at light saturation. Stomatal resistance was also recorded simultaneously, using a Li-Cor steady state porometer.

Chloroplast isolation: About 50 g of leaf tissue was homogenised for 30–45 s in a Sorvall Omni-mixer at full speed in 250 ml of buffer (0.33 M sucrose,

50 mM Tricine-KOH (pH 7.5), 10 mM NaCl, 1 mM EDTA, 3% PVP, mM sodium ascorbate and 1mM PMSF). The homogenate, after filtration was centrifuged at $100 \times g_{max}$ for 3 min and the supernatant was recentrifuged at $7000 \times g_{max}$ to collect the plastid pellet. The organelles were resuspended in the same buffer and layered on a 3-step sucrose gradient (20%, 45% and 60%). The gradients were centrifuged in a swing-out rotor at $7000 \times g_{max}$ for 30 min. The band at the interphase of 45–60%, containing intact chloroplasts was collected, washed twice with the above buffer (excluding PVP) and used immediately.

Isolation and estimation of CF_1 activity: The intact plastids were lysed and CF_1 was isolated according to the method of Younis *et al.* (3), omitting the column purification step. The preparation was resolved in a 6% polyacrylamide slab gel. One half of the gel was stained for protein by Coomassie blue and the other half for ATPase activity according to Delapelaire and Chua (4). The activity was quantified by scanning a photographic negative of the gel.

Estimation of RuBP carboxylase and CF_1 contents by rocket immunoelectrophoresis: Partially purified CF_1 preparations, described above, were also quantified immunologically by rocket electrophoresis and calculating the areas of the peaks.

The content of RuBP carboxylase was estimated from total soluble protein of leaf by the above method. Soluble protein was extracted from leaf tissue, by homogenisation in a buffer (1 g/10 ml) containing 100 mM Tris-HCl (pH 8), 200 mM NaCl and 1mM PMSF, and clarified by centrifugation at $15,000 \times g_{max}$ for 30 min.

The antibodies were raised against purified antigens (CF_1 and RuBP carboxylase) from *Vigna* chloroplasts in rabbits.

Estimation of newly synthesised CF_1 sub-units: About 1.5 g of leaf tissue was fed with 50 μCi of ^{35}S-methionine for 2 h and the crude CF_1 was prepared from this tissue as described above. The labelled protein was used for immunoprecipitation against anti-CF_1 and the precipitate was analysed by SDS-PAGE-fluorography.

Estimation of cytochrome f: The crude chloroplast preparation was resuspended in a buffer containing 0.4 M sucrose, 25 mM Tris-HCl (pH 7.5) and 0.1% Triton $\times-100$, and cyt f was assayed spectrophotometrically from HQ reduced minus ferricyanide oxidised difference spectra (5).

Cytochromes were also stained by heme-dependent peroxidase activity from total thylakoid proteins after SDS-PAGE analysis (6).

Chlorophyll fluorescence induction: *In vivo* and *in vitro* chlorophyll fluorescence curves were recorded at room temperature. The excitation light (25 μmol photon m^{-2} s^{-1}) was passed through a blue filter (Corning 4–72) and the signal was stored in a storage oscilloscope. Both fast and slow kinetics were recorded.

Results and Discussion

Field measurements of net photosynthetic rates (Table 2) showed that the yellow sectors completely lacked in CO_2 fixation. However, the photosynthetic rate of green sectors was nearly double (or more) that of the wild type leaves. Besides, the variegated leaves were nearly three times larger

TABLE 2

Rates of field photosynthesis (mg Co_2 m^{-2} s^{-1}) in wild type and mutant nerium leaves

Type of sample	5th July	19th August
Wild type	0.27	0.59
Green sector	0.64	0.77
Yellow leaf/sector (Respiration)	−0.22	−0.14

TABLE 3

Leaf morphological and anatomical characters in wild and mutant biotypes of nerium

Character	Wild type		Mutant		
	Light	Shade	Green sector	Yellow sector	Yellow leaf
Leaf thickness (mm)	0.48	—	0.58	0.50	—
Mesophyll thickness (mm)	0.09	—	0.18	0.15	—
Intercellular air space volume (%)	13.87	—	20.63	13.82	—
Specific fresh mass (mg/cm^2)	38.05	22.90	42.80	38.90	37.67
Specific dry mass (mg/cm^2)	13.84	9.44	14.23	9.60	8.49
Area (cm^2)	14.65	29.40	37.50 (variegated)		15.28
Water content (% of dry mass)	178.00	143.00	199.00	314.00	344.00

in their surface area than the wild type leaves (Table 3). Even if the green tissue were to account only for one-third of the total leaf area in a plant, the integrated P_N of this tissue would be equal to that in a wild type plant with similar leaf area. Actually, the green tissue represented more than one-half of the total leaf area in a mutant plant.

We do not have any information on the genotype of the wild and mutant biotypes. Even if they are not isolines, the large difference in their net photosynthesis per unit chlorophyll cannot be dismissed as plant to plant variation. We interpret that the thinning effect of yellow tissue imposed an increased demand for assimilates on the green sectors and this resulted in an enhancement in the net photosynthetic rate of the green tissue. Such changes in carbon fixation rates depending on the intensity of the sink demand either due to fruiting or defoliation are well documented in the literature (7, 8).

We wanted to examine whether the increase in CO_2 assimilation rate of green sectors was brought about by a direct increase in the enzymes of dark reactions or through more efficient photochemical reactions or both. RuBP carboxylase was estimated, as this is a rate limiting enzyme of the Calvin cycle and constitutes 50% of total leaf soluble protein. However, no difference could be noticed between the green sectors and the wild type leaves in the level of this enzyme protein (Fig 1), showing that there is no correlation of the enhanced rate of carbon assimilation with the level of this enzyme (assuming that there was no difference in its specific activity from the two sources). Then, it follows that modulations in the relative concentrations of electron transport carriers and ATP synthase of the thylakoid membranes, directly involved in the rate-limiting steps of

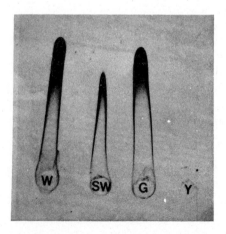

Fig. 1: Estimation on **RuBPC content by rocket immunoelectrophoresis.** Equal aliquots of soluble protein (1g/10 ml buffer) preparations were loaded in each well. Anti-RuBPC was raised against purified protein from *Vigna sinensis* leaves in a rabbit.

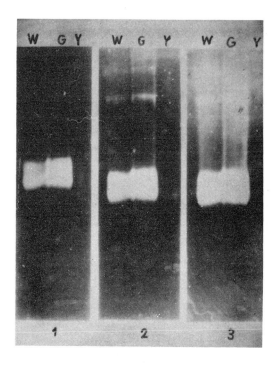

Fig. 2: (left) Coomassie blue staining of crude CF_1 preparations from wild type and mutant plastids after resolution in a 6% native acrylamide gel. Equal quantities of total thylakoid proteins were used for CF_1 extraction in all cases.

Fig. 3: (right) Same gel stained for ATPase activity. The gradual increase in intensities of different bands with time is shown.

photosynthesis, could correlate directly with the changes in overall photosynthetic capacity of intact leaves under lightsaturated conditions. Comparisons of the relative increases in thylakoid components demonstrated that the most substantial changes occurred in the coupling factor content whcih was almost trebled in the plastids of green sectors (Figs. 2 & 3). There was also a marked increase in cyt f content of these plastids and hence in the cyt b/f complex and its associated catalytic sites (Fig. 5). The bulk redox pool plastoquinone which delivers electrons to the cyt b/f complex and protons to the intrathylakoid space, was increased by about 50% (as estimated from the area above the chlorophyll fluorescence induction curves in the absence of DCMU at room temperature) in the plastids of green sectors (Fig. 6).

The'coupling factor was assayed both as activity and as concentration of the specific protein. There was a good correlation between the amount and activity of the preparations. The concentration was also estimated immunologically (Fig. 4) which confirmed the trends obtained by the other two methods (Figs. 2 and 3), namely, the thylakoid preparation from the

Fig. 4: Rocket immunoelectrophoretic estimation of CF_1 content from the above preparations. Antibody was raised in a rabbit against purified CF_1 from *Vigna sinensis* chloroplasts.

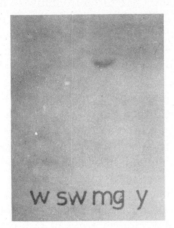

Fig. 5: Histochemical localisation of cytochromes in thylakoid polypeptides separated by SDS-PAGE. The major band is cytochrome f.

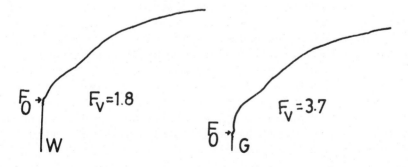

Fig. 6: Fast fluorescence kinetics of chloroplasts from wild type and green sectors (0.5, 5 mV).

green sectors of the mutant always had 2–3 times greater CF_1 content and activity per unit chlorophyll than the membranes of wild type leaves.

The cyt f content was assayed both spectrophotometrically as well as by heme-dependent peroxidase staining of thylakoid polypeptides. Both methods showed that the cyt f content per unit chlorophyll was more in the green sectors, but the activity staining showed a clearer trend. The percentage increase in PQ was less, but still substantial since PQ is the most abundant electron carrier in the thylakoid membrane. A greater number of PQ molecules would confer a greater capacity for electron transport from PS II to the cyt b/f complex. In summary, the greatest adaptive response seen in the multi-subunit protein complexes of green sector thylakoids on a chlorophyll basis was that of ATP synthase and the cyt b/f complex. Indeed, Wilhelm and Wild (9) observed previously that the cyt f content of *Chlorella* was changed in strict accordance with alterations of biomass productivity and photosynthetic capacity. Chow and Anderson (10) also reported that the increase in photosynthetic activity of pea leaves that had been transferred from low to high light were due to increases in PQ, cyt f and ATP synthase, only with a minor increase in PS II complex. Other reports also show that the differences we observed in the thylakoid composition of the green sectors resemble the adaptations of chloroplasts isolated from plants grown at high irradiances (11, 12, 13, 14 and 15). This was also exemplified by the Chl a/b ratios of these sectors. The chloroplasts from green sectors, like 'sun' adapted chloroplasts showed greater a/b ratio than the organelles from wild type leaves (see Table 1). The greater a/b ratio indicates lesser granal stacking and more stromal thylakoids. This is consistent with the observed increase in CF_1 content in green sector chloroplasts, since this protein complex is confined only to the non-appressed membranes (16). As the total chlorophyll concentration was similar in both types of leaves, the increased a/b ratio also suggests an alteration in the relative amounts of the Chl-protein complexes—namely an overall increase in the amount of the core Chl a—proteins of PS II and PS I and a concomitant decrease in the relative amounts of Chl a/b proteins of PS II and PS I (14). Assuming that the a/b ratio of LHC proteins is 1.3, a change in the Chl a/b ratio from 2.7 to 3.1 (from the wild type to the green sectors) implies a decrease from 62% to 55% of the total Chl present as LHC a/b proteins. To compensate for this in terms of total chlorophyll, a net increase in CP_a—the core complex of CP II (up to 3% of total Chl) can be expected. In fact, such changes were demonstrated to accompany the alterations in a/b ratio when plants were transferred to high light (10, 14). Thus, the chloroplasts from the green sectors, since they resembled the plastids from high light grown plants, should have the capacity for effective utilisation of high light. The observed increases in the concentrations of electron carriers such as PQ and cyt b/f (possibly with other mobile electron carriers such as PC and Fd) could lead to an increase in electron transport, as these components are involved in both non-cyclic and cyclic photophosphorylation and are among the rate-limiting factors of electron transport. The concomitant increase noticed in ATP synthetase should result in an increased capacity for ATP and NADPH formation and a prompt increase of the *in vivo* light saturated photosynthetic capacity.

At the gross morphological level, the variegated leaves or their green sectors in contrast to their 'sun' type chloroplasts resembled the shade-adapted leaves. They had higher surface area than leaves from wild type and resembled the shade leaves of wild type plants (Table 2). The increased surface area would compensate for the loss of light harvesting surface in the form of yellow sectors in these leaves. Further, their greater intercellular air space volume should facilitate better gaseous exchange as in the case of shade leaves. The variegated leaves were thicker than the wild type leaves due to additional layers of palisade cells (Table 2). This, however, is a 'sun' leaf characteristic, but obviously helps in better light harvesting per unit area of leaf surface. Thus, these multifarious adaptations at different levels of organisation help in increased light harvesting as well as effective light utilisation due to efficient electron transport, and thereby support greater photosynthetic activity. And, these modifications have mainly been confined to the thylakoid membrane, but for a few changes in leaf morphology. The other two important factors that influence the photosynthetic rate and are known to respond to increased

TABLE 4

Photosynthetic characters of green sectors

Character	Intensity with respect to wild type	Adaption	Advantage
Leaf area	Larger	Shade	Increased light Harvesting (compensates for yellow sectors/leaves)
Mesophyll thickness	Greater	Sun	Better light trapping
Intercellular	Greater	Shade	Better gaseous exchange
Chl a/b	Lower	*	Greater Photochemical activities
Cyt f/Chl	Greater	*	
CF1/Chl	Greater	* Sun	
		*	
		*	
RUBISCO/ g.Fr.wt	Equal	—	——

sink demand, namely, stomatal conductance and the carboxylation machinery were not modulated in this mutant. These results thus exemplify the adaptive and dynamic nature of the thylakoid membrane.

A close examination of the differences between the two types of tissues also reveal that a co-ordinated interplay of nuclear and chloroplast genomes is needed to effect these changes, especially as in the case of multisubunit proteins like the coupling factor. An attempt was made to address this question by analysing the rate of new synthesis of different subunits of this complex. Immunoprecipitation of radiolabelled CF_1, revealed that the stoichiometry of the nuclear coded (γ and β) subunits and plastid coded subunits (α, β and ϵ) was maintained in the newly synthesised CF_1. However, we have not analysed the other components of electron transport that also showed an increase in the thylakoid membrane of green sectors.

Conclusions

Modifications, proposed to have optimised the photosynthetic rates in the green sectors of the mutant, have mainly been confined to the thylakoid membrane, but for a few changes in leaf morphology. The other two important factors that influence the photosynthetic rate and are known to respond to increased sink demand, namely, stomatal conductance and the carboxylation machinery, were not modulated in this mutant. These results thus exemplify the adaptive and dynamic nature of the thylakoid membrane and also raise questions on the regulatory mechanisms—such as the nature of signals that cause differential gene activation, the mode of their transduction, co-ordination between nuclear and plastid genomes and the site of regulation during gene action. Our data may also be of agronomic significance, to be used in screening genotypes for superior photosynthetic performance.

Acknowledgements

This work was supported by a research grant (No. 21 (7) 84–STP–II) from the Department of Science and Technology.

References

1. Byott, G. S. (1976) *New Phytol.* 76, 295–299
2. Lichtenthaler, H. K. and Wellburn, A. R. (1983) *Biochem. Soc Transactions* 603, 591–592
3. Younis, H., Winget, G. D. and Racker, E. (1977) *J. Biol. Chem.* 252, 1814–1816

4. Delepelaire and Chua, N.-H. (1982) in Methods in Chloroplast Molecular Biology (Edelman, M., Hallick, R. B. and Chua, N. H. eds.) pp.000, Elsevier Biomedical Press, Amsterdam
5. Bendall, D. S., Davenport, H. E. and Hill, R. (1971) *Methods Enzymol.* 23, 327–344
6. Thomas, P. E., Ryan, D. and Levin, W. (1976) *Anal. Biochem.* 75, 168–176
7. Waering, P. F., Khalifa, M. M. and Treharne, K. J. (1968) *Nature* (Lond.) 220, 453–457
8. Gifford, R. M. and Evans, L. T. (1981) *Annu. Rev. Plant Physiol.* 32, 485–509
9. Wilhelm, C. and Wild, A. (1984) *J. Plant Physiol.* 115, 125–135
10. Chow, W. S. and Anderson, J. M. (1987) *Aust. J. Plant Physiol.* 14, 9–19
11. Bjorkman, O., Boardman, N. K., Anderson, J. M., Thorne, S. W., Goodchild, D. J. and Pyliotis, N. A. (1972) *Carengie Inst. Year Book* 71, 115–135
12. Grahl, H. and Wild, A. (1975) in *Environmental and Biological Control of Photosynthesis* (Marcelle, R., ed.) pp. 107–113, Dr. W. Junk Publishers, The Hague
13. Berzborn, R. J., Muller, D., Roos, P. and Anderson, B. (1981) in *Photosynthesis III. Structure and Molecular Organization of the Photosynthetic Apparatus* (Akoyunoglou, G., ed.) pp. 107–120, Balban Intern. Science Series, Philadelphia
14. Leong, T. Y. and Anderson, J. M. (1984) *Photosyn. Res.* 5, 117–128
15. Davies, E. C., Chow, W. S., LeFay, J. M. and Jordan, R. R. (1986) *J. Exp. Bot.* 37, 211–220
16. Anderson, J. M. (1986) *Annu. Rev. Plant Physiol.* 37, 93–136

Selective Alternations in Photosynthetic Pigment Characteristic and Photoelectron Transport during Senescence of Wheat Leaves

S. C. Sabat, Anil Grover and Prasanna Mohanty

School of Life Sciences,
Jawaharlal Nehru University
New Delhi 110067, INDIA

Summary

Aging induced selective changes were monitored both in intact and in detached primary leaves of wheat. We have shown that loss in senescence associated chlorophyll content in the case of wheat leaves, is linked to uniform loss in pigment in chloroplasts and not due to loss in chloroplast number. Analyses of loss of chloroplast absorption spectra both in detached and attached leaves indicate that far red absorbing forms of Chl a (Chl a 692, Chl a 700 and Chl a 708) are extremely sensitive to senescence induced losses. The shift in pH optimum of ferricyanide supported Hill activity of chloroplast during in vivo aging to the pH optimum of phenylenediamine supported Hill activity is indicative of a selective change in electron acceptance site of ferricyanide towards photosystem II. Thus *in vivo* aging seems to alter the oxidising side of photosystem II.

Introduction

In the final phase of leaf growth and development, technically termed leaf senescence, the cell physiology gets perturbed in several ways (1). A great deal of literature has accumulated on biochemical details of the senescing plant cells for processes such as photosynthesis, nitrogen metabolism, respiration etc. The current awareness points out that leaf senescence follows a well-defined and sequential cascade of metabolic alterations (1), and it is conceptually incorrect to consider leaf senescence as a phenomenon of merely deteriorative events (2). In support of this argument, Biswal and Mohanty (3) have earlier shown that the disorganisation of photosystem II (PS II) oxygen evolving complex follows an orderly scheme in senescing detached barley leaf. Their findings were based on assay of PSII electron transport activity in terms of ferricyanide (FeCN) reduction by isolated thylakoid membranes, and using water and diphenylcarbazide as electron donors.

We present in this paper further evidence that specific alterations in photochemical activity of chloroplasts mark the progress of leaf senescence. The work presented in this paper is a summary of the experiments done on senescing primary wheat leaf (4, 5, 6) and includes such aspects as relative sensitivity of various spectral forms of photosynthetic pigments to leaf senescence, age-induced alterations in electron transport chain and sequential loss of chloroplasts.

Materials and Methods

Wheat (*Triticum aestivum* L. var Kalyansona) seedlings were raised under laboratory conditions under continuous illumination (~20 watts m^{-2}) at 25 ± 1°C. The primary leaf was used in all experiments, the precise age and the conditions for induction of senescence (whether attached or detached) is mentioned along with the results in individual experiments. Broken chloroplast thylakoid membranes were prepared following standard procedure (6) and chlorophyll (Chl) was estimated following Arnon (7). Partial electron transport activities were measured polarographically under saturating light intensity (~480 watts m^{-2}) at 25°C. Details of the reaction conditions for electron transport assays are presented in the legends to the figures and tables. Chloroplast absorption spectra were recorded on a Hitachi 557 spectrophotometer. The opal surface of the quartz cuvettes was kept in the path of the light to minimise the scattering (8). The chloroplast number was counted using a haemocytometer (4).

Results

Relative sensitivity of various spectral forms of photosynthetic pigments to leaf senescence:

Both attached and detached leaves (16 d old seedlings) were used. One set of these seedlings was transferred to dark and the other set was maintained under continuous white light (~20 watt/m^2). The measurement of the spectral changes in chloroplasts isolated from the primary leaves of these seedlings was made after 6 d of incubation. The analysis of detached leaves was done after 3 d of incubation.

Both in attached and detached leaves, accompanying the overall loss of Chl content due to leaf senescence (data not shown), the absorbance at all the wavelengths of chloroplast absorption spectrum declined (Fig. 1A, 2A). Fig. 1B and 2B represent the percentage of decline in absorbance at various wavelengths. These values were calculated by considering the respective absorbance values observed with chloroplasts of 16 d old (control) primary leaves. The magnitude of decline in absorbance varied at different wavelengths of the absorption spectrum; particularly, Chl *a* 692

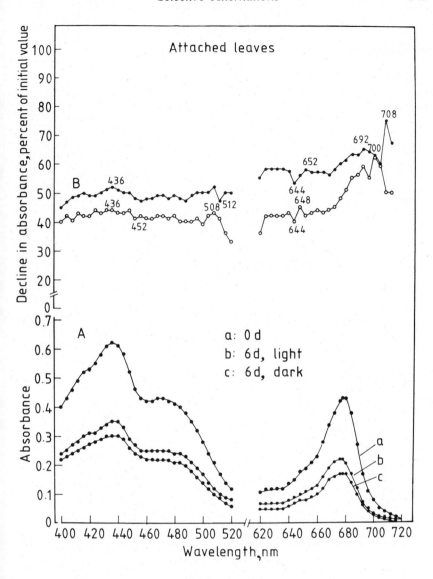

Fig. 1: Age dependent changes in absorption spectra of chloroplasts isolated from light and dark incubated attached 16-day-old primary wheat leaves.
[A] Absorption spectra of (a) control 0 day, (b) 6 days in continuous light (20 watt m²), (c) continuous dark 6 days.
[B] Percentage decline in absorbance as a function of wavelength. Control value was taken as 100 per cent.

was maximally affected due to senescence. In addition, higher extent of degradation were observed for Chl *a* 700 and Chl *a* 708.

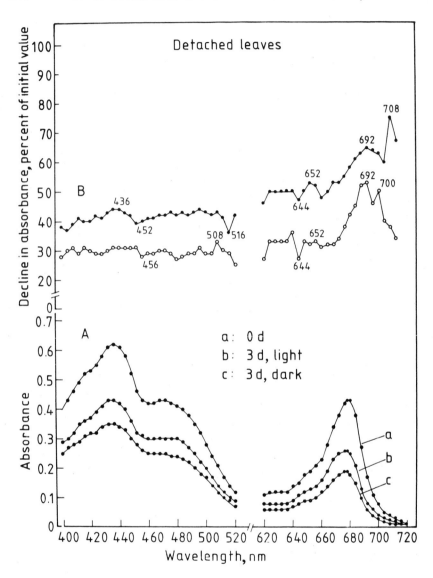

Fig. 2: Aging induced changes in absorption characteristic of light and dark incubated detached primary wheat leaves.

[A] Absorption spectra of chloroplasts (a) control 0 day, (b) 3 day light (c) 3 day dark incubated leaves. [B] The extent of loss as a function of wavelength is shown as percentage of control (100 per cent).

Aging-induced alteration in pH optima of oxygen evolution

Attached leaves were used for this experiment. The effect of various pH of the reaction medium on electron transport activity such as water to ferricyanide (with and without methylamine) and water to ferricyanide mediated by paraphenylenediamine (oxidised) was monitored. These assays were carried in chloroplasts sampled at three different stages (11, 15 and 19 days) of plant growth (Fig. 3, 4, 5). For 13–15 d the peak activity of FeCN Hill reaction as well as reduced catechol supported methyl viologen reduction was seen, and after 15 d a sharp decline was observed (Fig. 3). In the presence of methylamine, for 11–15-day-old leaf chloroplasts, the pH optimum for FeCN supported oxygen evolution was 8.0 and in aged (19 day old) leaf chloroplasts, this optimum indicated a shift of 0.5 pH unit of the acidic side (Fig. 4). However, for paraphenylenediamine mediated ferricyanide reduction, the pH optima remained unaltered during leaf again (Fig. 5). These results indicate that leaf age affects the site of electron acceptance by FeCN.

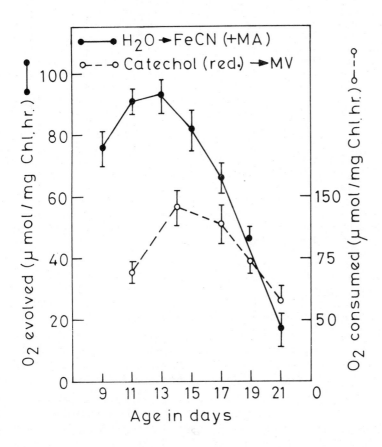

Fig. 3: Age dependent changes in chloroplast electron transport activity in primary wheat leaves. $H_2O \rightarrow FeCN$ and reduced catechol to MV assay were monitored polarographically. All assays were carried in three independent experiments. Bars represent S. D.

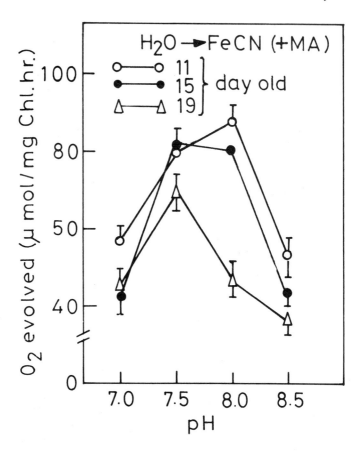

Fig. 4: Age induced shift in pH optimum of chloroplast. FeCN Hill activity was assayed in saturating light in the presence of 5 mM methylamine (MA); pH of reaction mixture, was adjusted with appropriate buffer.

Sequential loss of chloroplasts

Primary leaves of 16 d old seedlings were detached and were floated on water for induction of senescence. The pigment analysis and spectral studies were made after every 24 h, up to a total period of 72 h of incubation. One set of detached leaves was maintained in dark and the other set was kept under continuous white light. Chl content expressed on fresh weight basis declined continuously during the incubation period and Chl loss, as expected, was appreciably faster in dark-incubated leaves than in light-incubated ones (Table 1). Table 1 also reveals a close relation in the loss of Chl on unit fresh mass basis with the loss of absorbance at 678 nm on unit chloroplast number basis. The chloroplast number was adjusted to be nearly same for all assay intervals.

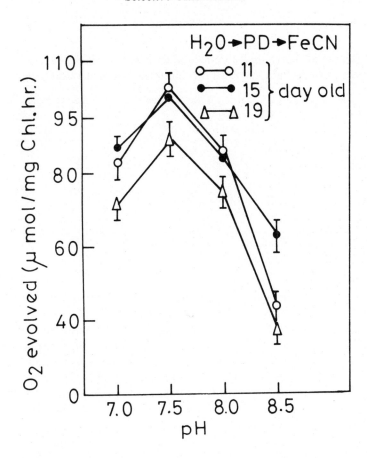

Fig. 5: Age dependent changes in photosystem II catalysed phenylenediamine (PD) supported O_2 evolution in primary leaf chloroplasts monitored as function of pH. Bars represent S. D. for three sets of independent experiments. Note no change in the pH optimum due to leaf again.

Discussion

Data presented in this paper provide further evidence for the contention that leaf senescence is an organised event (1). In case of photosynthetic pigments, it was noticed that carotenoids degrade to lesser extent than chlorophylls, and among chlorophylls Chl *b* is less sensitive than Chl *a* (Fig. 1, 2). Further analysis reveals that not all forms of Chl *a* degrade to a similar extent; Chl *a* 692, Chl *a* 700 and Chl *a* 708 represent forms which are extremely sensitive to leaf senescence stress. The results in attached and detached leaves in this respect followed nearly same trend. To ascertain the nature of alterations in PS II electron transport activity, we compared the effect of reaction pH on electron transport from water to ferricyanide,

TABLE 1

Loss in leaf total chlorophyll content and decrease in chloroplast red absorbance during light and dark incubation of detached wheat leaves

Time of incubation, h	Chl content mg/g fresh mass of leaf tissue		Chloroplast absorbance at 678 nm	
	Light	Dark	Light	Dark
0	1.57	1.57	0.48	0.48
24	1.39	1.16	0.42	0.37
48	0.99	0.78	0.34	0.25
72	0.98	0.51	0.34	0.25

16 d old leaves were incubated; for chloroplast 678 nm absorbance equal number of chloroplast were adjusted.

which accepts electrons at different sites viz., close to photosystem II and photosystem I of the electron transport chain (9). This was compared with oxidised paraphenylenediamine-mediated ferricyanide reduction which specifically accepts electron from a site close to photosystem II (10). A shift in the pH optimum of Hill activity in the presence of methylamine, an uncoupler, close to the pH optimum of paraphenylenediamine mediated electron transport activity (Fig. 3, 4 and 5) is suggestive of the fact that during senescence ferricyanide possibly accepts the electron largely from a site where PD (ox) accepts the electron from water. This site is known to be close to PS II (10). Based on experiments involving dibromothymoquinone, a plastoquinone antagonist, we suggest that shift in pH optimum during leaf senescence could favour the reoxidation of plastoquinone pool (data not shown). Experiments undertaken to analyse the empirical relationship of chlorophyll loss revealed that a close relation exists between loss of Chl on unit fresh weight basis and loss of absorbance at 678 nm on unit chloroplast number basis (Table 1). We therefore suggest that the similar extent of loss of Chl content and that of chloroplast number is due to the uniform extent of loss in Chl in all the chloroplasts. Otherwise, one would expect an imbalance in the pattern of Chl loss expressed on unit chloroplast number basis in case of senescence related Chl loss was only due to loss of certain fraction of chloroplast population.

The selected evidence presented above provides further support to the contention of Biswal and Mohanty (3) that photochemical activity-alternating in senescing leaves proceeds in a defined manner and this is possibly a general fact for all the senescence-related biochemical and physiological attributes (2).

Acknowledgements

Supported by the grant FG-IN-575; IN-SEA-170 to PM.

References

1. Stoddart, J. L. and Thomas, H. (1982) in *Encyclo Plant Physiology* (Boulter, D. and Parthier, B., eds.), pp. 592–636, Springer-Verlag, Berlin
2. Bonner, L. (1976) in *Plant Biochemistry* (Bonner, J. and Varner, J. E., eds.), pp 771–794, Academic Press, New York
3. Biswal, U. C. and Mohanty, P. (1976) *Plant Cell Physiol.* 17: 323–331
4. Grover, A., Sabat, S. C. and Mohanty, P. (1986) *Photosyn. Res.* 10: 223–229
5. Grover, A., Sabat, S. C. and Mohanty, P. (1987) *Biochem. Physiol. Pflanzen.* 182: 481–484
6. Sabat, S. C., Grover, A. and Mohanty, P. (1985) *Ind. J. Exp. Biol.* 23: 711–714
7. Arnon, D. I. (1949) Plant Physiol. 24: 1–15
8. Shibata, K. (1973) *Biochim. Biophys.* Acta 304: 249–259
9. Govindjee and Bazaz, M. (1967) *Photochem. Photobiol.* 6: 885–890
10. Saha, S., Ouitrakul, R., Izawa, S. and Good, N. E. (1971) *J. Biol. Chem.* 246: 3204–3209

PART 4
Molecular Structure—Functions, Relationships

New Results on the Mechanism of Photosynthetic Water Oxidation

G. Renger, H.–J. Eckert, R. Hagemann, B. Hanssum, H. Koike* and U. Wacker

Max-Volmer-Institut für Biophysikalische and Physikalische Chemie,
Technical University, Strasse des 17, Juni 135.
D 1000 BERLIN 12
*present address: RIKEN, WAKO, Saitama 351, Japan

Summary

Mechanistic aspects of photosynthetic water oxidation were analysed in dark-adapted PS II fragments from cyanobacteria (*Synechococcus vulcanus* Copeland) and higher plants (spinach) by detecting flash-induced absorption changes at 830 nm, at 350 nm and at 562 or 580 nm, reflecting the turnover of the photoactive chlorophyll a, P680, of the PS II reaction centre, of the univalent redox transitions $S_i \rightarrow S_{i+1}$ at the catalytic site of water oxidation and of external indicator dye response due to protolytic reactions, respectively. The possible role of hydrogen bonds for the structural integrity of the oxygen-evolving complex was investigated by measurements of the average oxygen yield per flash in thylakoids after heat treatment of suspensions in H_2O or D_2O.

   The results obtained led to the following conclusions: (a) regardless of the redox state of the catalytic site of water oxidation, S_i, P680$^+$ reduction is almost temperature- independent in the physiological range above $-1.5°C$, (b) in contrast to that, the activation parameters of the univalent oxidation steps $S_i \rightarrow S_{i+1}$ markedly depend on the redox state S_i. A characteristic break point is observed in cyanobacteria at about 16°C for S_3 oxidation leading to dioxygen formation, (c) about 6–7 hydrogen bonds play an important role for the structural integrity of the oxygen-evolving complex, and (d) the stoichiometry and kinetics of proton release exhibit a complex reaction pattern, depending on pH of the suspension.

   Based on these results and data taken from the literature a model is proposed for the mechanism of photosynthetic water oxidation to dioxygen.

Introduction

Photosynthetic water cleavage into dioxygen and metabolically bound hydrogen is of central relevance for the bioenergetics of the exploitation of solar radiation as the unique energy source of the biosphere (for a review see ref. 1). The key step of the overall process is water oxidation which takes place in system II via a four-step univalent redox reaction sequence. The kinetic pattern of photosynthetic water oxidation has been resolved (for recent review see ref. 2). It can be summarised by the scheme depicted in Fig. 1. Basically, three types of events can be distinguished: 1)

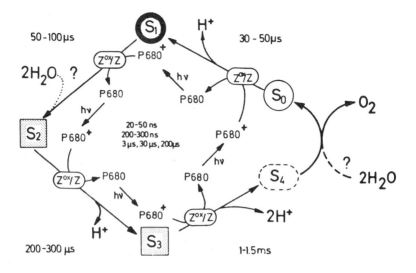

Fig. 1: Functional scheme of photosynthetic water oxidation to dioxygen. Oxidised redox equivalents are produced by the P680* photoreaction within few picoseconds, typical half transfer times via intermediate redox component Z (predominantly in the ns time domain) to the catalytic site (50 μs − 1.5 ms) are denoted inside and at the corners, respectively, of the scheme. The different properties of the five redox states of the catalytic site are characterised by different symbols (for details see text). Question marks indicate that the entry of water as substrate into the redox cycle (either at S_0 or at S_2/S_3) is a matter of debate. The deprotonation steps relate to the widely accepted extrinsic proton release pattern, the intrinsic pattern of protolytic reactions at the catalytic site is unknown (for details see text).

generation of sufficiently oxidising redox equivalents (holes) by photoox-idation of a special chlorophyll-a referred to as P680, 2) transfer of the holes via at least one redox group Z acting as one-electron component, and 3) stepwise univalent electron abstraction from the manganese-containing catalytic site of water oxidation leading to oxygen evolution after the accumulation of four holes.

   The functional groups participating in these reactions are incorporated into a protein complex containing at least seven polypeptides (3). The exact location has not yet been clarified.

   In this communication, new results are presented about: (i) the functional coupling between the catalytic site of water oxidation and $P680^+$, (ii) the activation parameters of the reactions leading to sequential oxidation of the catalytic site, (iii) the possible role of hydrogen bonds for the functional integrity of the system II complex, and (iv) the kinetics and stoichiometry of proton release coupled with the stepwise oxidation of the catalytic site of water oxidation. Based on these data and taking into account previous results the mechanism of photosynthetic water oxidation is discussed.

Materials and Methods

The preparation of thylakoids and PS II fragments from cyanobacteria (*Synechococcus vulcanus* Copeland) and higher plants (spinach) as well as the equipments for measuring flash-induced absorption changes and oxygen evolution have been described in detail in previous reports (4–9).

Results and Discussion

(A) The reduction kinetics of P680$^+$ as an indicator of the functional coupling between the catalytic site and the reaction centre

In samples with fully competent oxygen evolution the reduction of P680$^+$ is dominated by kinetics in the ns-domain (10–12). Based on measurements with the most powerful ADRY-agent (6), 2–(3–chloro–4–trifluoromethyl) anilino-3, 5–dinitrothiophene (ANT2p), the P680$^+$ reduction kinetics were shown to depend on the redox state of the catalytic site of water oxidation (13). More refined measurements at markedly higher time resolution confirmed this basic conclusion (14). The dependence of the ns-kinetics on the redox state S_1 was interpreted by electrostatic effects on the redox equilibrium between two components which are assumed to act as primary and secondary donor of P680$^+$ (14). Alternatively, the data could be also explained by structural changes of the states S_2 and S_3 that modify the electronic and nuclear coupling between P680$^+$ and its donor component (2) referred to as Z in the literature (15). In order to analyse this coupling more thoroughly, the temperature dependence of P680$^+$ reduction was measured by monitoring laser-flash-induced absorption changes at 830 nm in PS II membrane fragments from spinach. Fig. 2 shows traces obtained at +20°C and −1.5°C, respectively, under repetitive flash excitation on samples with intact oxygen evolution. The results reveal almost identical relaxation kinetics. This indicates that the overall P680$^+$ reduction is practically independent of temperature in the physiological range under steady state population of redox states S_i (i = 0,...3). For comparison Fig. 2 also shows 830 nm absorption changes in samples which are deprived of their oxygen evolution capacity due to incubation with 3 mM NH_2OH. In this case μs-kinetics arise at the expense of complete disappearance of the ns-kinetics (7). This phenomenon is indicative for a strong effect of the functional integrity of water oxidation on the P680$^+$ reduction kinetics. Therefore, it has been speculated that the extent of the ns-kinetics could be a measure of the oxygen evolution capacity (14). However, the latest detailed studies on the effects of trifluoperazine, diisothiocyanostilbene-2, 2-disulfonic acid (DIDS) and lauroylcholinechloride show this not to be the case (16). The data of Fig. 2 do not provide detailed information about the effect of temperature on P680$^+$ reduction as a function of the individual

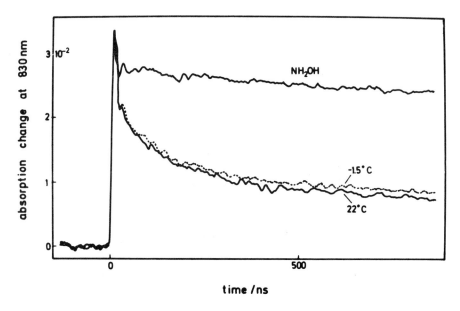

Fig. 2: Absorption changes at 830 nm as a function of time in O_2 evolving PS II membrane fragments from spinach under repetitive flash excitation (1 Hz) at − 1.5°C (dotted trace) and + 22°C (full trace). For comparison there is depicted the trace obtained in the presence of 3mM NH_2OH.
For experimental details see ref. 7, 16.

redox states S_i. To check this point, dark-adapted samples were illuminated with a laser flash train and the absorption changes caused by each flash were separately monitored. The results obtained are summarised in Fig. 3. An inspection of the data shows that there is almost no difference between the results at $-1.5°C$ and $+ 20°C$. A kinetic analysis of the data reveals that a satisfactory description can be achieved by the assumption of a triexponential decay of the form:

$$\Delta A_{830}(t) = \sum_{\mu=1}^{3} a_\mu(S_i) \cdot e^{-k_\mu \cdot t} \tag{1}$$

The amplitudes $a_\mu(S_i)$ depend on the redox state of the catalytic site as is clearly shown in Fig. 4 for the data obtained at $-1.5°C$. This dependence can be described by: $a_\mu(S_i) = \sum_{i=0}^{3} c_{\mu,i}[S_i]$, where $[S_i]$ is the probability of a catalytic site being in redox state S_i. A reasonable fit can be achieved with the following approximations:

$$c_{1,0} \approx c_{1,1} \text{ and } c_{1,2} \approx c_{1,3} \approx 0; \; c_{2,0} \approx c_{2,1} \approx 0 \text{ and } c_{2,2} \approx c_{2,3},$$
$$c_{3,0} \approx c_{3,1} \approx 0 \text{ and } c_{3,2} \approx c_{3,3}$$

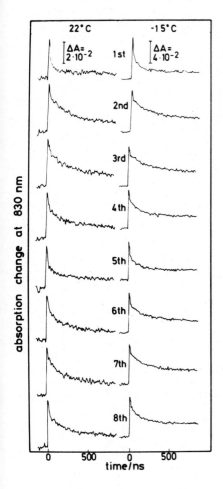

Fig. 3: Absorption changes at 830 nm as a function of flash N° and time in dark adapted O_2 evolving PS II membrane fragments from spinach at $-1.5°C$ (right side) and $+22°C$ (left side).

The measurement at $-1.5°C$ were performed with suspensions of doubled chlorophyll concentrations compared with those used at $+22°C$, for other details see ref. 16.

These relations are not exactly the same for $a_2(S_i)$ and $a_3(S_i)$ as reflected by small differences of the corresponding oscillation patterns depicted in Fig. 4. The origin of this phenomenon will be analyzed in a forthcoming study. The above-mentioned relations tacitly imply that the small contribution of μs-kinetics to the overall relaxation after the first flash is caused by a minor fraction of systems II which contain a damaged catalytic site of water oxidation. The oscillation at $-1.5°C$ closely resembles the pattern at $+20°C$ (data not shown). The relaxation kinetics at $-1.5°C$ are satisfactorily described by $k_1 = 1.7 \cdot 10^7$ s^{-1}, $k_2 = 3.10^6$ s^{-1} and $k_3 < 7.10^5$ s^{-1}. The same rate constants can be also used for the fit of the data obtained at $+20°C$, but in this case two different k_1-values of 5.10^7 s^{-1} (relaxation after the first and 5th flash) and of $1.7 \cdot 10^7$ s^{-1} have to be used. Except for this specific effect which will be analysed elsewhere, the present study shows that neither the kinetics nor the influence of the redox states of the

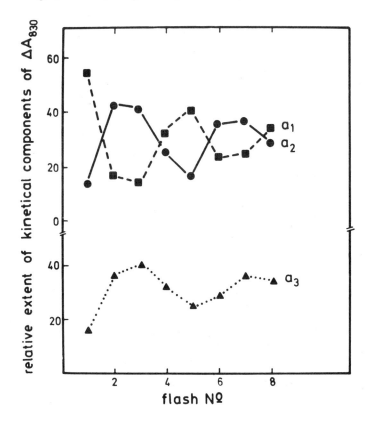

Fig. 4: Amplitudes of the decay kinetics in the ns-range as a function of flash N$^{\underline{o}}$ in dark-adapted O_2-evolving PS II membrane fragments at $-1.5°C$. The values were obtained by a triexponential fit of the data depicted in Fig. 3 (right side) with fixed rate constants of $k_1 = 1.7 \cdot 10^7 \ s^{-1}$, $k_2 = 3 \cdot 10^6 \ s^{-1}$, and $k_3 < 7 \cdot 10^5 \ s^{-1}$.

catalytic site exhibit a significant temperature dependence. The former phenomenon can be explained by a comparatively small activation energy of the electron transfer from Z to $P680^+$. The oscillation pattern of the $P680^+$ reduction kinetics could be interpreted either by the assumption that the structure of the catalytic site in S_0 and S_1 differs from that in S_2 and S_3, giving rise to markedly different coupling between $P680^+$ and Z, or that redox equilibria between electron transfer components responsible for the S_i state dependence of $P680^+$ reduction are only slightly enhanced in the range of $-1.5°C$ and $+20°C$. Direct experimental evidence is lacking for these types of equilibria. On the other hand, structural differences of S_2 and S_3 versus S_0 and S_1, respectively, are very likely as reflected by different Cl^- binding properties (17). Likewise, water substrate entry into the redox cycle is also discussed to occur at the level of S_2 (18). Therefore, regulation of $P680^+$ reduction via structural changes caused by the redox transition $S_1 \rightarrow S_2$ seems to be more attractive than a simple electrostatic point-charge effect. As a consequence of this idea, the temperature dependence of the oscillation pattern depicted in Fig. 4 indicates that the presumed conformational states regulating $P680^+$ reduction are practically the same at $-1.5°C$ and $+20°C$.

(B) Temperature dependence of the univalent redox steps at the catalytic site of water oxidation

The kinetics of the univalent oxidation steps $S_i \rightarrow S_{i+1}$ can be monitored by measurements of the rise-time of UV absorption changes (19,20). The oxidation rate depending on the redox level 'i' of the catalytic site of water oxidation kinetically coincides with the reduction of Z^{ox} (19–21). As these comprise protolytic reactions and structural modifications, the reaction coordinate of each individual process, $Z^{ox} S_i \rightarrow ZS_{i+1} + m_i H^+ + \delta_{i3} \cdot O_2$ (where $i = 0,...3$, with $i + 1 = 0$ for $i = 3$), is anticipated to depend on state S_i.

In order to resolve this problem, experiments were performed with PS II fragments from the thermophilic cyanobacterium *Synechococcus vulcanus* Copeland (5) which permit covering a broader temperature region. The results obtained are depicted in Fig. 5. The data at the left side, presenting 830 nm absorption changes induced by repetitive flash excitation at 0°C and 33°C, respectively, indicate that in PS II fragments from thermophilic cyanobacteria also the electron transport from Z to $P680^+$ is almost temperature independent. A markedly different behaviour is observed for the reactions $Z^{ox} S_i \rightarrow ZS_{i+1} + m_i H^+ + \delta_{i3} \cdot O_2$. The right side of Fig. 5 shows the reciprocal half-rise times of absorption changes at 350 nm as a function of the reciprocal temperature. As the kinetics of the transition $Z^{ox} S_o \rightarrow ZS_1 + H^+$ could not be resolved with sufficient precision (for details see 22), the temperature dependence of this reaction is omitted. The other three transitions exhibit remarkable differences of the corresponding reaction coordinates. Two findings are of special relevance: a) the activation energy of the reaction $Z^{ox} S_2 \rightarrow ZS_3 + H^+$ is significantly higher (26.8 kJ/mol) than that of $Z^{ox} S_1 \rightarrow ZS_2$ (9.6 kJ/mol), and b) the reaction $Z^{ox} S_3 \rightarrow ZS_o + O_2 + 2 H^+$ is characterized by a break point at about 16°C of its temperature dependence. The activation energy above the critical temperature, T_c (15.5 kJ/mol), is only about 25% of that observed below this point (59.4 kJ/mol) (22). Based on these data, some interesting conclusions can be drawn. If one assumes that the transmission coefficient σ of the reaction path is the same for both reactions, the transition state of $S_1 \rightarrow S_2$ requires a markedly higher activation entropy ΔS^* than that of $S_2 \rightarrow S_3$ (in the most simple case of $\sigma = 1$ one obtains $\Delta S^*_{2,1} = -140$ J/mol·K versus $\Delta S^*_{3,2} = -80$ J/mol·K). This phenomenon also supports the idea that S_2 is structurally distinctly different from S_1 (17), although on

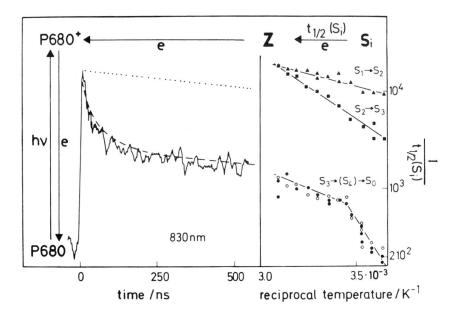

Fig. 5: Absorption changes at 830 nm (left side) as a function of time and reciprocal half-rise times of S_i oxidation (measured at 350 nm) as a function of reciprocal temperature (right side) in PS II fragments from *Synechococcus vulcanus* Copeland.

The signal on the left side represents a trace monitored at $+ 33°C$, the dashed curve symbolises the data at 0°C. The dotted curve was obtained in the presence of 3mM NH_2OH. Repetitive excitation (1 Hz) with laser flashes.

For experimental details see ref. 7, 16, 22.

the basis of EXAFS studies the microenvironment of the catalytic site itself was recently inferred to exhibit almost the same structure in S_1 and S_2 (23). The latter conclusion is also hardly reconcilable with the assumption that substrate water enters the redox cycle at state S_2 (18) rather than at S_0. Further studies are required to clarify this mechanistically crucial point. The break point in the temperature dependence of S_3 oxidation leading to dioxygen release probably reflects significant structural changes of S_3 around the critical temperature of 16°C in PS II fragments from thermophilic cyanobacteria. This idea raises questions about the nature of these changes. One possible explanation is based on the previous postulation that S_3 already satisfies the energy requirements of dioxygen bond formation at the redox level of a peroxide (24). Therefore, the break point could reflect a dramatic shift of the equilibrium between water substrate ligands and the oxidised manganese centre (22), i.e. above T_c the oxidation of S_3 could start at a peroxide-type level with a preformed 0–0 bond, whereas below T_c additional activation energy is required for peroxide formation. Alternatively, changes of the protolytic reactions should also be considered. The break point phenomenon at T_c implies a rather sharp and simultaneous change of ΔS^* and ΔH^* in the reaction path of S_3 oxidation. The origin of this "phase transition type" effect (structural changes of the whole membrane or "all or none" transition in the PS II polypeptide complexes) remains to be clarified.

The latest thermoluminescence data (25) provide additional interesting information about the reactivity of S_3 as a function of temperature. It was shown that in spinach thylakoids at subzero temperatures (between $-50°C$ and $-5°C$) electron abstraction from ZS_2 leads to a redox state S^*_3 which cannot be further oxidised into S_o and dioxygen. If one excludes the trivial possibility of S^*_3 being $Z^{ox}S_2$ (for discussion see ref. 25) one could consider this finding as further support of the idea that 0–0 bond preformation in S_3 is the key step of the reaction $n \cdot H_2O + Z^{ox}S_3 \rightarrow ZS_o + O_2 + 2 H^+$ (here $n = 0$ or 2 depending on the stage of water substrate entry into the redox cycle, see Fig. 1). This question of central mechanistic relevance for the mechanism of dioxygen formation remains to be answered by future studies.

(C) Effects of hydrogen bonds on the structural integrity of the oxygen evolution capacity

The oxygen evolution capacity can be selectively eliminated by a number of protein denaturing conditions. Among different interactions of charged and/or polarisable groups responsible for the structural integrity of a polypeptide, hydrogen bonds are of special interest. As an attempt to understand this effect in more detail, comparative experiments were performed with spinach thylakoids suspended either in H_2O or D_2O under denaturing conditions. The oxygen evolution capacity of spinach thylakoids was shown to be rather sensitive to a thermal shock (26). If this degradation process comprises the break of hydrogen bonds, then suspending of thylakoids in D_2O instead of H_2O is expected to provide some kind of protection. The number of systems II fully competent in oxygen evolution was determined by measurements of the average oxygen yield per flash, Y_{ss}, under repetitive excitation at sufficiently low frequencies (2 Hz) to exclude kinetical limitations. In order to separate effects of D_2O on thermal degradation from those of a possible influence on the activity, thylakoids were suspended in either D_2O or H_2O and subjected to a 3 min incubation of the desired temperature. After this treatment the samples were injected into a normal aqueous (H_2O) buffer solution and Y_{ss} was measured at 20°C. The data depicted in Fig. 6 reveal a significant stabilising effect of D_2O as reflected by the shift of thermal sensitivity by about 5 K towards higher temperatures. This result suggests that hydrogen bonds might play an important role in the native structure of the oxygen evolving complex. In the simplest case, the activation energies of the degradation process of samples incubated in either D_2O or H_2O differ only in the zero point energies of hydrogen bonds which are essential

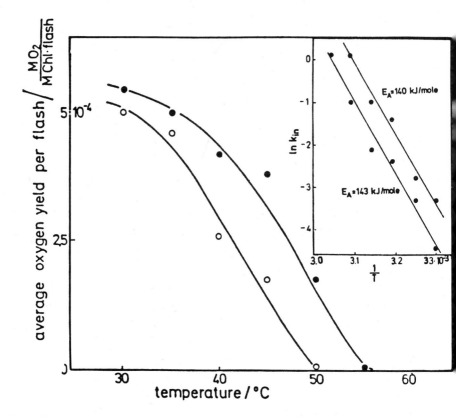

Fig. 6: Average oxygen yield per flash as a function of the temperature at a 3 min dark incubation.

The curves are computed for a single exponential degradation kinetics with A= 8.10^{20} s^{-1} and E_A = 140 kJ/mole (H_2O) and 143 kJ/mole (D_2O) (see insert).

Insert: ln k_{in} as a function of reciprocal incubation temperature, k_{in} is computed according to eq. (2) (see text).

for the native structure of the protein. As the detrimental effect is satisfactorily described by a first order kinetics (data not shown) the apparent degradation rate constant can be evaluated from the data of Fig. 6 by the equation

$$k_{in} = \frac{1}{t_{inc}} \cdot \ln \frac{Y_{ss}(0)}{Y_{ss}(t_{inc})} \tag{2}$$

where $Y_{ss}(0)$ = average oxygen yield per flash at t = 0 of thermal treatment and $Y_{ss}(t_{inc})$ = average oxygen yield per flash after incubation time of 3 min. The k_{in} values obtained from the data of Fig. 6 by applying eq. (2) are depicted as a function of T^{-1} in the insert of Fig. 6. The results can be fairly described by straight lines with activation energies of

$E_A(H_2O) = 140$ kJ/mole and $E_A(D_2O) = 143$ kJ/mole. If one assumes that deuteration affects only the reaction coordinate of the degradation process (i.e. the partition functions of all other internal modes remain invariant), the activation energy is expected to become changed solely due to the differences of the zero point energies of the hydrogen bonds $-H...Y-$ and $-D...Y-$ that are broken in the activated state. To calculate the order of magnitude of this effect, the number of hydrogen bonds undergoing cleavage and their zero point energies are required. Using 20–25 kJ/mole as data taken from the literature (27) for the average energy of hydrogen bonds in proteins, the experimental activation energy of 140–143 kJ/mole indicates a break of 6–7 hydrogen bonds in the activated state. Likewise, the activation entropy ΔS^* of denaturation is computed to be 190 J/K mole for thylakoids suspended in H_2O or D_2O. These values correspond with data reported for the denaturation of well-characterised isolated enzymes like trypsin or pancreatic proteinase (27). The difference of the zero point energy per hydrogen bond is given by $\Delta \varepsilon_o = 1/2h \cdot \nu_H (1 - 1/\sqrt{2})$, where ν_H = frequency of the normal hydrogen bond containing a proton and the difference in the reduced mass is about 2 due to the large mass of the protein compared with that of H and D, respectively. Typical wave numbers of hydrogen bonds are of the order of 200–300 cm^{-1} which gives rise to a difference of the activation energy of $(5–10) \cdot 10^{-22}$ J per hydrogen bond. Accordingly, for the cleavage of 6–7 hydrogen bonds, a total difference of the activation energy of $\Delta E_A = 2–4$ kJ/mole is obtained, in close agreement with the experimental findings. This supports the idea that some special hydrogen bonds are of crucial relevance for the native structure of the oxygen evolving complex. The localisation of these bonds and a possible functional involvement in protolytic reactions remain to be analysed in future studies.

(D) Kinetics and stoichiometry of proton release coupled with water oxidation

Water oxidation to dioxygen by holes produced at $P680^+$ is inevitably coupled with the total release of four protons into the lumenal space of thylakoids. Therefore, the stoichiometry of H^+ ejection from the catalytic site in each univalent redox step, $S_i \rightarrow S_{i+1}$, referred to as the **intrinsic** proton release pattern (24) should provide direct information about the deprotonation sequence of substrate water during the redox cycle. Results obtained with special pH-sensitive indicators in thylakoids (for review see ref. 28), inside-out vesicles (29) and PS II membrane fragments from spinach (9) in the physiological range of 6.0 <pH<7.5 can be satisfactorily explained by a stoichiometry of 1,0,1, 2 for the redox transitions $S_i \rightarrow S_{i+1}$, with i = 0,1,2,3, respectively (see Fig. 1). However, it has to be emphasised that, for several reasons, this experimentally detected **extrinsic** stoichiometry cannot be expected to reflect directly the protolytic state of the substrate water coordinated to the catalytic site at redox state S_i (for detailed discussion see ref. 9, 24, 30). The main source of ambiguity is obviously the participation of the apoprotein in the protolytic reactions due to pK-shifts of protonisable groups caused by the turnovers of Z and of the catalytic site itself (Bohr-type effect). If the apoprotein actually plays a key role for the protolytic reactions coupled with water oxidation, then changes of the stoichiometry and/or kinetics of proton release are anticipated to occur by variation of the environmental pH. In order to address this problem, experiments were performed with PS II membrane fragments in the range of $5.5 < \text{pH} < 9.0$. Flash-induced absorption changes of indicator dyes (bromocresol purple at $\text{pH} < 7.0$ and phenol red at $\text{pH} > 7.0$) caused by excitation of dark-adapted samples with a train of μs-flashes (for experimental details see ref. 9) are depicted in Fig. 7. The

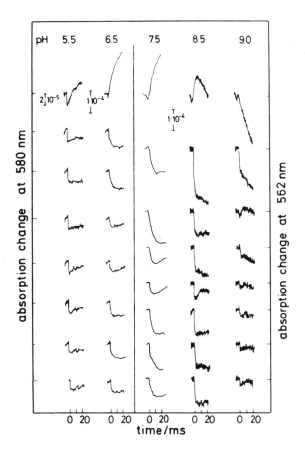

Fig. 7: Δ pH-indicating absorption changes at 562 and 580 nm as a function of time in O_2 evolving PS II membrane fragments from spinach at room temperature.

The traces at the left side were obtained with bromocresol purple, at the right side with phenol red. For experimental details see ref. 9.

data obtained reveal a significant dependence of H^+ release on flash No. 0. A pronounced H^+ uptake after the first flash is probably due to a single turnover of the high spin Fe^{2+} located between Q_A and Q_B at the PS II acceptor side. This phenomenon will not be discussed here (for details see ref. 9). The amplitudes of H^+ release exhibit characteristic oscillation patterns that are dependent on the external pH as is shown in Fig. 8. No significant changes (within the limit of experimental error) are observed in the physiological range of $6.0 < pH < 8.0$, but marked deviations arise outside this region. The results at alkaline pH are seriously affected by a decline of the oxygen evolution capacity in PS II membrane fragments (31) and therefore do not permit unequivocal conclusions. The interpretation of the other data within the framework of Kok's model (32) depends significantly on precise knowledge of the probability of misses (α) and

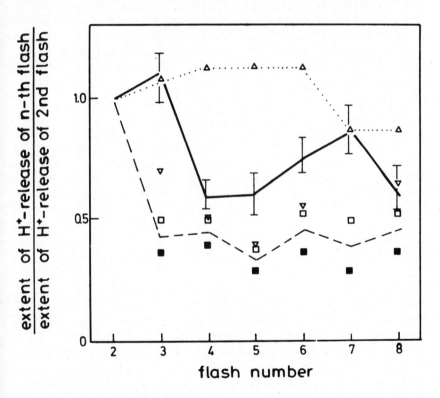

Fig. 8: Extent of proton release due to the n-th flash normalised to the extent due to the 2nd flash as a function of flash No. (n) in dark-adapted PS II membrane fragments from spinach at different pH.

Data are taken from Fig. 7, signals obtained in the presence of buffer were subtracted. For further details see ref. 9. The following symbols were used: Δ (pH = 5.5), ∇ (pH = 8.1), $\square$ (pH = 8.5) and $\blacksquare$ (pH = 8.9), the full curves represent data obtained in the range $6.0 < pH < 7.7$. The bars do not show error bars but reflect the variation of the data in the above-mentioned pH region.

double hits (β) and the apparent dark-state population $\{S_i\}_o$. A detailed analysis (Wacker, U. and Renger, G., unpublished results) however indicates that a 1,0,1,2 stoichiometry often does not provide the best fit. In many cases, stoichiometries of 0,0,2,2 and 0,0,1,3 permit a better description of the data. Regardless of the details, these findings support the idea that the apoprotein seriously affects the extrinsic proton release pattern. This conclusion implies that the intrinsic proton release is still an unresolved problem.

An inspection of the proton release kinetics reveals a multiphasic pattern. Generally, at least a four-exponential kinetics with half-life times of ≤ 100 μs, 200–300 μs, 500–800 μs and 1–5 ms is required for a satisfying fit. We were not able to find an unequivocal correlation between the amplitudes of the different kinetics and individual redox transitions $S_i \rightarrow S_{i+1}$, even within the physiological pH range. The most striking observation is the marked retardation of the proton release with increasing pH in the range of 5.5 up to 7.5. Preliminary studies did not show a clear kinetical correspondence of the rates of H^+ release and the $S_i \rightarrow S_{i+1}$ transitions in the above-mentioned pH range. Experiments are in progress to analyse this point.

(E) Mechanism of photosynthetic water oxidation

The understanding of the molecular mechanism of photosynthetic water oxidation to dioxygen requires detailed knowledge of the electronic structure and the nuclear geometry of the catalytic site in the different redox states S_i for $i = 0, ...4$. Different spectroscopic techniques (XANES, EXAFS, UV/VIS, IR, EPR, NMR) were used to resolve this problem (for recent reviews see ref. 2, 15, 30) but only few pieces of the huge puzzle were found. Despite lack of essential information (e.g. the ligands of the first coordination sphere at the catalytic manganese beyond substrate water, the existence and role of hydrogen bridges between substrate water and the apoprotein, dynamics of the apoenzyme and its effect on the reaction coordinate) an attempt will be made to discuss the mechanism of water oxidation within the framework of our current stage of limited knowledge. This discussion is based on two crucial postulates of our previous model (24,33): a) the central step of water oxidation, i.e. the 0–0 bond formation occurs at a binuclear manganese centre at the redox level of a peroxide in S_3, and b) there exists a special state or centre symbolised by M that undergoes a redox turnover at the steps $S_o \rightarrow S_1$ and $S_3 \rightarrow (S_4) \rightarrow S_o + O_2$ (protolytic reactions are omitted).

The former part of the first postulate is supported by the remarkable heterogeneity of the four manganeses which are present in each oxygen evolving complex (for review see ref. 2). Accordingly, two manganeses are assumed to form the functional core of water oxidation whereas the other two manganeses could act as further redox centres (redox buffer or electronic coupling with Z) or as an essential structural element (for detailed discussion see ref. 2,30,34). The second part of the first postulate appears to be mechanistically attractive because, in case of an intermediary peroxide formation, the electron abstraction from two water molecules takes place via two sequential 2-electron steps rather than by a more complex single 4-electron step. This idea has been accepted also for other models (see e.g. 35) but, in contrast to our original postulate (24), the peroxide level is assumed to occur only as a transient S'_4 state. However, it has to be emphasised that S_3 already satisfies the energy requirements for the formation of peroxidic state (based on experiments with $H_2^{18}O_2$ (36), even S_2 might be considered as a candidate). Recent mass spectroscopic

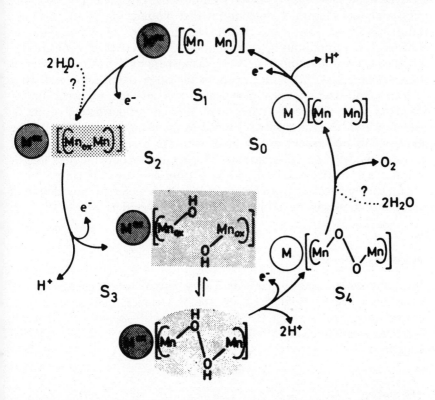

Fig. 9: Mechanism of photosynthetic water oxidation to dioxygen. Only the manganeses of the binuclear catalytic core are presented, for the nature of M see text. As the stage of water entry into the cycle is not unambiguously clarified (see text and Fig. 1) substrate ligation is only indicated for S_3 and S_4. The other ligands are symbolised by a half oval around Mn. Oxidised states are represented by grey areas. The possibility of a redox equilibrium between manganese and the water substrate ligands could lead to a peroxidic intermediate in S_3 as indicated in the scheme which tacitly implies that S_3 is oxidisable only in the peroxidic configuration. For the sake of simplicity, redox equilibria are omitted, which can occur between the binuclear manganese site and redox centres other than the substrate (for details see text). Likewise, each redox level (except that of S_3) is characterised by only one symbol.

studies (37,38) were used as argument against the formation of a peroxidic state prior to S_4. However, due to the very limited time resolution of this method, the data cannot be used as an unequivocal proof (for detailed discussion see 2,20). Furthermore, a nonexchangeable metastable S_3 state does exist in strictly dark-adapted *oscillatoria chalybea* (38). The latest results on the reactivity of S_3 as a function of temperature (see section B) might also support our postulate that 0–0 bond preformation occurs prior to S_4. The second postulate was primarily introduced in order to explain the extreme stability of S_1. Actually, it was later found that, in sufficiently dark-adapted samples, almost all catalytic sites attain the redox level S_1 (39). M is not yet identified. It could be either a special valence state of the catalytic manganese (40) or another redox-active component (2, 34). The latest and very elegant study confirms that the reaction $S_0{\rightarrow}S_1$ does not cause significant absorption changes in the range of 290–570nm, in contrast to the transitions $S_1 \rightarrow S_2$ and $S_2 \rightarrow S_3$ (41). Based on this result, it was concluded that, most likely, no manganese oxidation takes place at the transition $S_0 \rightarrow S_1$. With respect to the other redox steps, at least a 1-electron abstraction from manganese was inferred to occur for S_2 formation from S_1. On the other hand, it is not clear whether or not S_2 oxidation to S_3 is also mainly restricted to the manganese core (23). The above-mentioned results neatly fit in with the basic conclusions of our previous model (24). Different lines of evidence support the idea that S_1 contains a [Mn III) Mn(III)] and S_2 a [Mn(III) Mn(IV)] core.

As a summary of our fragmentary knowledge a molecular mechanism of water oxidation can be proposed which is depicted in Fig. 9.

Acknowledgements

The authors would like to thank Dr. T. Wydrzynski for critical reading and M. Müller and S. Hohm-Veit for skilful technical assistance and A. Bowe-Gräber for drawing the figures. Financial support by Deutsche Forschungsgemeinschaft (Sfb 312 and Re 354/8–1) is gratefully acknowledged.

References

1. Renger, G. (1983) in: *Biophysics* (Hoppe, W., Lohmann, W., Markl, H. and Ziegler, H., eds) 347–371, Springer, Berlin
2. Renger, G. (1987) *Angew. Chem. Int. Ed.* 26, 643–660
3. Ghanotakis, D. F., Waggoner, C. M., Bowlby, N. R., Demetriou, D. M., Babcock, G. T. and Yocum, C. F. (1987) *Photosynth. Res.* 14, 191–199
4. Völker, M., Ono, T., Inoue, Y. and Renger, G. (1985) *Biochim. Biophys. Acta* 806, 25–34
5. Koike, H. and Inoue, Y. (1985) *Biochim. Biopyhs. Acta* 807, 64–73
6. Renger, G. (1972) *Biochim. Biophys. Acta* 256, 428–439

7. Völker, M., Eckert, H. J. and Renger. G. (1987) *Biochim. Biophys.* Acta 890, 66-76
8. Weiss, W. and Renger, G. (1986) *Biochim. Biophys.* Acta 850, 173-183
9. Renger, G., Wacker, U. and Völker, M. (1987) *Photosynth. Res.* 13, 167-189
10. van Best, J. A. and Mathis, P. (1978) *Biochim, Biophys.* Acta 408, 153-163
11. Renger, G., Eckert, H. J. and Weiss, W. (1983) in: *The Oxygen Evolving System of Photosynthesis* (Inoue, Y., Crofts, A. R., Govindjee, Murata, N., Renger, G. and Satoh, K., eds) pp. 73-82, Academic Press Japan, Tokyo
12. Brettel, K. and Witt, H. T. (1983) *Photobiochem. Photobiophys.* 6, 253-260
13. Gläser, M., Wolff, C. and Renger, G.(1976) *Z. Naturforsch.* 31 c, 712-721
14. Schlodder, E., Brettel, K. and Witt, H. T. (1984) *Biochim, Biophys.* Acta 808, 123-131
15. Babcock, G. T. (1987) in: *Photosynthesis* (Amesz, J., ed) pp. 125-158, Elsevier, Amsterdam
16. Eckert, H. J., Wydrzynski, T. and Renger, G. (1988) *Biochim. Biophys Acta* 932, 240-249
17. Preston, C. and Pace, R. J. (1985) *Biochim. Biophys.* Acta 810, 388-391
18. Beck, W. F., de Paula, J. C. and Brudwig, G. W. (1986) *J. Am. Che. Soc.* 108, 4018-4022
19. Dekker, J. P., Plijter, J. J., Ouwehand, L. and van Gorkom, H. J. (1984) *Biochim. Biophys.* Acta 767, 176-179
20. Renger, G. and Weiss, W. (1986) *Biochem. Soc. Trans.* 14, 17-20
21. Babcock, G. T., Blankenship, R. E. and Sauer, K. (1976) *FEBS Lett.* 61, 286-289
22. Koike, H., Hanssum, B., Inoue, Y. and Renger, G. (1987) *Biochim. Biophys.* Acta 893, 524-533
23. Yachandra, V. K., Guiles, R. D., McDermott, A. E., Cole, J. L., Britt, R. D., Dexheimer, S. L., Sauer, K. and Klein, M. P. (1987) *Biochemistry* 26, 5974-5989
24. Renger, G. (1978) in: *Photosynthetic Water Oxidation* (Metzner, H., ed.) pp. 229-248, Academic Press, London
25. Koike, H. and Inoue, Y. (1987) *Biochim. Biophys.* Acta 894, 573-577
26. Kimimura, M. and Katoh, S. (1972) *Plant Cell Physiol.* 13, 287-298
27. Stearn, A. E. (1949) in: *Advanc. Enzymol.* 9, 25-74
28. Förster, V. and Junge, W. (1985) *Photochem. Photobiol.* 41, 183-190
29. Diedrich-Glaubitz, R., Völker, M., Gräber, P. and Renger, G. (1987) in: *Progr. in Photosynth. Res.* (Biggins, J., ed) Vol. 1, pp. 519-522, Martinus Nijhoff, Dordrecht
30. Renger, G. (1987) *Photosynthetica* 21, 203-224
31. Renger, G. and Hanssum, B. (1988) *Photosynth. Res.* (in press)
32. Kok, B., Forbush, B. and McGloin, M. (1970) *Photochem. Photobiol.* 11, 457-475
33. Renger, G. (1977) *FEBS Lett.* 81, 223-228
34. Renger, G. and Govindjee (1985) *Photosynth. Res.* 6, 33-55
35. Vincent, J. B. and Christou, G. (1987) *Inorg. Chim.* Acta 136, L41-L43
36. Mano, J., Takahashi, M. and Asada, K. (1987) *Biochemistry* 26, 2495-2501
37. Radmer, O. and Ollinger, (1986) *FEBS Lett.* 195, 285-289
38. Bader, K. P., Thibault, P. and Schmid, G. H. (1987) *Biochim. Biophys.* Acta 893, 564-571
39. Vermaas, W. J. F., Renger, G. and Dohnt, G. (1984) *Biochim. Biophys.* Acta 764, 194-202
40. Andreason, L. E., Hansson, Ö. and Vänngard, T. (1983) *Chemica Scripta* 21, 71-74
41. Lavergne, J. (1987) *Biochim. Biophys.* Acta 894, 91-107

Molecular and Cellular Biology of the Major Light-harvesting Pigment-protein (LHCII*b*) of Higher Plants

J. PHILIP THORNBER, GARY F. PETER, PARAG R. CHITNIS
AND ALEXANDER VAINSTEIN

Department of Biology and Molecular Biology Institute,
University of California at Los Angeles,
Los Angeles California 90024, USA

Summary

A review of the present state of knowledge regarding the distribution, composition, structure, function and biogenesis of the chlorophyll *a*/chlorophyll *b* containing light-harvesting complex of photosystem two (LHC-II) is given.

Introduction

The internal membranes of higher plant chloroplasts contain two multiprotein photosystems that are associated with chlorophyll (chl), carotenoid and the other cofactors necessary for performing light energy trapping and conversion. The chl *a* and *b*, carotene and xanthophyll molecules are coordinated with, but not covalently linked to, several specific hydrophobic proteins (see 1 for review). The proteins function to orient and space precisely their associated pigment molecules so that energy absorbed by any one of these often abundant pigment-proteins is transferred efficiently to the PS I or PS II reaction centres where it is converted into chemical potential. Each photosystem can be thought of as being composed of two distinct parts: (1) a core complex that contains those polypeptides and cofactors required for a plastid to have a stable and functional primary photochemical event, and (2) a light-harvesting complex (LHC) that houses most of the antenna pigments in the photosystem but is not essential for the structure and function of the core complex (1).

The light-harvesting complex of photosystem II (LHC II) is the most abundant of these pigmented complexes. It can be resolved into several pigment-protein components, termed by us LHC II*a, b* and *c*, of which LHC II*b* is the major component. LHC II*b*, was one of the first higher plant pigment-protein complexes to be described and it has been extensively investigated over the last twenty years (1). It has been an attractive component to study biochemically and biophysically since it is the depository for almost half of the chl and one-third of the protein in green plant thylakoids, and thus it provides a readily available water-insoluble pigment-protein in which to examine the molecular arrangement of pigments and energy transfer in a defined micro-environment. LHC IIb functions not only as an antenna but it and/or one of the other LHC II components also contributes to the stacking of the thylakoids to form grana (2). Furthermore, LHC IIb has been implicated in regulating, via its phosphorylation, the proportion of absorbed excitation energy directed to each of the two photosystems; i.e., the state I–state II transitions (see 3).

Earlier Studies of the Composition and Structure of LHC IIb

Preparations of LHC IIb have generally been reported to have a chl a/b ratio between 1.0 and 1.4 (4). A 1985 consensus composition of an LHC IIb monomer was 5–7 chl a, 4–6 chl b and 2–3 xanthophyll molecules per polypeptide (1); ratios of 6–13 chl molecules per polypeptide were reported (e.g., 4–6). There are multiple, typically two or three, apoproteins of 25–28 kDa in LHC IIb that apparently do not occur in a simple stoichiometry. The number of different LHC IIb apoproteins is seemingly not constant among different plants species. However, some caution should be exercised in automatically accepting some of the older data on this chl-protein. Previous LHC IIb preparations might have contained other LHC II pigment-proteins (see below); furthermore, older preparative procedures often removed substantial quantities of the pigment molecules originally bound to LHC IIb. This caroteno-chl-protein almost certainly occurs in an oligomeric form *in situ*, and although some evidence exists that it has a trimeric structure (7), it is likely that the 8 nm particle putatively identified as LHC II in freeze-fracture micrographs of thylakoid membranes (8), contains LHC IIb as an even higher oligomer. Biophysical data have indicated that three of the chl b molecules in an LHC IIb monomer are situated close enough to interact excitonically and are surrounded by chl a molecules (9, 10).

The molecular biology of this pigment-protein has received much attention during the last twelve years. As a consequence, characterisation of genes for the LHC IIb apoproteins and production of antibodies that react with these proteins are well advanced and have contributed greatly to our understanding of the biochemistry of this pigment-protein. This article aims to correlate the recent molecular genetic and biochemical studies to provide a more detailed picture of LHC IIb.

Genes Encoding LHC IIb Apoproteins

LHC IIb apoproteins are encoded by nuclear genes (11). Molecular genetic approaches have been used to isolate and sequence the genes encoding them; more than 20 genes from ten different species have now been sequenced (summarised in 12). These genes, often termed *cab* genes, have typical eukaryotic 5' and 3' flanking regions (TATA and CAAT boxes and polyadenylation sites). They are members of a multigene family containing between 3 (13) and 16 (14) genes, depending on the species. Each of the genes characterised so far codes for a precursor polypeptide (pre-LHC II*b* apoprotein) which has a transit peptide of 33–35 amino acids attached to the N-terminus of the mature polypeptide, which itself consists of approximately 233 amino acids. The transit peptide sequence deduced from different genes differ more from each other than do the sequences of the mature protein which are highly conserved among different species (12, 15).

The 3' and 5' untranslated sequences of these genes diverge from each other. Depending on this divergence, members of a gene family can be grouped into subfamilies; for example, tomato (16, 17) and petunia (14) genes have been divided into five subfamilies. The genes encoding LHC II*b* polypeptides can also be classified on the basis of the presence or absence of introns (18). Type I genes contain no intron, are more numerous in most plant species, code for a slightly longer polypeptide and have significantly different transit peptides as compared to type II genes that contain one intron (18, 19). The transit peptides of both types of genes are functional in import of pre-LHC II*b* apoproteins by plastids (20).

Regulation of Expression of *cab* Genes

The regulation of *cab* genes has been reported to occur at many different levels during gene expression. For example, phytochrome affects transcription of these genes (21); intermittent red light has an influence on the translation of LHC II*b* apoprotein mRNA (22); changes in light intensity affect post-translational modification (phosphorylation) (3, 23). The expression of *cab* genes is regulated in both a quantitative as well as a qualitative manner. Chimeric genes under the control of the 5'-flanking sequences of an AB80 gene from pea have been used to study tissue-specific and light-inducible expression of *cab* genes in transgenic tobacco plants (24). The results showed that about 400-base pairs of the 5'-flanking region are sufficient to give the tissue-specific pattern of expression observed for *cab* genes. The levels of expression are low however, indicating that some elements that affect LHC II*b* apoprotein transcription quantitatively are located still further upstream. A 247-base pair element from this 400-base pair region acts as a light-inducible enhancer and also as a tissue-specific silencer (25). Similarly, a region in the 5'—flanking sequence of a wheat *cab* gene is found to confer phytochrome responsiveness on the gene (26,27).

Biochemistry of the LHC II Components

Isolation: A recently devised electrophoretic system (28) can resolve the multiple pigment-proteins in thylakoids without displacing any of the chl, and probably carotenoid molecules from their association with protein. In this system, thylakoid membranes are solubilised by a glycosidic surfactant (e.g. octyl- or nonyl-glucoside, dodecyl-maltoside) and then fractionated by polyacrylamide gel electrophoresis (PAGE) using an upper reservoir buffer containing the surfactant, monosodium-N-lauryl beta-imino-diproprionate (Deriphat 160). Thus, for the first time, it is possible to describe unequivocally the pigment content of each green plant pigment-protein. In Fig. 1 the electrophoretic procedures has been applied to a photosystem II preparation. Four of the green bands obtained in this

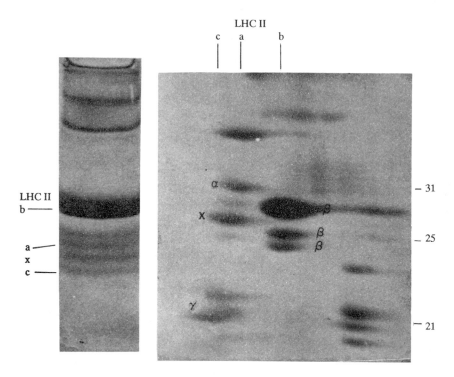

Figure 1. *Left,* Deriphat-PAGE of a barley photosystem II preparation. *Right,* two-dimensional PAGE shows the subunit composition of thylakoid pigment-proteins. A gel strip of the pigment-proteins separated by Deriphat-PAGE has been laid along the top of a fully denaturing SDS-PAGE so that migration of polypeptides out of it will be at right angles to the direction of the first electrophoresis. The positions of the LHC II pigment-proteins after the first dimension PAGE are marked at the top of the gel; note that LHC II*b* runs as an oligomer of 72kDa and LHC II*a* and *c* run as pigment-proteins of 35 and 24 kDa, respectively. The apoproteins associated with each pigment-protein are indicated by their respective Greek letters.

non-denaturing system are LHC II pigment-proteins: a 35 kDa LHC II*a*, also known as CP29 (29); a 72 kDa oligomeric form of LHC II*b*; a 25 kDa LHC II*c* component (CP24) (30); and a band at 30 kDa, which is either a mixture of LHC II*a* and *c* or a fourth chl-protein (CP27) (cp. 31, 32). The polypeptides associated with each LHC II component have been difficult to assign unequivocally because, not only can all four pigment-proteins migrate very near each other on non-denaturing PAGE, but so also can their apoproteins on fully denaturing PAGE. All LHC II*b* apoproteins have apparent sizes within 10kDa of each other (Table 1).

TABLE 1

Subunit composition of LHC II pigment-proteins of pea and barley

	LHC II*a*	LHC II*b*	LHC II*c*	LHC II*x*[1]
Pea	29	28,26,25	21	27
Barley	31	28,27,25	21	29,26

[1]We have not been able to establish whether these polypeptides bind pigment.

Note that since LHC II*b* migrates only as an oligomer under the conditions used, its apoproteins are well resolved from those of very similar sizes that are associated with the other LHC II components. In most plants we have examined there are three polypeptide subunits, the smallest of which is the least prevalent and the largest is generally the most prevalent.

Protein phosphorylation. We used standard conditions (2 min. in light with 200uM ATP) to label thylakoid proteins with ^{32}P-ATP which resulted in the 28 and 27kDa polypeptides of barley LHC II*b* being relatively strongly labelled, but the 25kDa LHC II*b* apoprotein was not phosphorylated.

LHC II*b* in a chl b-less mutant of barley. It was originally proposed that LHC II*b* was lacking in this mutant (e.g. 33). Later, others provided evidence for the presence of some LHC II *apoproteins* in the mutant (e.g. 34–37). Using our improved fractionation technique (Fig. 1) we examined which LHC II *pigment-proteins* are present in the mutant: LHC II*a* is present, but in significantly reduced amounts compared to those in the wild type; only chl *a* and xanthophylls are present in this complex, even in the wild-type plant. The 25kDa polypeptide component of LHC II*b* is also present but we are uncertain that pigment is associated with it. We do not detect the two major polypeptides of 28 and 27kDa that are associated with wild-type LHC II*b*.

The relationship between the different LHC IIb apoproteins. Three apoproteins (Table I) are readily separated by denaturing PAGE of the isolated pigment-protein. If urea (4M) is omitted from the separating gel and/or if the electrophoretic separation is done at low temperatures, then the smallest apoprotein co-migrates with the largest; i.e., only two apoproteins are readily seen (cf. 38). It is possible that further changes in the SDS-PAGE conditions might fractionate more than three apoproteins. Regardless of the exact number, we do not yet know how the apoproteins are related. The explanation that each apoprotein is solely a different gene product now seems less probable in light of the gene studies in *Arabidopsis thaliana* (13). The three strongly cross-hybridising members of this plant's LHC IIb gene family have been isolated and all three gene products are predicted to have identical amino acid sequences. Yet, LHC IIb in *A. thaliana* has a similar multiplicity of apoproteins as occurs in other plants that have as many as 15 cross-hybridising genes (14). The notion cannot be entirely abandoned, however, since there is a weakly cross-hybridising fourth gene in *A. thaliana* which has yet to be isolated and sequenced (13); nevertheless, it is more likely that the apoproteins differ in their post-translational modifications. Certainly the two slowest migrating forms are phosphorylatable whereas the fastest is not (see above). This difference, however, is probably not the complete explanation of LHC IIb apoprotein multiplicity, since whether or not an LHC IIb polypeptide can be phosphorylated may be indicative of whether or not proteolytic processing has occurred at the N-terminus. Also, LHC IIb has recently been found to have palmitylated polypeptides (39). A combination of modifications may, therefore, explain the variation in mobility of the apoproteins of PAGE, rather than being the result of different gene products.

Models for the Structure LHC IIb

Although crystals have been made of this hydrophobic chl-protein (7), none have yielded the high resolution X-ray data needed to have a detailed three-dimensional structure of this pigment-protein. In the absence of such data, we and others have resorted to modelling the complex on the basis of its predicted amino acid sequence (19,40,41). Using one of the deduced sequences, a model for the folding of this polypeptide with respect to the liquid bilayer (Fig. 2) has been predicted (19) and tested (42). Three membrane-spanning alpha-helical segments of ~ 23 residues were suggested from computer analyses of the sequence. Outside of the bi-layer little other alpha-helical structure was predicted except for two short amphiphilic helical stretches that would be located just prior to the first and after the third putative membrane helices. Several closely spaced beta-turns were predicted in the segment of protein connecting the putative second and third helices (cf. 19,42). The percentages of

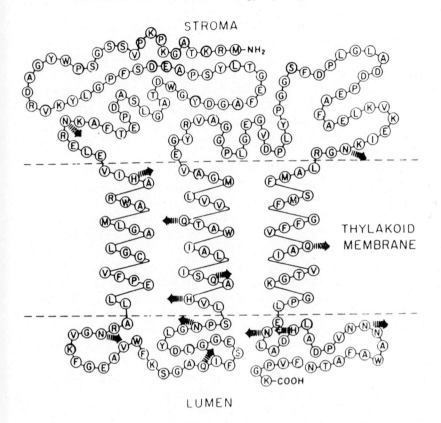

STROMA

THYLAKOID
MEMBRANE

LUMEN

Figure 2. Proposed model for the folding of the LHC II*b* polypeptide chain (cf. 19). The chl molecules are proposed (see text) to be co-ordinated to the polypeptide chain via ligation to the conserved histidine, glutamine and asparagine residues (marked by ▪◆).

alpha-helical and beta-structures, measured spectrally for this pigment-protein (43), agree well with the predicted amounts. The helices in the membrane are not aligned exactly perpendicular to the bilayer as shown in Fig. 2, but rather they make a net angle of 80° with respect to it (43). We proposed (19) that the charged residues within the membrane may form salt bridges either between helices of the same polypeptide, which would stabilise the folding of the monomer, or between monomers which would yield an oligomeric structure (see below). Since the N-terminus of the protein is firmly established to be in the stroma (2), and if the polypeptide chain spans the bilayer three times, its C-terminus will be in the lumen (Fig. 2). Support for the C-terminus being located there, and hence for a model in which the polypeptide spans the membrane three times, has come from tyrosine-labelling studies of the protein in right-side-out and inside-out thylakoid vesicles (25). Data from work with deletion mutants (42) are also consistent with three spans. Alternatively, work with a synthetic peptide antibody to a C-terminal sequence agglutinated thylakoids but not inside-out vesicles which suggests that the C-terminus is in the stroma and, therefore, that four membrane spans occur (41). Electron microscope data on two-dimensional arrays of LHC II*b* (5, 10) indicate that the polypeptide chain extends ~ 20A on one side of the membrane and ~ 7A on the other. If the three membrane-spanning helices are the only portions of the protein that are located in the lipid bilayer, then the folding of the remainder of the protein would be consistent with the size of such protrusions.

The location of the pigment molecules with respect to the folded polypeptide is of major interest. Direct determinations of their number per polypeptide have previously give equivocal results, and hence the best estimate of the ratio, at present, is probably that deduced from the ratio of chl *a*/*b* in the complex, which has been repeatedly determined to be 1.15 ± 0.02. Reasonable numbers that fit this ratio are 7 chl *a* and 6 chl *b* (however see 7), and hence, from the known ratio of chl/carotenoid in LHC II*b* 3 xanthophyll molecules per polypeptide Histidine residues have been clearly established to coordinate chl molecules to polypeptides in purple and green bacteria (cf. 44); however, there are too few of them in LHC II*b* to accommodate 13 chl molecules. Glutamine and asparagine residues are the next mostly likely candidates and, interestingly, the number of conserved His + Gln + Asn in the LHC II*b* sequence (cf. 12) equal the proposed number of chl molecules; furthermore, those His + Gln + Asn residues that do not occur in membrane-spanning helices are almost invariably located close to the end of a helix (Fig. 2), and thereby would be close to the lipid bilayer. It can be envisaged how either a chl *a* or chl *b* molecule can be selected to coordinate to a particular his, gln or asn residue so that when the helices are arranged in three dimensions the chl *b* molecules would be surrounded by chl *a* molecules and would give the relative positions indicated by biophysical data (9). The porphyrin rings of the chl molecules are oriented at right angles to the plane of the membrane (cf. 45).

The Form of LHC II*b* in Situ

In the model we have considered the structure of a monomer of LHC II*b*. Much evidence exists, however, that it occurs *in situ* in an oligomeric form(s): all of the complex can migrate on non-denaturing PAGE with a considerably larger apparent size than that of the supposedly monomeric form [e.g., Fig. 1; see also (40)], and electron diffraction data (5,7,10) are interpreted to indicate that LHC II*b* occurs as a trimer. We observe two oligomeric forms of LHC II*b* on Deriphat-PAGE of thylakoids or photosystem II particles, one having an apparent size of 72kDa and the other of 300kDa (28). Whether the two forms represent two different arrangements of LHC II*b* *in situ* or whether the smaller form is derived from dissociation of the larger one, has yet to be elucidated. Several studies have indicated that the basic structure of LHC II*b* is trimeric; however, our preliminary protein cross-linking studies on LHC II*b* reveal that it may be tetrameric: the major linked form we observe has the size of a dimer of the apoproteins but trimeric and tetrameric sizes of the cross-linked proteins are also obtained. Incidentally, no unequivocal evidence has yet been presented to show that the smallest pigmented form (30–35kDa) of LHC II*b* observed on PAGE is indeed monomeric. The interaction between subunits in the oligomer may involve one or more of the three charged residues in the putative membrane-spanning helices (Fig. 2) (47); salt bridges would permit more specific interactions than hydrophobic interactions alone.

Assembly of LHC II*b*

Since LHC II*b* subunits are particularly convenient markers for a thylakoid protein, they have been much used to study the intricacies of synthesis and assembly of a nuclear gene family coding for a membrane polypeptide. LHC II*b* apoproteins are synthesised on the cytoplasmic ribosomes as a higher molecular weight pre-LHC II*b* apoprotein(s). The assembly of LHC II*b* and other similar thylakoid proteins involves many intricate steps as well as interplay between cytoplasmic and chloroplastic products: (1) translocation of the pre-LHC II*b* apoprotein into thylakoid membranes; (2) processing to its mature form; (3) insertion into thylakoid membranes; (4) binding of chl and carotenoid molecules to it; and (5) association with other LHC II*b* apoprotein molecules to form an oligomeric and functional LHC II*b* (48). At some stage during the assembly process the water-soluble pre-LHC II*b* apoprotein changes to a water-insoluble processed product.

Several approaches have been used to dissect the different steps in the assembly of LHC II*b*. *In vitro* uptake of labelled polypeptides obtained by *in vitro* translation of total poly A-RNA by isolated intact plastids is the most popular one (49–53). Less equivocal data are obtained if only one particular precursor polypeptide is added to the plastids. Such a polypeptide can be made by using SP6 or T7 promoters in front of the gene of interest to synthesise the specific mRNA *in vitro* (42,54), which is then translated in a protein-synthesising system. The import of pre-LHC II*b* apoprotein into the plastids is post-translational (53), energy-dependent (51) and probably mediated through specific receptors in chloroplast envelopes (50). After the uptake of the *in vitro* synthesised pre-LHC II*b* apoprotein(s) by the isolated plastids, labelled mature polypeptide(s) can be detected in the oligomeric LHC II*b* (54). *In vitro*-import experiments using *in vitro*-mutated pre-LHC II*b* apoprotein of *L. gibba* have revealed the importance of the amino acid charge distribution in the membrane-spanning helices of the polypeptide for the stability of the newly imported

LHC II*b* apoprotein in the thylakoids and for its assembly into LHC II*b* (47). The pre-LHC II*b* apoprotein can be seen in the thylakoids and also in LHC II*b* under certain conditions: for example, when *in vitro* synthesised *L.gibba* pre-LHC II*b* apoprotein is incubated with either barley or maize plastids (54) but not with *L.gibba* etiochloroplasts (42), the mature *as well as* the precursor polypeptides are observed as integral thylakoid proteins. Both of these polypeptides migrate specifically with the LHC II*b* band in partially denaturing PAGE. The occurrence of both forms is also observed when *in vitro* synthesised barley pre-LHC II*b* apoprotein is imported into barley plastids isolated from etiolated plants which have been illuminated for less than 24 hours; when plastids from plants greened for 24 or more hours are used, only the mature LHC II*b* apoprotein is seen in the thylakoids (55). Thus the presence of pre-LHC II*b* apoprotein in thylakoids is not due solely to the heterologous nature of the system but is largely due to the influence of the developmental stage of plastids. Pulse-chase experiments showed that the pre-LHC II*b* apoprotein integrated into thylakoids of intact plastids can be processed to the mature LHC II*b* apoprotein (55).

Another way to study assembly is to separate and analyse the different steps *in vitro* and later reconstitute the entire assembly process. For example, pre-LHC II*b* apoprotein can be inserted into isolated thylakoids (56,57). Both Mg-ATP and a stromal factor are absolutely required for the insertion (56,58). The stromal factor is a protein (58). Although the exact way this factor helps in the insertion is not known, one possibility is that it is involved in attaching some hydrophobic moiety to the water-soluble pre-LHC II*b* apoprotein so that the modified pre-LHC II*b* apoprotein is then more compatible with hydrophobic environment of the thylakoid membranes. Such a moiety may be palmitic acid which is known to be attached to the mature LHC II*b* polypeptide (39). Alternatively, the stromal protein may change the conformation of the pre-LHC II*b* apoprotein to make its exterior surface hydrophobic. Evidence that membrane translocation is achieved by using some factor that apparently denatures/unfolds the polypeptide and permits its translocation has been found in *E.coli* (59) and yeast (60). The insertion of pre-LHC II*b* apoprotein into thylakoid membranes also depends on the stage of plastid development for both the appearance of the stromal factor that is required for its insertion into the thylakoid, and for the thylakoid membrane's receptivity for insertion (58).

Reconstitution experiments using the purified LHC II*b* apoprotein and chl molecules showed that chl molecules can bind to the apoprotein(s) *in vitro* and that carotenoids play an important role in permitting or stabilising the binding of chl and in the formation of LHC II*b* (61). Lastly, the assembly of purified LHC II*b* with PSII core complex in membranes of intermittent light-grown plants that lack LHC II*b*, has been demonstrated; however, it occurred with low efficiency (62,63).

LHC II*b* in Bundle Sheath and Mesophyll Cells of C4 Plants

Several laboratories have correlated the structural dimorphism in C4 plants with the absence, or with greatly diminished amounts, of photosystem II activity, and a dogma was established which equated agranal chloroplasts of bundle sheath cells with a lack of photosystem II activity (e.g. 64). To determine which components of photosystem II were actually missing from agranal bundle sheath chloroplasts, various laboratories have used enzymatic or mechanical means to separate mesophyll from bundle sheath cells. Some analyses of thylakoids in these separated tissues showed the presence, and some the absence, of LHC II*b* (see 65 for summary). Furthermore, when these tissues were investigated for the presence of the LHC II*b* mRNA, no single, definitive result was obtained. One can explain the equivocal data in a number of ways: (1) contamination of one cell type by the other: (2) the selective enrichment of a subpopulation of bundle sheath cells/chloroplasts during the separation procedure may be due to degradation of some bundle sheath components; or (3) different types of plants were used. The drawbacks of cell fractionation in C4 plant studies can be bypassed if *in situ* hybridisation techniques, which do not require tissue fractionation, are used. We detected the presence of the LHC II*b* transcripts in both mesophyll and bundle sheath cells in thin sections of intact maize leaves (65). A similar approach was used to look for the presence of the LHC II*b* apoprotein; namely, sections of maize leaves were immunodecorated with the MLH1 (35) monoclonal antibody to the LHC II*b* subunits. The data showed that bundle sheath plastids contain the LHC II polypeptide(s), albeit in apparently lesser amounts than found in mesophyll plastids; furthermore, the subunit composition of LHC II*b* differs slightly but significantly in bundle sheath and mesophyll thylakoids (65).

Both cell types possess the machinery for uptake, processing and incorporation of *in vitro* synthesised pre-LHC II*b* apoprotein into the pigmented complex. Incubation of [^{35}S]-pre-LHC II*b* apoprotein with bundle sheath chloroplasts resulted in the appearance of a processed form of LHC II*b* apoprotein in the thylakoids. This processed form was precipitated by antibodies specific for the LHC II*b* apoprotein(s) and shown to be assembled into LHC II*b*. Although the machinery for the introduction of pre-LHC II*b* apoprotein into the thylakoid LHC II complex was present in bundle sheath cells, it was less active in uptake than in mesophyll cells. Hence the lower levels of the LHC II*b* apoproteins in bundle sheath thylakoids could be due not only to transcriptional/post-transcriptional control, but also to a less active transport system for pre-LHC II*b* apoprotein in bundle sheath cells than in the mesophyll cells.

Concluding Remarks

The structure and composition of LHC II*b* has been studied for more than two decades; however, we are still a long way from knowing unequivocally how these polypeptides are folded with respect to the lipid bilayer, how pigments are associated with its polypeptides, what are its oligomeric forms, and how the monomer interacts with other monomers and with other photosystem II components to yield the entire structure. Some of the required information will come when its structure is determined by X-ray crystallography, but in the interim, and to obtain other unequivocal data, more protein chemical studies are required to advance our knowledge of this important antenna component.

Bibliography

1. Thornber, J. P. (1986) Biochemical characterisation and structure of pigment-proteins of photosynthetic organism. *Encyclopedia of Plant Physiol.*, New Series, 19:98–142
2. Mullet, J. E (1983) The amino acid sequence of the polypeptide which regulates membrane adhesion (grana stacking) in chloroplasts. *J. Biol. Chem.* 258:9941–9948
3. Bennett, J. (1983) Regulation of photosynthesis by reversible phosphorylation of the light-harvesting chlorophyll *a/b* protein. *Biochem J.* 212:1–13
4. Thornber, J. P., Markwell, J. P. and Reinman, S. (1979) Plant chlorophyll-protein complexes: recent advances. *Photochem. Photobiol.* 29:1205–1216
5. Kuhlbrandt, W. (1984) Three-dimensional structure of the light-harvesting chlorophyll *a/b*-protein complex. *Nature* 307:478–480
6. Ryrie, I. J., Anderson J. M. and Goodchild, D. J. (1980) The role of light harvesting chlorophyll *a/b* protein complex in chloroplast membrane stacking. Cation-induced aggregation of reconstituted proteoliposomes. *Eur. J. Biochem* 107:345–354
7. Kuhlbrandt, W. (1988) Structure of the light-harvesting chlorophyll *a/b*-protein complex from chloroplast membranes. In Scheer, H. and Schneider, S. (eds.) *Photosynthetic light-harvesting systems-Structure and Function* W. de Gruyter and Co. Berlin, 211–215
8. Staehelin, L. A. (1986) Chloroplast structure and supramolecular organziation of photosynthetic membranes. *Encyclopedia of Plant Physiol.*, New Ser. 19:1–84
9. Knox, R. S. and Van Metter, R. L. (1979) Fluorescence of light-harvesting chlorophyll *a/b*-protein complexes: implications for the photosynthetic unit. *CIBA* Foundation Symp 61:177–190
10. Li, J. (1985) Light-harvesting chlorophyll *a/b*-protein: three-dimensional structure of a reconstituted membrane lattice in negative stain. *Proc. Natl. Acad. Sci.* USA 82:386–390
11. Kung, S. D., Thornber, J. P. and Wildman, S. G. (1972) Nuclear DNA codes for the photosystem II chlorophyll-protein of chloroplast membranes. *FEBS Lett* 24:185–188.
12. Chitnis, ˙P. R. and Thornber, J. P. (1988) The major light-harvesting complex of photosystem II: Aspects of its moleculer and cell biology. *Photosyn. Res.*, in press
13. Leutwiler, L. S., Meyerowitz, E. M. and Tobin, E. M. (1986) Structure and expression of three light-harvesting chlorophyll *a/b*-binding protein genes in *Arabidopsis thaliana. Nucleic Acids Res.* 14:4051–4064
14. Dunsmuir, P. (1985) The petunia chlorophyll *a/b* binding protein genes: a comparison of Cab genes from different gene families. *Nucleic Acids Res.* 13:2503–2518
15. Karlin-Neumann, G. A. and Tobin, E. M. (1986) Transit peptides of nuclear-encoded chloroplast proteins share a common amino acid framework. *EMBO J.*5:9–13

16. Piechulla, B., Pichersky, E., Cashmore, A. R. and Gruissem, W. (1986) Expression of nuclear and plastid genes for photosynthesis-specific proteins during tomato fruit development and ripening. *Plant Mol. Biol.* 7:367–376

17. Pichersky, E., Bernatsky,R., Tanksley, S. D., Malik, V. S. and Cashmore, A. R. (1987) Genomic organization and evolution of the *rbc* and *cab* gene families in tomato and other higher plants. *Proc. Tomato Biotechnology Symp.*, in press

18. Stayton, M. M., Black, M., Bedbrook, J. and Dunsmuir, P. (1986) A novel chlorophyll *a/b* binding (Cab) protein gene from petunia which encodes the lower molecular weight Cab precursor protein. *Nacl. Acids Res.* 14:9781–9796

19. Karlin-Neumann, G. A., Kohorn, B. D., Thornber, J. P. and Tobin E. M. (1985) A chlorophyll *a/b*-protein encoded by a gene containing an intron with characteristics of a transposable element. *J. Mol. Appl. Genet.* 3:45–61

20. Kohorn, B. D. and Tobin, E. M. (1986) Chloroplast import of light-harvesting chlorophyll *a/b*-proteins with different amino termini and transit peptides. *Plant Physiol.* 82:1172–1174

21. Silverthorne, J. and Tobin, E. M. (1984) Demonstration of transcriptional regulation of specific genes by phytochrome action. *Proc. Natl Acad. Sci.* USA 81:1112–1116

22. Slovin, J. P. and Tobin, E. M. (1982) Synthesis and turnover of the light-harvesting chlorophyll *a/b*-protein in *Lemna gibba* grown with intermittent red light: possible translational control *Planta* 154:465–472

23. Bennett, J., Markwell, J. P., Skrdla M. P. and Thornber, J. P. (1981) Higher plant chlorophyll *a/b*-protein complexes: studies on the phosphorylated apoproteins. *FEBS Lett* 131:325–330

24. Simpson, J., Van Montagu, M. and Herrera-Estrella, L. (1986) Photosynthesis-associated gene families: Differences in response to tissue-specific and environmental factors. *Science* 233:34–38

25. Simpson, J. Schell, J., Van Montagu, M. and Herrera-Estrella, L. (1986) Light-inducible and tissue-specific pea *lhcp* gene expression involves an upstream element combining enhancer-and silencer like properties. *nature* 323:551–554

26. Nagy, F., Fluhr, R., Kuhlemeir, C., Kay, S., Boutry, M., Green P., Poulsen, C. and Chua, N.-H. (1986) Cis-acting elements for selective expression of two photosynthetic genes in transgeneic plants. *Phil. Trans. R. Soc. Lond* B314:493–500

27. Nagy, F., Kay, S. A., Boutry,M., Hsu, M. Y. and Chua, N. H. (1986) Phytochrome-controlled expression of a wheat *cab* gene in transgenic tobacco seedings. *EMBI J.* 5:1119–1124

28. Peter, G. F., and Thornber, J. P., (1988) Identification and characterization of higher plant photosynthic pigment-proteins of photosystem I and photosystem II. *J. Biol. Chem.* submitted

29. Camm, E. L., and Green, B. R. (1980) Fractionation of thylakoid membranes with the nonionic detergent octyl-beta-D-glucopyranoside. *Plant Physiol.* 66:428–432

30. Dunahay, T. G., and Staehelin, L. A. (1986) Isolation and characterization of a new minor chlorophyll *a/b*-protein complex (CP24) from spinach. *Plant physiol.* 80:429- 434

31. Bassi, R., Hoyer-Hansen, G., Barbato, R., Giacometti, G. M., and Simpson, D. J., (1987) Chlorophyll-proteins of the photosystem II antenna system. *J. Biol. Chem.* 262:13333–13341

32. Dunahay, T. G., Schuster, G. and Staehelin, L. A. (1987) Phosphorylation of spinach chlorophyll-protein complexes. CPII but not CP29, CP27 or CP24, is phosphorylated *in vitro FEBS Lett* 215:25–30

33. Thornber, J. P., and Highkin, H. R., (1974) Composition of the photosynthetic apparatus of normal barley leaves and a mutant lacking chlorophyll *b*. *Eur. J. Biochem.* 41:109–116

34. Burke, J. J., Steinback, K. E. and Arntzen, C. J. (1979) Analysis of the light-harvesting pigment-protein complex of wild type and a chlorophyll *b*-less mutant of barley. *Plant Physiol.* 63:237–243

35. Darr, S. C., Somerville, S. C. and Arntzen, C. J. (1986) Monoclonal antibodies to the light-harvesting chlorophyll a/b-protein complex of photosystem II. J. Cell. Biol. 103:733–7410

36. White, M. J. and Green B. R. (1987) Polypeptides belonging to each of the three major chlorophyll a + b protein complexes are present in the chlorophyll b/less barley mutant. Eur. J. Biochem. 165:531–535

37. Ryrie, I. J. (1983) Immunological evidence for the apoproteins of the light-harvesting chlorophyll protein complex in a mutant of barley lacking chlorophyll b. Eur. J. Biochem 131:149–155

38. Thornber, J. P., Peter, G. F., Nechushtai, R., Chitnis, P. R., Hunter, F. A. and Tobin, E. M. (1986) Electrophoretic separation of chlorophyll-protein complexes and their apoproteins. In Regulation of Chloroplast Differentiation, Alan Liss, N. Y. 249–258

39. Mattoo, A. K. and Edelman, M. (1987) Intramembrane translocation and posttranslational palmitoylation ofthe chloroplast 32-kDa herbicide-binding protein. Proc. Natl. Acad. Sci. USA 84:1497–1501

40. Burgi, R., Suter, F. and Zuber, H. (1987) Arrangement of the light-harvesting chlorophyll-a chlorophyll-b protein complex in the thylakoid membrane. Biochim Biophys. Acta 890:346–51

41. Anderson, J. M. and Goodchild, D. J. (1987) Transbilayer organization of the main chlorophyll a/b-protein of photosystem II of thylakoid membranes. FEBS Lett 213:29–33

42. Kohorn, B. D., Harel, E., Chitnis, P. R., Thornber, J. P. and Tobin, E. M. (1986) Functional and mutational analysis of the light-harvesting chlorophyll a/b protein of thylakoid membranes. J. Cell. Biol. 102:972–981

43. Nabedryk, E., Andriaambinistsoa, S. and Breton, J. (1984) Transmembrane orientation of the alpha-helices in the thylakoid membrane and in the light-harvesting complex. A polarized infrared spectroscopy study. Biochim. Biophys. Acta 765:380–387

44. Deisenhofer, J., Eppo, O., Miki, O., Huber, R. and Michel, H., (1985) Structure of the protein subunits in the photosynthetic reaction center of Rhodopseudomonas viridis at 3A resolution. Nature 318:618–624

45. Breton, J. (1986) Molecular orientation of the pigments and the problem of energy trapping in photosynthesis. Encl. plant Physio., New Ser. 19:319–36

46. Dunkley, P. R. and Anderson, J. M. (1979) The light-harvesting chlorophyll a/b-protein complex from barley thylakoid membranes. Polypeptide composition and the characterization of an oligomer. Biochim. Biophys. Acta 545:175–187

47. Kohorn, B. D. and Tobin, E. M. (1987) Amino acid charge distribution influences the assembly of apoprotein into light-harvesting complex II. J. Biol. Chem. 262:12897–12899

48. Schmidt, G. W. and Mishkind, M. L. (1986) The transport of proteins into chloroplasts. Annu. Rev. Biochem. 55:879–912

49. Chua, N.-H. and Schmidt, G. W. (1979) Transport of proteins into mitochondria and chloroplasts. J. Cell. Biol. 81: 461–483

50. Cline, K., Werner-Washbourne, M., Lubben, T. H. and Keegstra, K. (1985) Precursors to two nuclear-encoded chloroplast proteins bind to the outer envelope membrane before being imported into chloroplasts. J. Biol. Chem. 260:3691–3696

51. Grossman, A. Bartlett, S. and Chua, N.-H. (1980) Energy-dependent uptake of cytoplasmically synthesized polypeptides by chloroplasts. Nature 285:625–628

52. Mullet, J. E. and Chua, N.-H. (1983) In vitro reconstitution of synthesis, uptake, and assembly of cytoplasmically synthesized chloroplast proteins. Methods Enzymol. 97:502–509

53. Schmidt, G. W., Bartlett, S. G., Grossman, A. R., Cashmore, A. R. and Chua, N.-H. (1981) Biosynthetic patheays of two polypeptide subunits of the light-harvesting chlorophyll a/b protein complex. J. Cell Biol. 91:468–478

54. Chitnis, P. R., Harel, E., Kohorn, B. D., Tobin,E M. and Thornber, J. P. (1986). Assembly of the precursor and processed light-harvesting chlorophyll a/b protein of

Lemna into the light-harvesting complex II of barley etiochloroplasts. *J. Cell Biol.* 102:982–988

55. Chitnis, P. R., Morishige, D., Nechushtai, R. and Thornber, J. P. (1988) Assembly of the barley light-harvesting chlorophyll *a/b*-proteins in barley etiochloroplasts involves processing of the precursor on thylakoid. *Plant Mol. Biol.* in press.
56. Cline, K. (1986) Import of proteins into chloroplasts. Membrane integration of a thylakoid precursor protein reconstituted in chloroplast lysates. *J. Biol. Chem.* 261:14804–14810
57. Chitnis, P. R., Nechushtai, R., Harel, E. and Thornber, J. P. (1987) Some requirements for the insertion of the precursor of apoproteins of *Lemna* light–harvesting complex II into barley thylakoids. In Biggins, J. (ed) *Progress in Photosynthesis Research* 4:573–576 The Hague: Martinus Nijhoff/Junk

Transport of Proteins into Chloroplasts

KENNETH KEEGSTRA, CYNTHIA BAUERLE, ALAN FRIEDMAN, THOMAS LUBBEN, LAURA OLSEN AND STEVEN THEG

Botany Department,
University of Wisconsin,
Madison, Wisconsin 53706, USA

Summary

The transport of cytoplasmically synthesised precursor proteins into chloroplasts can be divided into several steps. The first is binding of precursors to the surface of chloroplasts. This binding is thought to be mediated by proteins located on the outer envelope membrane. Subsequent to binding, the precursors are translocated across the two envelope membranes via an ATP-dependent process. For certain proteins, such as plastocyanin, an additional step is needed to transport plastocyanin across the thylakoid membrane to the thylakoid lumen.

Introduction

Most chloroplastic proteins are encoded in the nucleus and synthesised as higher molecular weight precursors in the cytoplasm. The extra amino acids that are present at the amino terminus of precursors are called transit peptides. Work in the past few years has demonstrated that a transit peptide is necessary for transport of precursors into chloroplasts (1,2,3). Moreover, it has been demonstrated in several cases that a transit peptide is sufficient to direct the transport of foreign passenger proteins into chloroplasts (4,5,6). However, the mechanisms responsible for decoding the targeting information present in the transit peptide are less well characterised.

A key development that allowed important advances in understanding protein transport into chloroplasts was the demonstration that import could be reconstituted in vitro (7,8). In these early studies a mixture of radioactive precursor proteins, obtained by in vitro translation of polyadenylated mRNA, was incubated with isolated intact chloroplasts. The precursor to a small subunit of ribulose-1,5-bisphosphate carboxylase (prSS) was imported into chloroplasts, processed to its proper size and assembled with large subunit to yield holoenzyme (7,8). This in vitro assay has subsequently been used to examine the import of numerous precursor proteins (9) including precursors to light-harvesting chlorophyll *a/b* protein (LHCP) (10), plastocyanin (PC) (11), ferredoxin (Fd) (11), and 5-enolpyruvylshikimate-3-phosphate (EPSP) synthase (12).

The utility of the *in vitro* import assay was further enhanced by the development of *in vitro* transcription systems (13). These systems allow the synthesis of precursor proteins from cloned genes via sequential *in vitro* transcription and translation reactions (see (1) for example). One advantage of this technique is that it allows experiments to be performed with radiochemically pure precursors, rather than with the mixture of precursors that results from the translation of total polyadenylated mRNA. This not only eliminates the possibility of competition between precursors, but also simplifies data analysis by eliminating the need for immunoprecipitation of imported proteins. An additional and probably more significant advantage is that it allows the relatively rapid production of altered precursor proteins via modification of precursor genes by recombinant DNA techniques. The ability of the modified precursor proteins to be imported and properly localised can be examined using reconstituted import assays (see for example 11, 14, 15).

Studies using *in vitro* assays have led to several important insights into the import process that would have been difficult or impossible to obtain from *in vivo* experiments. For example, the *in vitro* assay has been used to demonstrate that import occurs after synthesis of the precursor protein is complete (7). A second important insight derived from *in vitro* studies is that the import process can be divided into several discrete steps. The first step is the binding of precursor proteins to the chloroplast surface. It has been suggested that this binding is mediated by a receptor protein present in the outer envelope membrane (16), although evidence for this suggestion remains incomplete (see discussion below). The second step in the import process is translocation of the precursor protein across the two envelope membranes. This step requires energy in the form of ATP. A third step, which occurs during or immediately after translocation, is proteolytic processing of the precursor to remove the transit peptide. For many proteins, additional steps are required to reach their final location and achieve their active form. For example, the small subunit (SS) of ribulose-1, 5-bisphosphate carboxylase must assemble with the chloroplast-synthesised large subunit to form the active holoenzyme. Other imported proteins must have prosthetic groups or metal ions while still others must be inserted into or across another membrane before reaching their final location. In this brief review we will focus primarily on the binding and translocation events.

Binding of Precursors to Chloroplasts

The first step in protein import is the binding of precursors to the surface of chloroplasts. Cline *et al.* (17) demonstrated that precursor binding can be assayed separately from translocation across the envelope membranes. They showed that binding of either prSS or prLHCP without transport could be achieved by incubating the chloroplasts with the precursors in the absence of ATP (see below). Chloroplasts treated in this manner could be washed to remove unbound precursor, and then assayed for transport of the bound species upon addition of exogenous ATP.

The initial characterisation of the binding of prSS to chloroplasts was extended by Friedman and Keegstra (manuscript in preparation). They found that binding was both saturable and specific. Under the conditions used for the assays, binding approached saturation at a precursor concentration of 5–10 nM, with approximately three thousand precursor molecules bound to each chloroplast. The binding was specific in that prSS was bound, but SS was not. These results are consistent with the idea that prSS binding is mediated by specific receptor proteins located in the outer envelope membrane.

The early studies of Cline *et al.* on precursor binding assumed that the binding step did not require energy because binding could be measured at levels of ATP that did not support import (17). This assumption has proven to be incorrect. Recent studies, using precursor preparations with ATP completely removed by gel filtration, have demonstrated that precursor binding requires ATP (Olsen, Theg, Selman and Keegstra, manuscript in preparation). Binding is maximal at an ATP concentration of approximately 100 µM, a level which cannot support protein translocation. The function of ATP in precursor binding is not known, but non-hydrolysable analogs will not substitute for ATP, which suggests that ATP hydrolysis is needed.

Membrane Receptors for Precursor Binding

A number of attempts have been made to indentify the receptor(s) involved in protein import. Pfisterer *et al.* (18) examined the binding of a mixture of precursor proteins to isolated chloroplastic envelope membranes, and more recently, Bitsch and Kloppstech (19) provided evidence that precursors could be bound to solubilised envelope proteins reconstituted into lipid vesicles. The latter authors also attempted to cross-link prSS to proteins on the surface of chloroplasts (20), but no clear receptor protein emerged from that study. Cornwell and Keegstra (21) also used cross-linkers in an effort to identify the receptor for prSS. Utilising a heterobifunctional, photoactivatable cross-linking reagent, they found a specific interaction of prSS with a 66 kD protein on the surface of chloroplasts. This protein was postulated to be the receptor, or part of the receptor complex, that mediates the binding of prSS to the chloroplastic surface.

Recently Pain *et al.* used a very different approach to identify a putative receptor protein (22). They prepared anti-idiotypic antibodies that act as an analogue of prSS and bind to the putative receptor, thereby blocking the import of prSS (22). These antibodies were used in several types of studies. One was to demonstrate that the putative receptor, identified by this antibody preparation, was localised at contact sites where the two envelope membranes are tightly oppressed. They also used antibodies for Western blotting in an effort to identify the putative receptor. The antibodies reacted with two chloroplastic proteins. The first was identified as the large subunit of ribulose-1, 5-bisphosphate carboxylase. The reason for a positive reaction with this protein remains unexplained. The second protein which reacted with the antibody was the major 30 kD protein of the chloroplastic envelope membrane. They concluded that this protein is the putative receptor protein.

If one assumes that the level of precursor binding at saturation gives an accurate estimate of the amount of receptor protein, then it is possible to calculate the relative abundance of the putative receptor protein. Assuming 3000 receptors per chloroplast (see above) and a molecular weight of 66 kD, the receptor protein would be approximately 0.003% of the total chloroplast protein. If one assumes that envelope proteins constitute approximately 1% of the total chloroplast protein, then the receptor protein would be approximately 0.3% of the total envelope protein. Thus the putative receptor protein should be a relatively minor protein in envelope polypeptide profile. This conclusion is in conflict with the observations of Pain *et al.* who identified the major 30 kD protein of the envelope membrane as the putative receptor (22).

Protein Translocation Across the Envelope Membrane

Energy Requirements for Import into the Stroma

The energy requirement for the import of nuclear-encoded proteins into chloroplasts was first addressed by Grossman *et al.* (23). They demonstrated that chloroplastic protein import was stimulated in the light over that observed in the dark, and that uncouplers were effective inhibitors of that stimulation. DCMU, which inhibits linear, but not cyclic, photophosphorylation, did not eliminate protein import. Exogenously added ATP was able to support protein import into dark-incubated chloroplasts, either in the absence or presence of uncouplers. These experiments led the authors to propose that protein import into isolated chloroplasts requires ATP as the sole energy source. As mentioned above, the latter approach, wherein exogenously added ATP was used to overcome the uncoupler-induced inhibition of import, was later employed by Cline *et al.* (17) to separate the binding and translocation steps.

Several additional reports in the energy requirements for protein import into chloroplasts have appeared recently (24–27). The hydrolysis of ATP was found to be necessary for import of prSS into spinach chloroplasts (25), pea chloroplasts (26, 27) and castor bean leucoplasts (24). We have recently extended these observations to demonstrate that ATP is required for import of the precursors of ferredoxin and plastocyanin into pea chloroplasts (Theg, Bauerle, Olsen, Selman and Keegstra, manuscript in preparation). Other nucleotides (GTP, CTP, UTP and AMP) were found to be ineffective substitutes for ATP in supporting protein uptake by plastids in the dark (25, 27), as were a number of non-hydrolysable ATP analogs (25–27). Moreover, it was found that the ATP-dependent import of precursors into dark-incubated chloroplasts was unaffected by agents which collapse either or both the pH and electrical components of the protonmotive force (25–27). Thus the consensus from all these studies is that protein translocation across the envelope membranes requires ATP, but not an electrochemical potential.

Another important question is the location of ATP utilisation during protein import. One strategy for examining this problem is to generate ATP inside plastids by the addition of glycolytic or Calvin–Bensen cycle intermediates. Import of prSS into chloroplasts is supported by internally-generated ATP (25–27). These results are not unexpected since light can stimulate import and ATP produced by light is also generated in the stroma. Although internally generated ATP supported protein import into chloroplasts, it was unknown whether this ATP was actually utilised internally or after it had been transported to the external medium. To address this question, ATP was generated in the stroma while external ATP was consumed by exogenous membrane-impermeable ATP traps. Presumably, ATP-mediated protein import would occur under these conditions only if it was driven by ATP inside the chloroplasts. However, the results of the experiments seem to depend upon the choice of enzyme used to consume ATP. Pain and Blobel used either glucose/hexokinase or apyrase and observed that import of prSS into pea chloroplasts occurred normally, leading them to conclude that ATP was hydrolysed in the stroma. Schindler *et al.* (27) performed a similar experiment with glucose/hexokinase, and observed that protein import was inhibited. They concluded that ATP was required externally for protein import. Flügge and Hinz (25) performed the most careful study of this question, measuring the actual ATP levels in the chloroplast under conditions of the import assays. Furthermore, they reduced possible complications which could result from rapid transport of ATP across the envelope by using spinach chloroplasts, which display lower adenylate translocator activity than do pea chloroplasts. These authors found that when either glucose/hexokinase or fructose-6-phosphate/fructose-6-phosphate kinase were used as external ATP consuming systems, the import of prSS was not inhibited. However, when alkaline phosphatase was used as the external ATP trap, the import of prSS was inhibited and correlated with the external, rather than the internal ATP concentration. They concluded that ATP outside the stroma appeared to be required for import, although an alternative explanation could involve the participation of an externally accessible phosphorylated intermediate, if this intermediate was dephosphorylated by the alkaline phosphatase.

We have recently re-examined the problem of the location of ATP utilisation using pea chloroplasts and several different precursors. Our strategy was similar to that employed by the other groups, except that we examined the rate of precursor import under the various conditions. Because pea chloroplasts have an active adenylate translocator, thereby allowing internal and external pools of ATP to rapidly equilibrate, it was possible that external ATP traps could deplete internal ATP levels (27). This emphasises the importance of measuring the ATP levels produced by various treatments. However, by monitoring the ATP levels and correlating them with the rates of protein import, we have tentatively concluded that the ATP utilised for protein translocation is needed inside chloroplasts (Theg, Bauerle, Olsen, Selman and Keegstra, manuscript in preparation).

The recent evidence that ATP is also needed for precursor binding may complicate the issue of the location of ATP utilisation during import. At present we have not examined the location of the ATP requirement for protein binding, which is approximately one order of magnitude lower than that needed to support protein translocation. However, it seems reasonable to speculate that this ATP is needed outside chloroplasts. If correct, this leads to the conclusion that ATP is required on both sides of the envelope membranes; on the outside to support precursor binding and on the inside to support protein translocation. This tentative conclusion needs further experimental support.

Although initial studies indicated that ATP was required for protein import into mitochondria (28), subsequent studies indicated that instead, the electrical component of the proton electrochemical potential gradient was required (29,30). More recently, the energy requirements of protein import into mitochondria were reinvestigated, and a requirement for ATP was observed (31,32). Thus the emerging consensus is that both ATP and a transmembrane electric field are necessary to drive the import of proteins into mitochondria. A requirement for ATP hydrolysis may be common to all protein translocation systems; such a requirement has also been reported for the transport of proteins across the endoplasmic reticulum membrane (33–36), movement of secretory proteins through the Golgi apparatus (37,38), and for protein secretion from E. coli (39,40). In the last case a protonmotive force also appears to be involved.

Mechanism of Protein Translocation

The actual mechanism by which proteins are translocated across the two envelope membranes remains obscure. Eilers and Schatz (41) addressed this question for mitochondria and have proposed that protein unfolding must precede translocation. This hypothesis is based largely on the observation that the import of a chimeric precursor containing the transit peptide from subunit IV of cytochrome oxidase fused to dihydrofolate reductase was inhibited by the folate antagonist, methotrexate. Methotrexate also imparted some protection against dihydrofolate reductase degradation by thermolysin. Additional evidence for the requirement of unfolding prior to translocation was recently provided by Verner and Schatz (42), who showed that a prematurely terminated chimeric precursor, which presumably was incompletely folded, was imported into mitochondria without the usual ATP requirement. While these results are consistent with the notion the protein unfolding occurs during import into mitochondria, other interpretations of these data cannot be rigorously excluded.

In chloroplasts, recent evidence suggests that, if proteins are unfolded during import, they are not exposed to a hydrophobic environment during translocation. Lubben *et al.* (43) fused two different hydrophobic stretches of amino acids that function as stop-transfer sequences (44) to the carboxyl terminus of prSS. Neither chimera was halted during translocation by an interaction of the hydrophobic stop transfer region with the membranes, which indicates that the regions were probably never exposed to the lipid layer. These results argue against the threading of an unfolded protein directly through the lipid bilayer or other hydrophobic environment. They also suggest that if translocator proteins are a universal feature of protein transport, then a putative chloroplast translocator protein must be sufficiently different from the endoplasmic reticulum translocator protein such that it could not recognise the stop-transfer domains of either chimera.

Although such studies begin to provide some insight into the mechanism of protein translocation across the envelope membranes, a great deal of work still needs to be done. At least two fundamental but related questions need to be answered. First, are there translocator proteins present in the chloroplast envelope membranes that mediate the movement of proteins across the two membranes? Such a hypothesis has been described in detail most recently by Singer *et al.* (45). The second question has to do with the conformation of the precursor during translocation. Specifically, is the protein unfolded during transport so that it can be threaded across the membrane or is the protein translocated in a folded state?

Protein Translocation Across the Thylakoid Membrane

The internal thylakoid membrane system of chloroplasts contains several proteins that originate in the cytoplasm. These proteins not only must be transported across the envelope membranes, but also must be inserted into or transported across the thylakoid membrane. The transport of PC, a thylakoid lumen protein, has been used as a model system to study this pathway. The precursor to plastocyanin has a complex transit sequence that consists of two structural domains (46). Smeekens *et al.* provided evidence that the transport of plastocyanin occurs in two steps (11). The first step, transport of prPC across the two envelope membranes, is similar in many respects to the import of stromal proteins discussed above. This step is mediated by the amino terminal domain of the transit sequence. After import, this region of the transit sequence is removed by a stromal processing protease to form an intermediate-sized form of PC. The second step, transport of the intermediate across the thylakoid membrane, is thought to be mediated by the carboxyl terminal domain of the transit sequence. At present many details of the thylakoid translocation process are still unknown. The best way to investigate this step will be to reconstitute it *in vitro* using isolated thylakoid membranes and either prPC or intermediate PC. Efforts to accomplish this *in vitro* reconstitution are under way in our laboratory. In the interim, studies utilising intact chloroplasts have yielded some insights. Experiments with ionophores have demonstrated that an electrochemical potential across the thylakoid membrane is not needed for PC transport to the lumen (Theg, Bauerle, Olsen, Selman and Keegstra, manuscript in preparation). The tentative conclusion at present is that transport across the thylakoid membrane requires ATP, just as transport across the envelope membranes does. However, until the second transport step can be studied independent of the first, it will be difficult to confirm the exact energy requirement. Many other questions also remain unanswered. For example, are stromal proteins or other factors required to initiate translocation across the thylakoid membrane, as has been demonstrated for insertion of LHCP into the thylakoid membrane (47)? Does the thylakoid membrane have receptors or other proteins which mediate the translocation of proteins? These questions indicate that this will be a fruitful area of investigation for many years.

Acknowledgements

The authors thank Willow Ealy for assistance in preparing this manuscript. Work from the authors' laboratory was supported by grants from the NSF, the Division of Biological Energy Research at DOE and the Competitive Research Grants Office at USDA.

References

1. Anderson, S. and Smith, S. M. (1986) *Biochem. J.* 240, 709–715
2. Mishkind, M. L., Wessler, S. R. and Schmidt, G. W. (1985) *J. Cell Biol.* 100, 226–234
3. Schmidt, G. W. and Mishkind, M. L. (1986) *Annu. Rev. Biochem.* 55, 879–912

4. Cashmore, A., Szabo, L., Timko, M., Kausch, A., Van den Broeck, G., Schreier, P., Bohbnert, H., Herrera-Estrella, L., Van Montagu, M. and Schell, J. (1985) *Bio/Tech* 3, 803–808
5. Della-Cioppa, G., Kishore, G. M., Beachy, R. N. and Fraley, R. T. (1987b) *Pl. Physiol.* 84, 965–968
6. Schreier, P. H. and Schell, J. (1986) *Phil. Trans. R. Soc. Lond.* B313, 429–432
7. Chua, N. -H. and Schmidt, G. W. (1978) *Proc. Natl Acad. Sci. USA* 75– 6110–6114
8. Highfield, P. E. and Ellis, R. J. (1978) *Nature* 271, 420–424
9. Grossman, A. R., Bartlett, S. G., Schmidt, G. W., Mullet, J. E. and Chua, N. -H. (1982) *J. Biol. Chem.* 257, 1558–1563
10. Schmidt, G. W., Bartlett, S. G., Grossman, A. R., Cashmore, A. R. and Chua, N. -H. (1981) *J. Cell Biol.* 91, 468–478
11. Smeekens, S., Bauerle, C., Hageman, J., Keegstra, K. and Weisbeek, P. (1986) *Cell* 46, 365–375
12. della-Cioppa, G., Bauer, S. C., Klein, B. K., Shah, D. M., Fraley, R. T. and Kishore, G. M. (1986) *Proc. Natl Acad. Sci. USA* 83, 6873–6877
13. Krieg, P. A. and Melton, D. A. (1984) *Nucl. Acids Res.* 12, 7057–7071
14. Lubben, T. H. and Keegstra, K. (1986) *Proc. Natl Acad. Sci. USA* 83, 5502–5506
15. Reiss, B., Wasmann, C. C. and Bohnert, H. J. (1987) *Mol. Gen. Genet.* 209, 116–121
16. Chua, N. -H. and Schmidt, G. W. (1979) *J. Cell Biol.* 81, 461–483
17. Cline, K., Werner-Washburne, M., Lubben, T. H. and Keegstra, K. (1985) *J. Biol. Chem.* 260, 3691–3696
18. Pfisterer, J., Lachmann, P. and Kloppstech, K. (1982) *Eur. J. Biochem.* 120, 143–148
19. Bitsch, A. and Kloppstech, K. (1986) *Eur. J. Cell Biol.* 40, 160–166
20. Kloppstech, K. and Bitsch, A. (1986) In: *Regulation of Chloroplast Differentiation*, 235–240. Alan R. Liss, Inc.
21. Cornwell, K. L. and Keegstra, K. (1987) *Plant Physiol.* 85, 780–785
22. Pain, D., Kanwar, Y. and Blobel, G. (1988) *Nature* 331, 232–237
23. Grossman, A., Bartlett, S. and Chua, N. -H. (1980) *Nature* 285, 625–628
24. Boyle, S. A., Hemmingsen, S. M. and Dennis, D. T. (1986) *Plant Physiol* 81, 817–822
25. Flügge, U. I. and Hinz, G. (1986) *Eur. J. Biochem.* 160, 563–570
26. Pain, D. and Blobel, G. (1987) *Proc. Natl. Acad. Sci. USA* 84, 3288–3292
27. Schindler, C., Hracky, R. and Soll, J. (1987) *Z. Naturforsch* 42c, 103–108
28. Nelson, N. and Schatz, G. (1979) *Biochemistry* 76, 4365–4369
29. Gasser, S. M., Daum, G. and Schatz, G. (1982) *J. Biol. Chem.* 257, 13034–13041
30. Schleyer, M., Schmidt, B. and Neupert, W. (1982) *Eur. J. Biochem.* 125, 109–116
31. Eilers, M., Oppliger, W. and Schatz, G. (1987) *EMBO J.* 6, 1073–1077
32. Pfanner, N. and Neupert, W. (1986) *FEBS* 209, 152-156
33. Hansen, W., Garcia, P. D. and Walter, P. (1986) *Cell* 45, 397–406
34. Rothblatt, J. A. and Meyer, D. I. (1986) *EMBO J.* 5, 1031–1036
35. Waters, M. G. and Blobel, G. (1986) *J. Cell Biol.* 102, 1543–1550
36. Schlenstedt, G. and Zimmermann, R. (1987) *EMBO J.* 6, 699–703
37. Balch, W. E., Elliott, M. M. and Keller, D. S. (1986) *J. Biol. Chem.* 261, 14681–14689
38. Balch, W. E. and Keller, D. S. (1986) *J. Bio. Chem.* 261, 14690–14696
39. Chen, L. and Tai, P. C. (1986) *J. Bact.* 168, 828-832
40. Yamane, K., Ichihara, S. and Mizushima, S. (1987) *J. Biol Chem.* 262, 2358–2362
41. Eilers, M. and Schatz, G. (1986) *Nature* 322, 228–232
42. Verner, K. and Schatz, G. (1987) *EMBO J.* 6, 2449–2456
43. Lubben, T., Bansberg, J. and Keegstra, K. (1987) *Science* 283, 1112–1114
44. Blobel, G. (1980) *Proc. Natl Acad. Sci. USA* 77, 1496–1500
45. Singer, S. J., Maher, P. A. and Yaffe, M. P. (1987) *Proc. Natl Acad. Sci. USA* 84, 1015–1019
46. Smeekens, S., de Groot, M., van Binsbergen, J. and Weisbeek, P. (1985) *Nature* 317, 456–458
47. Cline, K. (1986) *J. Bio. Chem.* 261, 14804–14810

The Role of Lipid–Protein Interactions in the Structure and Function of Photosynthetic Membranes

DENIS J. MURPHY[1], LI GANG[2], PETER F. KNOWLES[2] AND
IKUO NISHIDA[3]

[1]Department of Botany,
University of Durham,
Durham DHI 3LE, U.K.

[2]Department of Biophysics,
University of Leeds,
Leeds LS2 9JT, U.K.

[3]National Institute for Basic Biology,
Okazaki, Japan,

Summary

The mechanism and functional role of lipid–protein interactions in photosynthetic membranes has been examined. The efficacy of various acyl lipids in the restoration of energy and electron transfer between solubilised chlorophyll–protein complexes was studied. The best reconstituting agent was an impure phosphatidylcholine preparation which is rarely found in native photosynthetic membranes. Other purified photosynthetic membrane lipids were relatively poor reconstituting agents. Spectroscopic techniques such as electron spin resonance (ESR) and saturation transfer ESR were employed to study more directly the mechanisms of lipid–protein interactions. Nitroxide-labelled acyl lipids were introduced into thylakoid membranes and sub-membrane preparations and the resultant ESR spectra were analysed under various conditions. There was a pronounced selectivity of thylakoid protein for different lipids as reflected by increasing lipid motion restriction from monogalactosyldiacyl-glycerol <phosphatidylcholine <phosphatidylglycerol. Photosystem II preparations showed considerably more spin label motion restriction than photosystem I preparations or thylakoid membranes. Chilling sensitive plants, such as cucumber or tomato, also showed more spin label motion restriction than chilling tolerant plants, such as spinach and pea. In all of the above cases the extent of spin label immobilisation was correlated with the protein:lipid ratio of the membranes. Rotational correlation times, deduced from ST-ESR spectra, indicated that different lipids bind at different sites on proteins. These data are discussed in the context of the specificity of lipid-protein interactions in photosynthetic membranes.

Introduction

Biological membranes are composed of varying proportions of lipids and proteins. In general, the more metabolically active a membrane is, the higher the proportion of protein. Very active membranes such as mitochondrial inner membranes (1), *E. coli* inner membranes (2), chloroplast thylakoid membranes (3), and purple bacterial membranes (4) contain 70–85% protein by weight. These membranes are, therefore, largely made up of planar proteinaceous assemblies separated by relatively small stretches of lipid bilayer.

There are many thousands of bilayer lipids found in biological membranes. These lipids may differ in their headgroup, acyl chain length and unsaturation. Despite all the possible variations, the lipid composition of a given membrane, such as the chloroplast thylakoid, is normally as invariant as its protein composition. This in itself suggests that the lipid population probably plays an important role in defining the structure and function of the membrane via interactions with the membrane proteins.

Several types of lipid–protein interaction are found in biological membranes. Firstly, in some rare cases, lipids or lipid derivatives may be covalently bound to proteins. A second and more common type of strong lipid–protein binding is found in the pigment-protein complexes of photosynthetic membranes. These proteins can be completely denuded of their acyl lipid complement by either detergent or solvent treatment while the pigments remain firmly bound in the interior of the protein molecules (5,6). It is the third and commonest type of lipid–protein binding which is the subject of this article. Here the bilayer lipids in the vicinity of membrane proteins interact relatively weakly with the latter, but in a way which is yet to be satisfactorily described. It has been calculated that, in protein-rich membranes, as many as 50–75% of the total lipid population may be immediately adjacent to a protein surface at a given instant (6). There is, therefore, no doubt that lipid–protein interactions are ubiquitous in such membranes. What is less certain is whether particular lipids specifically interact with particular proteins or particular domains on the protein surface. We also know relatively little about the effects of such interactions upon the physical and biochemical properties of the participating lipid and protein molecules and their consequences at the level of membrane structure and function.

Techniques

The techniques used in the study of lipid–protein interactions fall into two broad groups. Firstly there are the physical methods, such as spectroscopy, where lipid–protein interactions can be directly monitored in either native or reconstituted membranes. Secondly there are the biochemical methods where an isolated membrane protein is re-inserted into different lipid milieux and the effects on its biological activity are measured. We have employed both techniques in the study of photosynthetic membranes and such a complementary approach is probably the most useful.

Reconstitution of Biological Activity

There have been many studies on the reconstitution of membrane proteins in a wide variety of biological systems, as reviewed in (7). There are also recent reviews on the use of such techniques for the study of photosynthetic membranes (6,8,9). As yet there is little consensus about which lipids or lipid mixtures are optimal for the functioning of any given photosynthetic membrane protein. For example, it has been claimed that the presence of monogalactosyldiacylglycerol (MGD) will enhance the activities of reconstituted ATP synthetase (9), LHCII–PSII and LHCI -PSI (10,11), but this has been disputed (12,13). Other thylakoid lipids were able to restore the activities of reconstituted cytochrome b_6-f (14). PSII oxygen evolution (15), high potential cytochrome b_{559} (16,17), PSII PSI electron transport (13) and PSII→PSI energy transfer (13,18).

In Fig. 1 the results of such an experiment are shown. Native thylakoid membranes exhibit three fluorescence emission peaks, at 685, 693 and 735nm, which originate respectively from LHCII, the PSII reaction centre, and PSI. These peaks reflect excitation energy transfer between the various protein complexes. Solubilisation of the membranes with Triton X-100 results in the dissociation of these complexes and the consequent loss of inter-complex energy transfer. The result is a single fluorescence emission peak at 685nm. When the dissociated complexes are reconstituted into membranes containing different thylakoid lipid populations, varying extents of activity restoration are observed. The two galactolipids, MGD and DGD, which together comprise about 75% total thylakoid acyl lipid, were very poor reconstituting agents. In contrast, the relatively scarce phospholipids PG and PC were good reconstituting agents. The best reconstituting agent, PC, is a very minor thylakoid lipid component (<5% w/w) and may even, as some workers claim, be absent altogether from such membranes (19). Even better than purified PC, was an impure form containing 60% impurities such as lysolipids and other phospholipids, i.e. Sigma type IIS azolectin. The conclusion from such a study is that the ability of a lipid to reconstitute biological activity *in vitro* may not necessarily reflect lipid–protein interactions *in vivo*. Other factors such as the relative liposome forming capacity or the ability of a liposome to incorporate a solubilised protein may be the real determinants of the efficacy of a particular lipid or lipid mixture as a reconstituting agent.

It is noteworthy that, in over a decade of similar studies, very little evidence has been accrued of a specific lipid requirement for the restoration of activity in isolated membrane proteins. Indeed, it has been the experience of the present authors (12,13) and others (reviewed in reference 20) that lipid mixtures, and even detergents, are far more potent reconstituting agents than single lipid species. Whether specific lipid mixtures are required for optimal protein function and whether these truly reflect *in vivo* interactions remains the subject of some controversy (6,13).

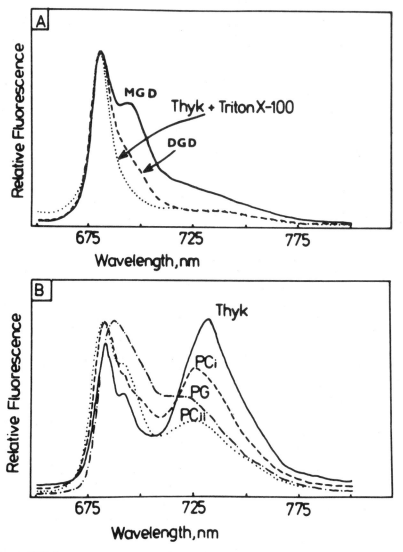

Fig. 1: 77K fluorescence-emission spectra of (A) solubilised thylakoids compared with solubilised thylakoids subject to detergent-dialysis in the presence of monogalactosyldiacylglycerol (MGD) or digalactosyldiacylglycerol (DGD), (B) intact thylakoids (Thyk) compare with solubilised thylakoids subject to detergent-dialysis in the presence of phosphatidylglycerol (PG), pure β-oleoyl, ζ-palmitoyl phosphatidylcholine (PCi) or the less pure phosphatidylcholine Sigma Type IIS preparation (PCii).

Spectroscopic Methods

A number of spectroscopic techniques are available for studying the dynamics of lipid–protein interactions. The most commonly used methods are electron spin resonance (ESR), nuclear magnetic resonance (NMR) and fluorescence spectroscopy. The useful timescale of ESR and fluorescence spectroscopy is about 3×10^{-8}S whereas that of NMR is 10^{-4}S. The slower timescale of NMR spectroscopy results in spectral averaging of the signals due to lipid–protein in interactions with those due to lateral diffusion of lipids in fluid bilayer regions. This limits its usefulness for the study of lipid–protein interactions. Fluorescence spectroscopy can be usefully employed to study lipid–protein interactions, even at relatively high temperatures ($>40°$), but has the drawback that the relatively bulky probe moiety may itself perturb the membrane environment. With ESR spectroscopy, a relatively small probe moiety renders the technique less likely to generate artefactual perturbations and this has been the method routinely used in our investigations.

Electron Spin Resonance Spectroscopy

This technique involves the specific labelling of part of a lipid molecule with a reporter group sensitive to molecular motion on an appropriate time scale. In most ESR studies of membranes, one of the acyl chains of a polar lipid bears a paramagnetic probe and these 'spin-labels' are then introduced into the chosen membrane. The difference in mobility between lipids in the bulk bilayer phase and those adjacent to protein surfaces will give rise to different ESR spectra for spin-labels in these two regions. The spin-label ESR technique is well suited for the study of lipid–protein interactions in membranes because the time scale of the interactions is of the same order as the maximum sensitivity of the spectrum to molecular motion, i.e. approximately 10^{-8}S.

A criticism levelled against the spin-label method is the potential perturbing effect of the nitroxide probe. The mobility of the spin probe may be less than for the parent lipid and also there could be specific interactions between the spin probe and membrane proteins. These criticisms have been addressed in recent reviews (Marsh 1983, 1985) and the conclusion has been drawn that such perturbating effects are minor.

ESR spectra from spin-labelling of biological membranes are always two-component (21). The two components arise from the fluid bilayer lipid and the motionally-restricted lipid in the vicinity of a protein. Single component spectra can be obtained from the spin-label in lipid only (for the fluid bilayer component) and from a completely immobilised spin probe (for the motionally restricted component). The proportion of the two components in a biological membrane can then be determined by computeraided spectral subtraction methods. This simple method can only be used when the exchange rate between bilayer and motionally-restricted lipid is appreciably slower than the time scale of the ESR technique, i.e. $< 3 \times 10^{-8}$S. The lipid exchange rate has been estimated to be $< 10^->$ at 20°C (22), which indicates that our spectral subtraction procedure is valid.

ESR Studies on Thylakoid Membranes

There have been very few published reports of spin-label studies on complex biological membranes. We have reported previously on the high degree of motional restriction in thylakoid membranes (23) and have now extended our observations to the membrane sub-fractions as well as to different plant species and physiological conditions.

The lipid composition of thylakoid membranes is 70–75% by weight galactolipids, 10–15% phosphatidyl glycerol and 0–5% phosphatidylcholine (6,8). For any systematic spin-label study of thylakoid membranes, it is essential to use spin probes which reflect the behaviour of all membrane constituents. For this reason, we have used both galactolipid and phospholipid spin-labels in our studies.

Selectivity by membrane proteins for certain lipid classes has been observed in several model and biological membranes (21). In Figure 2, it can be seen that there is a similar selectivity by thylakoid membrane proteins. The extent of lipid motional restriction can be seen qualitatively in Figure 2 as a variation in the proportion of the immobilised spectral component. This can be quantified according to the method described by Marsh (21). The results show that, for each lipid class, the proportion of motionally restricted lipid in freshly prepared thylakoid membranes is as follows: 14-PGSL, 43%; 14-PCSL, 31%; 12-PCSL, 34%; 12-MGDSL, 26%. This indicates the following selectivity of thylakoid protein for lipid: PG < PC < MGD.

Selectivity could arise from differences in the relative binding constants for the different classes of polar lipid or from differences in the number of binding sites for the different lipids. It has been argued that for model and biological membranes containing a single protein species, specificity is probably due to differences in binding strength (24). However, the diversity of proteins present in the thylakoid membrane prevents such a conclusion being drawn on the origin of selectivity in this case.

The interaction of lipid spin probes with different thylakoid subfractions is shown in Figure 3. This demonstrates that, in the absence of chlorophyll and proteins, there is little motional restriction. The inclusion of chlorophyll at the physiological polar lipid chlorophyll ratio causes the appearance of some motional restriction in the spectrum. This may indicate an interaction between lipid acyl chains and the phytyl chain of chlorophyll as suggested previously (6). A greater degree of motional restriction is seen in thylakoids where, as discussed above, 26–43% of a given lipid is motionally restricted. This obviously results from lipid–protein interactions. Even more extensive motional restriction is seen (Figure 3) with PS 2 fractions prepared using the detergent procedure (25). Spectral subtraction shows that with 14-PGSL, 86% of the spin–label is motionally restricted compared with 75% for the corresponding PC and MGD spin–labels. The data imply extensive lipid–protein interactions in

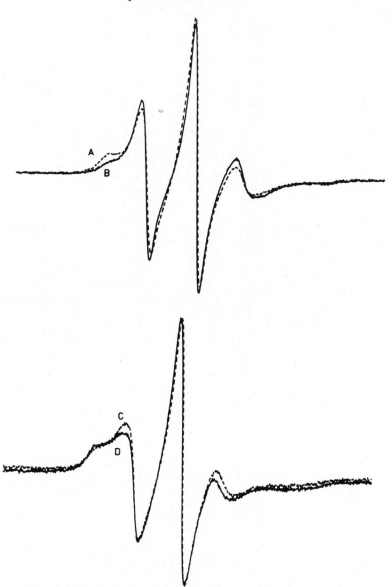

Fig. 2: ESR spectra of various acy lipid spin-labels incorporated into pea leaf thylakoids. (A) Phosphatidylglycerol spin-label, 14-PGSL, – – – –, (B) phosphatidylcholine spin-labels, 14 PCSL, – – – – (C) phosphatidylcholine spin-label, 12 PCSL – – – – – (D) monogalactosy-Idiacylglycerol spin-label, 12-MGDSL, – – – – . The numbers refer to the acyl carbon atom to which the spin-label is attached. Spectra were recorded at 20°C.

PS 2 preparations. These preparations, which are active with regard to PS 2 electron transfer and water splitting, are composed of inverted bithyla-koids and are depleted in polar lipids (3,6,26). It is probable that the low

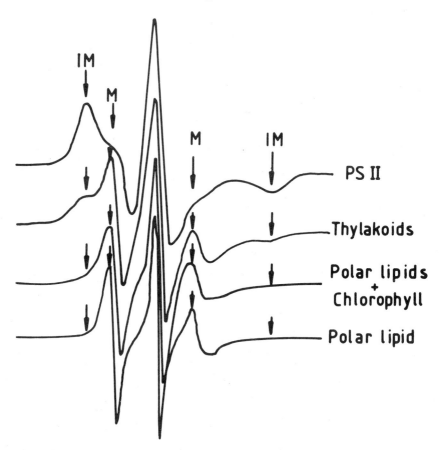

Fig. 3: ESR spectra of the phosphatidylglycerol spin-label 14-PGSL incorporated respectively into photosystem II preparations (PSII), intact thylakoids, vesicles of total thylakoid lipid + pigment extracts and vesicles of total thylakoid polar lipids. Spectra were recorded at 20°C.

Note the increasing magnitude of the immobile spectral components (IM) and decreasing magnitude of the mobile components (M) from the bottom to the top spectra.

lipid: protein ratio in these membranes gives rise to the extensive motional restriction (6).

Lipid mobility in PS 2 preparations can be studied further using saturation-transfer ESR techniques. ST-ESR extends the useful time scale of ESR to $> 10^{-4}$S and can be used to determine the rate of motion of lipids in the motionally restricted environment (27).

The rotational correlation time (Tc) deduced from the ST-ESR spectra is inversely proportional to the rate of rotation of the spin probe. For the PS 2 preparations, the following values for Tc at 20°C have been calculated: 14PGSL, 1.5×10^{-5}S; 14-PCSL, 3.6×10^{-5}S; 12-PCSL, 2.8×10^{-5}S; 12-MGDSL, 1.5×10^{-5}S. It can be seen that Tc values for PCSL are greater than for corresponding PG or MGD spin-labels.

These data are potentially important since they indicate that the different spin-labels bind at different sites, the site for PCSL binding being the most motionally restricted. The most attractive explanation for these effects is that PCSL is associated with a larger protein complex than either PGSL or MGDGSL and therefore tumbles more slowly. However, location of the spin-labels between protein aggregates or in gel-phase lipid pools cannot be ruled out as alternative explanations.

The above ESR results were obtained using pea membranes. We have also used other crop plants in order to study species-specific differences in membrane organisation. In Fig. 4, the ESR spectra of thylakoids labelled

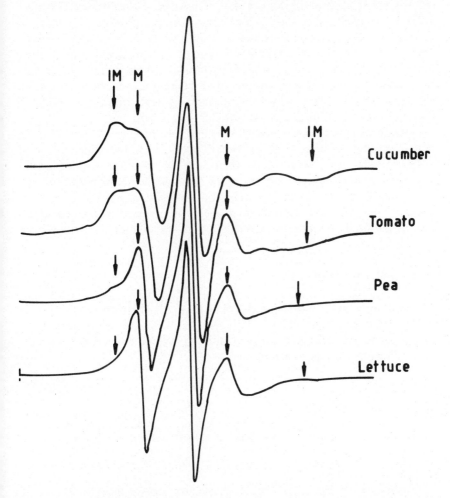

Fig. 4: ESR spectra of the galactolipid spin-label 12-MGDSL incorporated into thylakoid membranes of different plants. Spectra were recorded at 28°C.

Note the increasing magnitude of the immobile spectral components (IM) from the most (lettuce) to the least (cucumber) chilling-resistant species.

with 14 PGSL and 14 MGDSL are shown for pea, tomato, cucumber and lettuce. There is a clear difference in the extent of lipid motion restriction between the different species with cucumber > tomato > pea > lettuce. The extent of lipid motion restriction is correlated with the susceptibility of the plant to chilling injury. However, compositional data suggest that this is a secondary effect since the more highly motionally restricted species also have the lowest lipid:protein ratios and the most extensive granal stacking.

In general, there are two conclusions that can be drawn from our studies: (i) the extent of motion-restriction of a lipid spin-probe in a membrane reflects the lipid–protein ratio of that membrane. Hence there is little or no motion-restriction in the absence of protein, but progressively more in protein-rich thylakoid membranes and lipid-depleted PSII preparations. This is also true for different plant species with different lipid:protein ratios, (ii) the extent of motion-restriction of a lipid spin-probe is also influenced by the identity of the lipid headgroup. In both thylakoid membranes and PSII enriched membrane fragments there is a degree of specificity in the interaction of proteins and lipids.

Despite its complexities in terms of lipid and protein composition, we have been able to apply ESR and other techniques to probe the nature of lipid–protein interactions. It is important to elucidate the molecular dynamics of such membranes in order to understand them fully at the functional level. Membrane proteins do not exist alone and both their structure and function can only be properly appreciated by the consideration of the lipid populations with which they are so intimately associated.

Acknowledgements

We gratefully thank Dr D. Marsh and Dr L. J. Horvath for the use of STESR facilities and valuable discussions. G. Li is supported by a postgraduate studentship from the Chinese Government. DJM thanks the Royal Society for its assistance.

References

1. Colbeau, A., Nachbaur, J. & Vignais, P. M. (1971) *Biochim. Biophys.* Acta 299, 462–492
2. Burnell, E., Van Alphen, L., Verkheij, A. & De Kruijff, B. (1980) *Biochim. Biophys.* Acta 597, 492–501
3. Murphy, D. J. & Woodrow, I. E. (1983) *Biochim. Biophys.* Acta 725 104–112
4. Kushwaha, S. C., Kates, M. & Martin, W. G. (1975) *Can. J. Bot.* 53, 284–292
5. Murphy, D. J. & Frinsley, R. T. (1985) *Biochem. J.* 229, 31–37
6. Murphy, D. J. (1986) *Biochim. Biophys.* Acta 864, 33–94
7. Sanderman, H. (1978) *Biochim. Biophys.* Acta 515, 209–237

8. Murphy, D. J. (1986) in *Encyclopedia of Plant Physiology*, New Series, Vol 19 (Staehelin, L. A. & Anrtzen, C. J. ed) 713–725, Springer, Berlin
9. Ryrie, I. J. (1986) *Encyclopedia of Plant Physiology*, New Series, Vol 19 (Staehelin, L. A. & Anrtzen, C. J. eds) 675–682, Springer, Berlin
10. Siefermann-Harms, D., Ross, J. W., Kaneshiro, K. H. & Yamamoto, H. Y. (1982) *FEBS Lett.* 149, 191–196
11. Siefermann-Harms, D., Ninnemann, H., Ross, J. W. & Yamamoto, H. Y. (1984) in *Advances in Photosynthesis Research*, Vol 1 (Sybesma, C. ed) 741–744, Nijhoff/Junk, Amsterdam
12. Murphy, D. J. (1984) in *Structure, Function & Metabolism of Plant Lipids* (Siegenthaler, P. A. & Eichenberger, W. eds) 465–470, Elsevier, Amsterdam.
13. Murphy, D. J. (1985) *Photosynth. Res.* 8, 219–233.
14. Chain, R. K. (1985) *FEBS Lett.* 180, 321–325
15. Gounaris, K., Whitford, D. & Barber, J. (1983) *FEBS Lett.* 163, 230–234
16. Matsuda, H. & Butler, W. L. (1983) *Biochim. Biophys* Acta 724, 123–127
17. Matsuda, H. & Butler, W. L. (1983) *Biochim. Biophys* Acta 725, 320–324
18. Murphy, D. J., Crowther, D. & Woodrow, I. E. (1984) *FEBS Lett.* 165, 151–155
19. Douce, R. & Joyard, J. (1980) in *The Biochemistry of Plants*, Vol. 4 (Stumpf, P. K. & Conn, E. E. eds) 321–363, Academic Press, New York
20. Devaux, P. F. & Seigneuret, M. (1985) *Biochim. Biophys.* Acta 822, 63–125.
21. Marsh, D. (1985) in *Progress in Protein–Lipid Interactions* (Watts, A. & De Pont eds) 143–172, Elsevier, Amsterdam
22. Marsh, D., Watts, A., Pates, R. D., Uhl, R., Knowles, P. F. & Esmann, M. (1982) *Biophys. J.* 37, 265–274.
23. Murphy & Knowles (1984) in *Structure, Function and Metabolism of Plant Lipids* (Siegenthaler, P. A. & Eichenberger, W. eds) 425–428, Elsevier, Amsterdam
24. Knowles, P. F., Watts, A. & Marsh, D. (1981) *Biochemistry* 21, 5888–5893
25. Berthold, D. A. Babcock, G. T. & Yocum, C. F. (1981) *FEBS Lett.* 134, 231–234
26. Dunahay *et al.* (1984) *BBA* 764, 179–193
27. Horvarth, L. I. & Marsh, D. (1983) *J. Mag. Res.* 54, 363–369

Introduction to Triplet-minus-Singlet Absorbance Difference Spectroscopy Monitored with Absorbance-detected Magnetic Resonance

A. J. HOFF

Department of Biophysics,
Huygens Laboratory of the State University P. O. Box 9504,
2300 RA Leiden (The Netherlands)

Summary

The principle of triplet-minus-singlet (T–S) spectroscopy monitored via absorbance-detected magnetic resonance (ADMR) is briefly introduced. An extension to the recording of linear dichroic T–S spectra is described. In the companion research paper the method is applied to *Rhodobacter sphaeroides* R26.
Key words: ADMR, triplet-minus-singlet spectroscopy.

1. Why T–S Spectroscopy

The conversion of the photon energy of light into free chemical energy by plants and certain bacteria is a process of enormous complexity spanning a time range of some ten decades (10^{-13}–10^{-3}s). After capture of a photon by antenna pigments, the excitation is shuttled to a photochemical trap, the reaction centre. Here, the excitation energy is within a few picoseconds, converted into separated charges, which are stabilised by dark electron transport in several steps lasting a few hundred picoseconds to microseconds. This fundamental process takes place in a complicated array of pigments (cofactors), of which the detailed configuration is only now emerging form X-ray crystallography (1). In spite of the structural knowledge so obtained, to understand the intricacies of the charge separation and electron transport reactions one needs various forms of spectroscopy, foremost among which is kinetic optical spectroscopy.

The interpretation of (laser) flash spectroscopy necessitates a detailed unravelling of the optical absorption spectra of the reaction centre. One way to do this is to perturb the pigment constellation in the reaction centre

---

Abbreviations: T–S, triplet-minus-singlet; LD, linear dichroism; ADMR, absorbance-detected magnetic resonance.

and to study its effects on the absorbance spectrum. Most often this has been done by oxidation or reduction of selected cofactors, either by light or by chemical agents. These procedures generate charged species, which affect the optical properties of the neighbouring pigments by their electric fields (the Stark effect), thus considerably influencing the spectrum in a way that is only partially understood.

A more gentle perturbation of the array of electron transport components is the generation of a triplet state. Just as for oxidised or reduced cofactors, the electronic structure of a triplet state is different from that of the unperturbed ground state, but it does not carry a charge and is therefore free from the complications arising from the Stark effect.

In bacterial and plant reaction centres the triplet state of the primary donor is easily generated by blocking forward electron transport by reducing or deleting the secondary (quinone) acceptor. Charge separation still takes place but the only decay channel of the charges is now that of recombination, which at low temperatures generates almost exclusively the primary donor triplet state, ^3P. This state has at cryogenic temperatures a lifetime of about 0.1 to 1 ms, depending on the photosystem studied. Absorbance difference spectroscopy can now be performed by subtracting the singlet ground state 'normal' absorption spectrum from that of the reaction centre containing the triplet state. In the past this has been done by generating ^3P by a saturating flash of light and quickly measuring the flash-induced difference in absorption at a number of wavelengths across the absorption spectrum (2-4). This method is not very sensitive, and therefore rather time consuming. Recently, a completely different way of recording triplet-minus-singlet (T−S) absorbance difference spectra has been developed (5). Here, the steady state triplet concentration that is generated by continuous illumination is changed through electron paramagnetic resonance by a resonant microwave field. The concomitant change in the optical absorption spectrum is then recorded as a function of wavelength. A full T−S spectrum takes only a few minutes to complete, and the method allows simultaneous recording of the linear-dichroic (LD) T−S spectrum.

In the following, a brief explanation will be given of the new technique. Its use for the investigation of photosynthetic reaction centres will be illustrated by a study of reaction centres of the photosynthetic bacterium *Rhodobacter sphaeroides* R26, both native and modified by borohydride treatment.

2. Physics of the Triplet State

Normally, aromatic molecules are in the singlet ground state, where all electrons are paired. What does that mean? All electrons possess a magnetic angular momentum, called 'spin'; they can be viewed as little bar magnets with a strength of $\frac{1}{2}$ in the appropriate units. The bar magnet can have two different positions in space, 'up' and 'down'. Every electronic orbital can accomodate two and not more than two electrons, one with spin up, one with spin down (the so-called Pauli principle). For every orbital the sum of the magnetic angular momentum of the two electrons is then zero (compare the field of two bar magnets that are lying side-to-side, the north pole of one next to the south pole of the other). The total spin angular momentum of the molecule, which we shall label S, is then also zero: S=0. This is called a singlet state.

If one now photoexcites the molecule (for simplicity we will only consider aromatic molecules) one electron from the highest occupied molecular (electronic) orbital, the HOMO, is excited to the lowest unoccupied molecular orbital, the LUMO. Although the two electrons in the HOMO and LUMO are now in different orbitals, and therefore need not obey the Pauli principle, they are still paired because of the law of conservation of angular momentum. Thus, the molecule is still in a singlet state. During the lifetime of this state, there is a finite probability that the electron in one of the orbitals inverts its spin, from up to down or vice versa. Spin angular momentum is then borrowed from, or taken up by, the orbital angular momentum of the electrons. The result is that now the total spin angular momentum S does not sum to zero but to $S=\frac{1}{2}+\frac{1}{2}=1$. The molecule has now two unpaired electrons; it is paramagnetic.

The above process of spin inversion is called Intersystem Crossing (ISC), its efficiency is very much dependent on molecular properties. Another process by which a spin may invert is radical recombination. In photosynthetic reaction centres charge separation limits the lifetime of the excited state of the primary donor to less than 10 ps. This is much too short for ISC to occur. During the lifetime of the separated charges is blocked reaction centres, the two spins may become unpaired (total spin S=1) and upon recombination the triplet state ^3P is formed.

One can show, by applying the laws of quantum mechanics, that a molecule with spin S has 2S + 1 magnetic energy levels. In a spherical symmetric environment these levels have the same energy (they are 'degenerate') but in an applied magnetic field they split into 2S + 1 distinct levels separated by the so-called Zeeman energy. (In a magnetic field one has to apply force to a bar magnet to turn it from an antiparallel to a parallel position, hence these two positions then differ in energy.) A singlet state, with S = 0 (all electrons paired), has only one magnetic energy level. When there are two unpaired electrons, however, S = 1, the molecule has three magnetic energy levels and is in the (excited) triplet state.

If the molecule is not spherically symmetric (and almost all molecules are not) then even without external magnetic field the three triplet levels do not have the same energy. They are non-degenerate because of the magnetic dipole–dipole interaction between the two tiny electronic bar magnets. The energy separations between the three levels, commonly labelled x, y, z, depend on the strength and the symmetry of the dipole–dipole interaction. For rhombic symmetry all levels have different energy. The energy separation between any two levels is given by $|D| + |E|$, $|D| - |E|$ and $2|E|$ where D and E are two parameters characterising the dipole–dipole interaction ($|D| \geqslant 3\,|E|$). For axial symmetry, $E = 0$ and two levels (conventionally the x and y levels) are degenerate. For planar molecules, e.g. chlorophylls, one expects D>0, whereas the sign and magnitude of E depends on the symmetry in the plane of the molecule. In Fig. 1a the three triplet energy levels are depicted for D>0 and E<0.

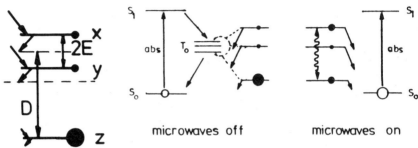

microwaves off   microwaves on

Fig. 1a: The triplet sublevels in zero magnetic field. Arrows pointing to the right: population probabilities, to the left: decay rates.

Fig. 1b: Principle of absorbance-detected magnetic resonance (ADMR). Filled circles denote relative equilibrium populations of the triplet sublevels, open circles that of the singlet ground state. Corrugated arrow denotes microwave transition.

Under equilibrium conditions, e.g. continuous illumination of a sample of aromatic molecules, and at low temperatures, the relative population of the three triplet sublevels n_x, n_y, n_z, will depend on the population probabilities p_x, p_y, p_z and the decay rates k_x, k_y, k_z via $n_i = p_i/k_i$ ($i = x,y,z$). Generally the p_i's and k_i's are not equal, and the equilibrium populations of the three levels differ from each other. At high temperatures, spin-lattice relaxation will distribute the populations over the three sublevels according to the Boltzmann distribution. For ^3P formed by recombination in reaction centres, the p_i's are almost equal, and $k_{x,y} \gg k_z$, making $n_z \gg n_x$, n_y.

If two sublevels have different equilibrium populations one may transfer population from one level to the other, less populated, level by applying an electro-magnetic field whose photon energy is resonant with the energy separation between the two levels. This is called electron paramagnetic resonance (EPR), because the magnetic vector of the field acts on the magnetic moment of the electrons and elicits a transition between magnetic energy levels. It is the analogon of optical absorption spectroscopy, where the electric vector of the field acts on the electronic charge. Normally, EPR is carried out on molecules that have, for one reason or another, only *one* unpaired electron ($S=\frac{1}{2}$, two magnetic energy levels, a 'doublet' state). One then needs an external magnetic field to split the levels and to generate unequal populations. For the triplet state, however, an external magnetic field is not needed, as the energy levels are almost always already split by the dipolar interaction. Thus, we can do an EPR experiment in zero field, for which the three possible transitions lie in the microwave range. Often, it is advantageous to detect the transition of population from one level to another not by detecting energy absorption from the electromagnetic field, but by monitoring an optical characteristic of the molecule. This is called optically detected magnetic resonance (ODMR) to which we now turn.

3. Absorbance Detected Electron Paramagnetic Resonance (ADMR) of Triplet States

ADMR is a variant of Optically Detected Magnetic Resonance (ODMR) of triplet states, reviewed in (6). The equilibrium population n_i of two sublevels of a triplet state generated by cw illumination is altered by the application of a resonant microwave field. If the molecular decay rates of these two sublevels are unequal, this results in a change in the overall equilibrium triplet concentration because the microwave field opens up a new decay channel for the slower decaying sublevel. This change in the triplet concentration entails a change of equal magnitude and of opposite sign in the singlet ground state population. This translates in a change in the fluorescence ($S_1 \leftarrow S_o$) or in the absorbance (both $S_n \leftarrow S_o$ and $T_n \leftarrow T_o$ transitions) of the sample (Fig. 1). The change in fluorescence, $\triangle F$, or absorbance, $\triangle A$, can be used to trace an ODMR spectrum by plotting $\triangle F$ or $\triangle A$ at a particular wavelength against the frequency of the microwave field. Such a spectrum is similar to an EPR spectrum. Conversely, keeping this frequency constant at a resonance value, one may plot $\triangle F$ or $\triangle A$ versus the detection wavelength, resulting in microwave induced ffluorescence (MIF) or triplet-minus-singlet (T–S) absorbance difference spectra (7).

4. Triplet-minus-Singlet Spectra Monitored by ADMR

The recording of MIF or T–S spectra is facilitated by modulating the amplitude of the microwave field at a low frequency (about 330 Hz) and detecting $\triangle F$ or $\triangle A$ with a lock-in detector. Simply scanning the wavelength via a monochromator and recording the lock-in output then immediately gives the difference spectrum. The experiment has to be carried out at low temperatures, commonly below 2.1 K, although recently higher temperatures, up to 50 K, have also been used (8, 9 and E. J. Lous, unpublished experiments). The low temperature ensures that spin relaxation, which competes with the microwave absorption, is inhibited.

5. Linear Dichroic T−S Spectra

Linear dichroic T−S spectra are obtained by inserting a photoelastic modulator (PEM) and a polarisation analyser between sample and detector. The impinging unpolarised (horizontal) light beam is elliptically polarised in the sample by the directionality of the microwave field in a specially designed cavity (10). In the randomly oriented sample only those triplet molecules are affected by the (vertical) microwave field that have their microwave transition moment (close to) parallel to the field direction. This means that the distribution of triplet states is made anisotropic by the resonant microwave field, so that in directions parallel and perpendicular to this field the optical absorption will be slightly different (assuming that at most wavelengths the triplet state's absorbance is different from that of the singlet ground state). The ellipticity of the transmitted light is converted by the PEM and the analyser into an intensity modulation at a frequency of 100 kHz. This modulation is lock-in detected via the same photodiode as the 330 Hz intensity modulation due to the amplitude modulation of the microwave field. Thus, by the double modulation and demodulation technique, T−S and linear dichroic T−S spectra are simultaneously recorded. (Note that the LD-(T−S) spectrum is really the difference between the T−S spectrum measured with the analyser parallel and perpendicular to the direction of the microwave field; it is a (T−S)″−(T−S), or △(T−S)″ spectrum.) A special calibration method (11) permits one to calibrate the amplitude of the LD-(T−S) spectrum relative to that of the T−S spectrum.

6. Information on Orientation

From the above it will be clear that the ellipticity of the transmitted light, i.e. the amplitude of the LD-(T−S) spectrum, at a particular wave-length λ depends on the angle α that the optical transition moment at λ makes with the direction of the microwave transition moment of the triplet state. The position of the latter in the molecular coordinate frame is usually not, or only approximately, known. One can, however, derive that the moments of the three possible ODMR transitions between two out of the three triplet sublevels are mutually perpendicularly oriented (they span a Cartesian coordinate frame). Usually, molecular symmetry considerations will give some idea of the orientation of the triplet transition moments in the molecule. More precise information is obtained from a simulation of the LD-(T–) spectra recorded for two different ADMR transitions (that are per force perpendicular to each other).

For an isolated absorption band at wavelength λ the angle α defined above follows from the relative amplitude of the T–S and LD-(T–S) spectrum (11–13):

$$R_i = \left| \frac{\text{LD}-(\text{T}-\text{S})}{\text{T}-\text{S}} \right|^i = \frac{3 \cos^2\alpha_i - 1}{3 + \cos^2\alpha_i} \tag{1}$$

where i refers to one of the possible ADMR transitions (usually labelled x, y or z. The x transition is between the y and z sublevel, and so on). It follows that from two ratios R, e.g. R_x and R_y, α_λ can be unambiguously determined in the x,y,z frame. For different isolated absorption bands, this then also gives the mutual angles between their respective transition moments. Unfortunately, most often the absorption bands in the T−S and LD-(T−S) spectra overlap considerably and the sought-for information must be obtained by computer simulation of the spectra, varying stepwise the angles α and the orientation of the x,y,z frame relative to the molecular frame.

Another complication arises when there are aggregates of molecules present, each absorbing in the same wavelength region. This is for example the case for bacterial photosynthetic reaction centres, which (for purple bacteria) comprise two BChl's making up the primary electron donor, two accessory BChl's, and two BPh's. The absorbance transition moments of all these pigments will electrostatically interact, the so-called exciton interaction, which gives rise to considerable band shifts and changes of intensity and polarisation relative to the *in vitro*, isolated monomer values. The simulation of the T−S and LD-(T−S) spectra then involves the use of an appropriate exciton model [11,14]. Fortunately, visual comparison of T−S and LD-(T−S) spectra often gives already valuable information on the pigment configuration, without extensive computer simulations.

7. Instrumentation

A block diagram of an ADMR LD-(T−S) spectrometer is shown in Fig. 2 (10, 13). Essentially, the instrument consists of a single-beam absorption spectrophotometer. The light source is a tungsten−iodine lamp which serves both as excitation and as probe light. The sample is located inside a helium bath cryostat in which helium under reduced pressure is in the superfluid state (below 2.1 K). Surrounding the sample is a split-ring cavity (15) which contains the vertically polarised amplitude-modulated microwave field (frequencies ranging between a few hundred MHz to one GHz).

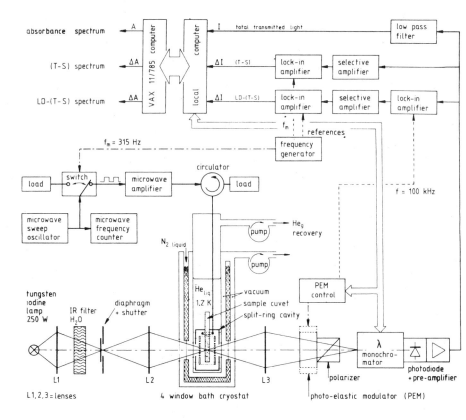

Fig. 2: Block schematic of the LD-(T−S) set-up.
Courtesy Dr. E. J. Lous.

The sample consists of reaction centres (or photomembrane particles) in a glassy bufffer-ethyleneglycol mixture, and is illuminated by white light. Behind the sample and cryostat is a PEM and an analyser, which together give rise to a 100 kHz intensity modulation of the light that is superimposed on the ~330 Hz modulation effected by the microwave field. After passing a monochromator, these modulations $\triangle I$ are detected by a Si photodiode and demodulated in three lock-in amplifiers. One is tuned to 330 Hz; its output gives the T−S spectrum. The other two are tuned to 330 Hz and 100 kHz, respectively, and give the LD-(T−S) spectrum. Both spectra are recorded simultaneously. To correct for the variation in lamp output, monochromator throughput and detector efficiency as a function of wavelength the $\triangle I$ signals are divided by the intensity I falling on the photodiode. The value of $\triangle I/I$ is low, of the order of 10^{-4} to 10^{-6}. The T−S and LD-(T−S) detection channels are calibrated relative to each other with a LED at the place of the sample. Data collection, monochromator scan and correction procedures are all automated via a dedicated mirocomputer. Spectra are stored on tape and can be read into a VAX laboratory computer for further analysis.

8. Applications to Photosynthetic Reaction Centres

T–S and LD-(T–S) spectroscopy monitored via ADMR have been applied to a variety of reaction centres from plant and bacterial origin (7,10,11,16–26), reviewed in (7,13,27,28). Simulations of T–S and LD-(T–S) spectra with exciton theory have been carried out in (11,14). Detailed information concerning the geometrical arrangement of reaction centre transition moments and triplet axes has been extracted from the spectra. The richness of these data makes T–S spectroscopy eminently suited to monitor differences in cofactor configuration betwcen reaction centres of the plant photosystems and those of bacterial origin, and between the reaction centres of different bacterial species. As an illustration, in the companion paper (29) T–S and LD-(T–S) spectra of native reaction centres of *Rb. sphaeroides* R-26 are compared with the spectra of modified reaction centres, which were treated with sodiumborohydride to remove one of the accessory bacteriochlorophylls. It is anticipated that T–S spectroscopy will be an important tool to study the changes brought about by site-directed mutagenesis (this volume, chapters oooo).

Acknowledgements

The author is indebted to Drs H. J. den Blanken, E J. Lous, R. F. Meiburg and H. Vasmel for their enthusiasm and innovative contributions when working as graduate students in the Department of Biophysics in Leiden. This work was supported by the Netherlands Foundation for Chemical Research (SON), financed by the Netherlands Organization for Scientific Research (NWO).

References

1. Deisenhofer, J., Epp, O., Miki, K., Huber, R. and Michel, H. (1984) *J. Mol. Biol.* 180, 385–395
2. Cogdell, R. J., Monger, T. G. and Parson, W. W. (1975) *Biochim. Biophys.* Acta 408, 189–199
3. Shuvalov, V. A. and Parson, W. W. (1981) *Proc. Natl Acad. Sci. USA* 78, 957–961
4. Shuvalov, V. A. and Parson, W. W. (1981) *Biochim. Biophys.* Acta 638, 50–59
5. Den Blanken, H. J. and Hoff, A. J. (1982) *Biochim. Biophys.* Acta 681, 365–374
6. Clarke, R. H. (ed.) (1982) *Triplet State ODMR Spectroscopy*, John Wiley Inc., New York

7. Hoff, A. J. (1986) in: *Photosynthesis* III. *Photosynthetic Membranes and Light Harvesting Systems* (Staehelin, L. A. and Arntzen, C. J., eds.), Vol. 19 *Encyclopedia of Plant Physiology*, New Series (Pirson, A. and Zimmerman, M. H., eds.), pp. 400–421, Springer-Verlag, Berlin
8. Ullrich, J., Angerhofer, A., von Schütz, J. U. and Wolf, H. C. (1987) *Chem. Phys. Lett.* 140, 416–420
9. Lous, E. J. and Hoff, A. J. (1988) in: *Studies of the Photosynthetic Reaction Center* (Breton, J. and Vermeglio, A., eds.) in the press
10. Den Blanken, H. J., Meiburg, R. F and Hoff, A. J. (1984) *Chem. Phys. Lett.* 105, 336–342
11. Lous, E. J. and Hoff, A. J. (1987) *Proc. Natl Acad. Sci. USA* 84, 6147–6151
12. Meiburg, R. F. (1985) Doctoral thesis, University of Leiden
13. Hoff, A J. (1985) in: *Antennas and Reaction Centers of Photosynthetic Bacteria. Structure, Interaction and Dynamics* (Michel-Beyerle, M. E., ed.), pp. 150–163, Springer-Verlag, Berlin
14. Knapp, E. W., Scherer, P. O. J. and Fischer, S. F. (1986) *Biochim. Biophys. Acta* 852, 295–305
15. Hardy, W. N. and Whitehead, L. A. (1981) *Rev. Sci. Instrum.* 57, 213–216
16. Den Blanken, H. J., van der Zwet, G. P. and Hoff, A. J. (1982) *Chem. Phys. Lett.* 85, 335–338
17. Den Blanken, H. J. and Hoff, A. J. (1983) *Biochim. Biophys. Acta* 724, 52–61
18. Den Blanken, H. J., Hoff, A. J., Jongenelis, A. P. J. M. and Diner, B. A. (1983) *FEBS Lett.* 157, 21–27
19. Den Blanken, H. J. and Hoff, A. J. (1983) *Chem. Phys. Lett.* 98,. 255–262
20. Den Blanken, H. J., Vasmel, H., Jongenelis, A. P. J. M., Hoff, A. J. and Amesz, J. (1983) *FEBS Lett.* 161, 185–189
21. Den Blanken, H. J., Jongenelis, A. P. J. M. and Hoff, A. J. (1983) *Biochim. Biophys. Acta* 725, 472–482
22. Vasmel, H., den Blanken, H. J., Dijkman, J. A., Hoff, A. J. and Amesz, J. (1984) *Biochim. Biophys.* Acta 767, 200–208
23. Hoff, A. J., den Blanken, H. J., Vasmel, H. and Meiburg, R. F. (1985) *Biochim. Biophys.* Acta 806, 389–397
24. Lous, E. J. and Hoff, A. J. (1987) in: *Progress in Photosynthesis Research* (Biggins, J., ed.), Vol. 1, pp. 399–402, Martinus Nijhoff Publishers, Dordrecht
25. Lous, E. J. and Hoff, A. J. (1987) *Chem. Phys. Lett.* 140, 620–625
26. Beese, D., Steiner, R., Scheer, H., Angerhofer, A., Robert, B. and Lutz, M. (1988) *Photochem. Photobiol.* 47, 293–304
27. Hoff, A. J. (1985) in: *Optical Properties and Structure of Tetrapyrroles* (Blauer, G. and Sund, H., eds.), pp. 453–473, Walter de Gruyter & Co., Berlin
28. Hoff, A. J. (1986) in: *Light Emission by Plants and Bacteria* (Govindjee, Amesz, J. and Fork, D. C., eds.), pp. 225–265, Academic Press, Orlando, FL (USA)
29. Angerhofer, A., Beese, D., Hoff, A. J., Lous, E. J. and Scheer, this volume, pp. 197–204

Photosystem I Particles from Sorghum: Isolation and Characterisation of Chlorophyll-Protein Complexes under Differing Growth Light Regimes

A. AMRUTHAVALLI, A. RAMACHANDRA REDDY AND
V. S. RAMA DAS

Department of Botany,
Sri Venkateswara University,
Tirupati 517502, INDIA

Summary

Photosystem I (PS I) particles were isolated from sorghum thylakoids by Triton-X-100 solubilisation. The purified photosystem I particles possessed a chlorophyll/P700 ratio of 180 and a chlorophyll a/b ratio of 10. The polypeptide composition of the isolated PS I complex revealed a dominant 68kD polypeptide, four polypeptides in the range of 18-25kD associated with LHC-I and six minor polypeptides in the range of 8-24kD. The chlorophyll-protein (chl-protein) complexes of PS I were isolated from the purified particles by centrifugation in SDS sucrose density gradients. The complexes were designated as CP I, CP Ia and LHC-I. Sucrose density gradient ultracentrifugation resulted in two chlorophyll protein complexes associated with LHC-I-730 which were identified by their fluorescence emission spectra. The changes in the structural organisation of PS I complex under varying levels of growth irradiance were investigated. The CP I complex remained stable while the proportion of CP Ia and LHC-I increased under low growth light. The PS I complex exhibited two fluorescence peaks at 730 and 688 nm associated with core antenna of reaction centre and LHC-I respectively. The enhanced F 688 indicates the increased size of LHC-I. Further, the two complexes of LHC-I, namely LHC-I-730 (inner antenna) and LHC-I-685 (peripheral antenna), exhibited marked variations in their fluorescence characteristics. The results suggest the indispensable role of the inner and peripheral antenna of PS I of sorghum in the efficient harvesting of available photon flux density under light limited conditions.

Introduction

Higher plants respond to the amount of incident light available during growth with specific changes in the composition, structure and function of chloroplast thylakoids (1,2,3). The modulations in chl-protein content resulting from different growth light intensity has been markedly shown to influence the light harvesting assemblies of PS II and PS I (4). An increase in light irradiance during plant growth causes substantial increases in PS II reaction centres and a slight increase in PS I reaction centres on a chlorophyll basis (5,6,7). The Q/P700 ratios in grain sorghum have increased with increased growth light intensity (8).

Although the effects of varying light intensity on the structure and function of thylakoid membranes are well known, very little information is available about the changes in the organisation of individual photosystems. Particularly, there is an obvious lack of information about the changes in the organisation of PS I in the chloroplast membranes as affected by light intensity. Recently, our knowledge on the antenna chl a/b protein of PS I has been considerably advanced (9, 10). Mullet *et al.* (9) have first attributed several polypeptides of 22 to 25 kD present in a 'native' PS I complex to a peripheral antenna. A LHC-I preparation which contains four polypeptides in the 21 to 23.5 kD range has been isolated from PS I-110. However, the changes in the number and structure of chl-protein complexes associated with PS I reaction centre complex as a function of light intensity has not been studied. In this paper an attempt was made to examine the changes in the chl-protein complexes of PS I in the thylakoids of grain sorghum grown at two different light regimes.

Materials and Methods

Growth conditions

Grain *Sorghum (Sorghum bicolor* L. CSH-5) plants were grown in plots of 45 cm width and the seedlings were thinned to approximately 16 plants per sq. m. Seed beds received nitrogen and phosphorous at the rate of 50 kg and 40 kg per hectare respectively. After 30 days another dose of 40 kg nitrogen was applied to the crop. Artificial shading was provided on the experimental plots with wooden frames made up of wooden reapers. The number and size of holes in the frames were altered to provide the low light conditions of 450 μmol m^{-2}s^{-1} (normal sunlight, 2000 μmol m^{-2}s^{-1}). Light intensity was measured with a lambda meter using a quantum sensor (LI-COR/Li 70). Temperatures were 38/28°C day/night with an approximately 12 h photoperiod. Studies were made with the leaves from 45-day-old plants.

Isolation and purification of PS I particles

Photosystem–I particles were isolated and purified according to Mullet *et al.* (9) with slight modifications. The leaves were homogenised in a buffer containing 0.4M sorbitol, 0.05M tricine-KOH pH 7.8, and the slurry was filtered through four layers of cheese cloth and centrifuged at 1,000 × g for 5 min. The pellet was re-suspended in a medium containing 0.05M sorbitol, 5 mM EDTA-NaOH, pH 7.8 and centrifuged at 10,000 × g for 5 min. The pellet was suspended with distilled water to give a chl concentration of 0.8 mg/ml. The de-stacked thylakoids (0.8 mg chl/ml) were solubilised by an addition of 20% triton X-100 to a final concentration of 0.6–0.7% and incubated for 30 min at 20°C with constant stirring. The contents were centrifuged at 42,000 x_g for 30 min in a Beckman VTi 70 rotor. 8 ml of the supernatant was loaded on to the top of 0.1M to 1.0M linear sucrose gradient containing 5 mM tricine-KOH pH 7.8 and 0.02% triton X-100, underlaid with 2M sucrose cushion and centrifuged at 27,000 rpm for 16 h at 4° in a Beckman SW 28.1 rotor.

P700 content

P700 content was determined by difference spectra by using 25 μM $K_3Fe(CN)_6$ and 0.5 mM sodium ascorbate. Samples of equal chl concentration were placed in identical cuvettes. After a base line was recorded, 25 μM $K_3Fe(CN)_6$ was added to sample cuvette and 0.5 mM sodium ascorbate to the reference cuvette and allowed to equilibrate prior to recording the spectra. Reversible absorbance changes were again recorded by re-reducing the oxidised sample. P700 was calculated using an absorption coefficient of 64 mm^{-1} cm^{-1}.

Analysis of PS I complex polypeptides

Analysis of polypeptides of PS I particles by SDS-PAGE was performed in slab-gel apparatus using 7.5–15% gradient polyacrylamide running gel and 5% stacking gel in a discontinuous buffer system as described by Laemmli (11). The submembrane fractions of thylakoids and PS I particles were solubilised in a buffer containing 0.062 M tris-HCl pH 6.8 containing 2% (w/v) SDS, 2% (w/v) β-mercaptoethanol and 10% glycerol. Electrophoresis was carried out at a constant current of 30 mamps for 10–12 h. Gels were scanned at 550 nm in Shimadzu UV-240 spectrophotometer.

Separation of Chl-protein complexes of PS I complex by ND-SDS-PAGE

The chl-protein complexes were analysed under non-denaturing conditions at 4°C (12). The PS I particles were suspended in a buffer containing 0.06 2M tris-HCl pH 6.8, 10% glycerol and 1% SDS to give a final concentration of 2 mg chl/ml. Electrophoresis was carried out immediately after solubilisation in the dark at 4°C for 2 h by using 10% (w/v) acrylamide separating gel and 4% (w/v) stacking gel as described by Henriques and Park (13). Gels were scanned at 675 nm with a Shimadzu UV-240 Spectrophotometer.

Separation of chl protein complexes of PS I complex by sucrose density gradient ultracentrifugation

Chl-protein complexes of PS I were separated by sucrose density gradient ultracentrifugation as described by Bassi and Simpson (10) with some modifications. The dark green band of PS I resolved at the junction of 1M and 2M sucrose was collected with a syringe and dialysed overnight in a buffer containing 50 mM sorbitol, 5 mM tricine-KOH pH 7.8 and centrifuged at 42,000 x_g for 1 h at 4°C to pellet purified PS I particles. PS I particles were solubilised in 150 mM NaCl to give 0.3 mg chl/ml. Triton X-100 and SDS were added in the concentrations to give chl: SDS: Triton X-100 in a ratio of 1:4:1. Aliquots of this solubilised solution were loaded onto 0.1M to 1M sucrose gradient containing 5 mM tricine-KOH pH 7.8 and 0.2% triton X-100 and centrifuged at 27,000 rpm in a Beckman SW 28.1 rotor for 24 h.

Chlorophyll fluorescence measurements

Room temperature chlorophyll fluorescence emission spectra were recorded on a Hitachi-654-IOS spectrofluorometer. Samples were diluted in 50 mM tricine-KOH pH 7.8, 0.3M sucrose to a final concentration of 5 μg chl/ml. Fluorescence emission spectra were recorded by providing excitation at 488 nm with 10 nm band width.

Chlorophyll content was measured by the method of Arnon (14).

Results

Isolated PS I particles from *Sorghum* were characterised by the pigment content, protein composition and their photo-chemical properties. The PS I complex was obtained as a heavy green band at 1M sucrose. The contents of P700 and chlorophylls of isolated PS I were compared with those of total thylakoids (Table 1). The purified PS I particles exhibited 45% recovery of total chlorophyll, chl a/b ratio of 10 and chl-P700 ratio of 180. The contents

TABLE 1

Characterisation of PS I complex from Sorghum thylakoids

Experiment	Thylakoid membranes		PS I complex	
	Normal light	*Low light*	*Normal light*	*Low light*
% Chl recovery	100	100	40	43
Chl a/b ratio	4	3	10	8
Chl/P700	450	410	180	189
P700 content (nmol/mg chl)	180	167	719	681

of P700 and the ratio of chl a/b decreased while the chl/P700 ratio increased under low growth light intensity (Table 1).

SDS-PAGE analysis of PS I complex was compared with de-stacked thylakoids (Fig. 1 and 2). PS I complex was resolved into 11 bands with a prominent reaction centre apoprotein of 68 kD. Two polypeptides of 24 kD and 22 kD, two heavily stained bands of 20 kD and 18 kD and six low molecular, weight polypeptides in the range 8–16 kD were noticed. Variations in the protein composition of PS I, associated with low light conditions are shown in Figure 2. The band intensity of 68 kD and the number of low molecular weight polypeptides in the range of 8–16 kD and their protein content decreased under low growth light. An increase in the intensity of 20 kD and 18 kD polypeptides was observed under low growth light intensity (Figure 2). Non-denatured SDS-PAGE analysis of PS I complex showed three chlorophyll protein complexes: CP I, CP Ia, and LHC-I (Fig. 3). The sizes of CP Ia (core antenna of PS I) and LHC-1 were significantly increased under low growth light intensity.

The functional characteristics of the chl-protein complexes of PS I were

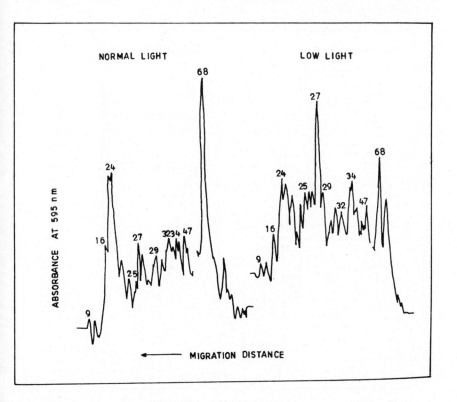

Fig. 1: Densitograms of sorghum thylakoids polypeptides under normal and low growth light conditions.

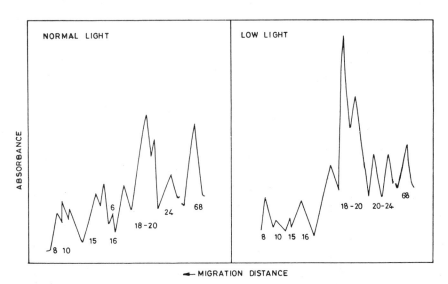

Fig. 2: Densitograms of isolated PS I complex polypeptides of sorghum under normal and low light conditions.

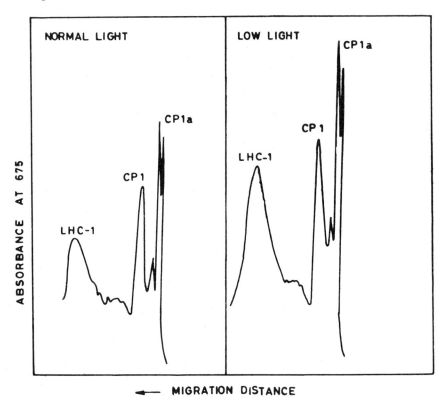

Fig. 3: Densitograms of PS I complex of sorghum separated by non-denaturing SDS-PAGE.

investigated by fluorescence spectroscopy. The sucrose density gradient ultracentrifugation revealed four major chlorophyll bands which were designated as CP I, CP Ia and two bands associated with LHC-I. The chl-protein complexes resolved by sucrose density gradient ultra-centrifugation were dialysed against 10 mM tricine-KOH pH 7.8 for 6 h to remove the detergent. Fluorescence emission spectra of PS I complex showed a major peak at 688 nm and a shoulder at 732 nm (Fig. 4). The relative fluorescence emission of PS I at 688 nm increased under low growth light. A shift of F688 to F690 and F730 to F725 was also observed under low light conditions (Fig. 4).

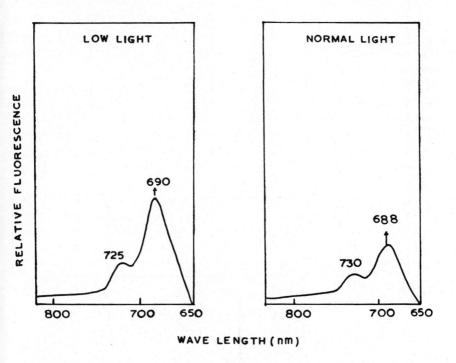

Fig. 4: Fluorescence emission spectra of isolated PS I complex of sorghum grown under normal and low light.

The two bands associated with LHC-I were further characterised by fluorescence emission spectra. The lighter band had a fluorescence emission maximum at 685 nm and designated as LHC-I-685 (Fig. 5). The heavier band exhibited fluorescence emission peaks at 688 and 727 nm and was designated as LHC-I-730 (Fig. 6). In low light F685 increased significantly with 5 nm shift towards far red (Fig. 5). The emission spectra of LHC-I-730 indicated a decreased F688 under low light (Fig. 6).

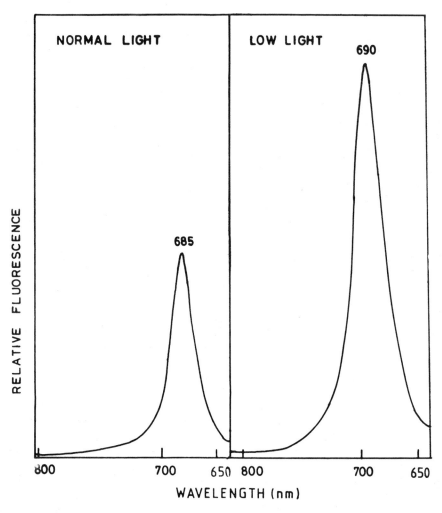

Fig. 5: Fluorescence emission spectra of isolated LHC-680 of sorghum under normal and low growth light conditions.

Discussion

Isolated PS I particles from Sorghum leaves had a chl a/b ratio of 10 and chl/P700 ratio of 180. These characteristics of PS I were slightly different from those reported by others (9,15,16). Native PS I complex of *Pisum sativum* exhibited a chl a/b ratio of 14–18 and chl/P700 ratio of 110 (9) while Haworth *et al.* (15) reported a chl a/b ratio of 7–8 for isolated PS I particles. Our results for sorghum show that a chl a/b ratio of 10 indicate a more association of LHC-I with isolated PS I with a concomitant increase

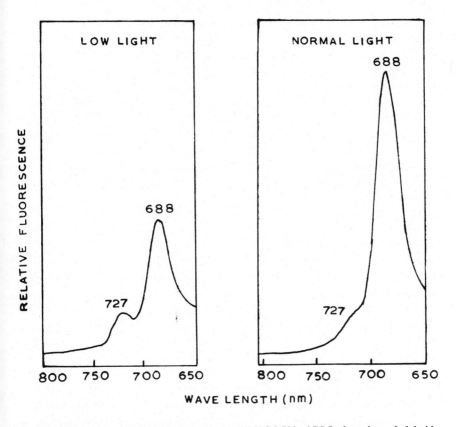

Fig. 6: Fluorescence emission spectra of isolated LHC-I-730 of PS I of sorghum thylakoids under normal and low growth light conditions.

in chl/P700 ratio. The isolated PS I preparation was highly enriched in P700 content as evidenced by the values in Table 1. The lowered chl a/b ratio in low light grown plants indicates more association of chl b while the increased chl/P700 ratio suggest an increase in the size of LHC-I in low light. P700 content was only slightly reduced in low light.

The polypeptide profiles of purified PS I complex were compared with Triton-X-100 solubilised thylakoids. The reaction centre apoprotein of 68 kD was associated with a LHC-I in the range of 18–24 kD along with some low molecular weight polypeptides. The polypeptides of sorghum PS I were similar to those reported for other plants (6,9,15,16,17). The results described here provide for the first time evidence for the changes in the structure and function of chl-protein complexes of PS I with varied growth light intensity.

The increase in the band intensities of polypeptides in the range 18–20 kD in low light clearly indicates an increased antenna size of PS I (Figs. 1 and 2). The modulation of protein composition of reaction centre complex has been evidenced by a decrease in the reaction centre apoprotein as well as lowered number of low molecular weight polypeptides and their protein content. Our present study has shown that isolated PS I particles of sorghum has three distinct chlorophyll-protein complexes: CP I, CP Ia and LHC-I (Fig. 3). The increased CP Ia content and an increase in the absorbance of LHC-I in low light suggests an increased size of core antenna and LHC-I. The fluorescence spectroscopy of isolated PS I complex indicates that the chlorophyll antenna of PS I complex is constituted by two groups of chl-protein complexes: each associated with core antenna of reaction centre and LHC–I, having fluorescence emission peaks at 730 and 688 nm respectively (Fig. 5). The enhanced F688 in low light indicates an increase in size of LHC-I.

The sucrose density gradient ultracentrifugation of PS I complex resulted in two chlorophyll protein bands associated with LHC-I in addition to CP I and CP Ia. These two chl-protein complexes of LHC-I were further characterised by fluorescence studies and designated as LHC-I 685 and LHC-I-730, which clearly indicate that LHC-I of sorghum PS I has two antenna groups of which LHC-I 730 corresponds to inner antenna and LHC-I-685 is the peripheral antenna (10). Fluorescence emission spectra of LHC-I-730 showed a major peak at 688 nm and a shoulder at 727 nm analogous to total PS I complex confirming that this complex acts as an inner antenna for PS I reaction centre. The fluorescence emission of LHC-I-685 which is similar to the other light harvesting systems indicates that this complex functions as peripheral antenna.

Enhancement of F685 and its shift in fluorescence towards far red of LHC-I 685 complex in low light (Fig. 6), indicates a significant increase in the size of peripheral antenna which would facilitate more harvesting of the available photon flux density. In addition, the decrease in the F688 of LHC-I-730 in low light clearly suggests that the inner antenna of LHC-I plays an indispensable role in transferring the excitation energy from LHC-I-685 to the reaction centre of PS I.

References

1. Boardman, N. K. (1977) *Ann. Rev. Plant Physiol.* 28, 355–357
2. Thornber, J. P., Cogdell, R. J., Pierson, B. K. and Seftor, R. E. B. (1979) *Photochem. Photobiol.* 29, 1205–1216
3. Anderson, J. M. (1986) *Ann. Rev. Plant Physiol.* 37, 193–196
4. Leong, T. Y. and Anderson, J. M. (1983) *Biochim. Biophys.* Acta 723, 391–399
5. Melis, A. and Harvey, G. W. (1981) *Biochim. Biophys.* Acta 637, 138–145
6. Malkin, R. (1982) *Ann. Rev. Plant Physiol.* 33, 455–479
7. Leong, T. Y. and Anderson, J. M. (1984) *Photosynth. Res.* 5, 105–115
8. Bhaskar, C. V. S. and Rama Das, V. S. (1987) *Photosynthetica* 21, 566–571
9. Mullet, J. E., Burke, J. J. and Arntzen, C. J. (1980) *Plant Physiol.* 65, 814–822
10. Bassi, R. and Simpson, D. (1987) *Eur. J. Biochem.* 163, 221–230
11. Laemmli, U. K. (1970) *Nature* 227, 680–685

12. Anderson, J. M. (1980) *Biochim. Biophys*. Acta 591. 113–126
13. Henriques, F. and Park, R. B. (1978) *Plant Physiol,* 62, 856–860
14. Arnon, D. I. (1949) *Plant Physiol*. 24, 1–15
15. Haworth, P., Watson, J. I. and Arntzen, C. J. (1983) *Biochim. Biophys*. Acta 724, 151–158
16. Lam, E., Ortiz, W., Mayfield, S. and Malkin, R. (1984) *Plant Physiol*. 74, 650–655
17. Nelson, N. and Notsani, B. E. (1977) in: *Bioenergetics of membranes* (Packer, L. Papageorigiou, G. C. and Trebst, A., eds.) pp. 233–244, Elsevier, Amsterdam

Index